Lecture Notes in Mathematics

Editors:
A. Dold, Heidelberg
F. Takens, Groningen

Subseries: Fondazione C. I. M. E., Firenze
Advisor: Roberto Conti

Springer
*Berlin
Heidelberg
New York
Barcelona
Budapest
Hong Kong
London
Milan
Paris
Santa Clara
Singapore
Tokyo*

M. Bardi   M. G. Crandall   L. C. Evans
H. M. Soner   P. E. Souganidis

Viscosity Solutions and Applications

Lectures given at the 2nd Session of the
Centro Internazionale Matematico Estivo
(C.I.M.E.) held in Montecatini Terme, Italy,
June 12–20, 1995

Editors: I. Capuzzo Dolcetta, P. L. Lions

Fondazione
C.I.M.E.

Springer

Authors

Martino Bardi
Dipartimento di Matematica
Università di Padova
Via Belzoni, 7
I-35131 Padova, Italy

Michael G. Crandall
Department of Mathematics
University of California
Santa Barbara, CA 93106, USA

Lawrence C. Evans
Department of Mathematics
University of California
Berkeley, CA 94720, USA

Halil Mete Soner
Department of Mathematics
Carnegie Mellon University
Schenley Park
Pittsburg, PA 15213, USA

Panagiotis E. Souganidis
Department of Mathematics
University of Wisconsin-Madison
480 Lincoln Drive
Madison, WI 53706, USA

Editors

Italo Capuzzo Dolcetta
Dipartimento di Matematica
Università di Roma "La Sapienza"
Piazzale Aldo Moro, 5
I-00185 Roma, Italy

Pierre Louis Lions
CEREMADE
Université Paris-Dauphine
Place du Maréchal de Lattre de Tassigny
F-75775 Paris Cedex 16, France

Cataloging-in-Publication Data applied for

Die Deutsche Bibliothek – CIP-Einheitsaufnahme

Viscosity solutions and applications: held in Montecatini Terme, Italy, June 12–20, 1995 / M. Bardi
... Ed.: I. Capuzzo Dolcetta;P. L. Lions. – Berlin; Heidelberg; New York; Barcelona; Budapest; Hong
Kong; London; Milan; Paris; Santa Clara; Singapore; Tokyo: Springer, 1997 (Lectures given at the ...
session of the Centro Internazionale Matematico Estivo (CIME) ... ; 1995.2)
(Lecture notes in mathematics; Vol. 1660 : Subseries: Fondazione CIME)
ISBN 3-540-62910-6 kart.
Centro Internazionale Matematico Estivo <Firenze>: Lectures given at the ... session of the Centro
Internazionale Matematico Estivo (CIME) ... – Berlin; Heidelberg; New York; London; Paris; Tokyo;
Hong Kong: Springer
Früher Schriftenreihe. – Früher angezeigt u.d.T.: Centro Internazionale Matematico Estivo: Proceedings
of the ... session of the Centro Internazionale Matematico Estivo (CIME)
1995,2. Viscosity solutions and applications. – 1997

Mathematics Subject Classification (1991): 35B37, 35J60, 49J15, 49K15, 49L25

ISSN 0075-8434
ISBN 3-540-62910-6 Springer-Verlag Berlin Heidelberg New York

This work is subject to copyright. All rights are reserved, whether the whole or part of the material is
concerned, specifically the rights of translation, reprinting, re-use of illustrations, recitation, broad-
casting, reproduction on microfilms or in any other way, and storage in data banks. Duplication of this
publication or parts thereof is permitted only under the provisions of the German Copyright Law of
September 9, 1965, in its current version, and permission for use must always be obtained from
Springer-Verlag. Violations are liable for prosecution under the German Copyright Law.

© Springer-Verlag Berlin Heidelberg 1997
Printed in Germany

The use of general descriptive names, registered names, trademarks, etc. in this publication does not imply,
even in the absence of a specific statement, that such names are exempt from the relevant protective laws
and regulations and therefore free for general use.

Typesetting: Camera-ready T$_E$X output by the author/editors
SPIN: 10520395   46/3142-543210 - Printed on acid-free paper

Preface

The C.I.M.E. School on Viscosity Solutions and Applications held in Monte-catini from June 12 to June 20, 1995 was designed with the aim to provide a rather comprehensive and up-to-date account of the theory of viscosity solutions and some of its applications.
The School comprised the following five series of lectures:

M.G. Crandall: General Theory of Viscosity Solutions

M. Bardi: Some Applications of Viscosity Solutions to Optimal Control and Differential Games

L.C. Evans: Regularity for Fully Nonlinear Elliptic Equations and Motion by Mean Curvature

M.H. Soner: Controlled Markov Processes, Viscosity Solutions and Applications to Mathematical Finance

P.E. Souganidis: Front Propagation: Theory and Applications

as well as seminars by:

L. Ambrosio, M. Arisawa, G. Bellettini, P. Cannarsa, M. Falcone, S. Koike, G. Kossioris, M. Motta, A. Siconolfi and A. Tourin.

The present volume is a record of the material presented in the above listed courses. It is our belief that it will serve as a useful reference for researchers in the fields of fully nonlinear partial differential equations, optimal control, propagation of fronts and mathematical finance.

It is our pleasure here to thank the invited lecturers, the colleagues who contributed seminars, all the participants for their active contribution to the success of the School and the Fondazione CIME for the support in the organization.

I. Capuzzo Dolcetta, P.L. Lions

Contents

VISCOSITY SOLUTIONS: A PRIMER

by

Michael G. Crandall[†]

Department of Mathematics

University of California, Santa Barbara

Santa Barbara, CA 93106

0. Introduction

These lectures present the most basic theory of "viscosity solutions" of fully nonlinear scalar partial differential equations of first and second order. Other contributions to this volume develop some of the amazing range of applications in which viscosity solutions play an essential role and various refinements of this basic material.

In this introductory section we describe the class of equations which are treated within the theory and then our plan of presentation.

The theory applies to scalar second order partial differential equations

$$\text{(PDE)} \qquad F(x, u, Du, D^2u) = 0$$

on open sets $\Omega \subset \mathbb{R}^N$. The unknown function $u : \Omega \to \mathbb{R}$ is real-valued, Du corresponds to the gradient $(u_{x_1}, \ldots, u_{x_N})$ of u and D^2u corresponds to the Hessian matrix (u_{x_i,x_j}) of second derivatives of u. Consistently, F is a mapping

$$F : \Omega \times \mathbb{R} \times \mathbb{R}^N \times \mathcal{S}(N) \to \mathbb{R}$$

where $\mathcal{S}(N)$ is the set of real symmetric $N \times N$ matrices. We say that Du (D^2u) "corresponds" to the gradient (respectively, the Hessian) because, as we shall see, solutions u may not be differentiable, let alone twice differentiable, and still "solve" (PDE). We write $F(x, r, p, X)$ to indicate the value of F at $(x, r, p, X) \in \Omega \times \mathbb{R} \times \mathbb{R}^N \times \mathcal{S}(N)$. (PDE) is said to be *fully nonlinear* to emphasize that $F(x, r, p, X)$ need not be linear in any argument, including the X in the second derivative slot.

F is called *degenerate elliptic* if it is nonincreasing in its matrix argument:

$$F(x, r, p, X) \le F(x, r, p, Y) \quad \text{for} \quad Y \le X.$$

The usual ordering is used on $\mathcal{S}(N)$; that is $Y \le X$ means

$$\langle X\xi, \xi \rangle \le \langle Y\xi, \xi \rangle \quad \text{for} \quad \xi \in \mathbb{R}^N$$

where $\langle \cdot, \cdot \rangle$ is the Euclidean inner product. If F is degenerate elliptic, we say that it is *proper* if it is also nondecreasing in r. That is, F is proper if

$$F(x, s, p, X) \le F(x, r, p, Y) \quad \text{for} \quad Y \le X, \quad s \le r.$$

[†] Supported in part by NSF Grant DMS93-02995 and in part by the author's appointment as a Miller Professor at the University of California, Berkeley for Fall 1996.

As a first example, F might be of first order

$$F(x, r, p, X) = H(x, r, p);$$

every first order F is obviously (very) degenerate elliptic, and then proper if it is nondecreasing in r. For an explicit example, the equation $u_t + (u_x)^2 = 0$ with $(t, x) \in \mathbb{R}^2$ is a proper equation (we are thinking of (t, x) as (x_1, x_2) above). On the other hand, the Burger's equation $u_t + uu_x = 0$ is not proper, for it is not monotone in u. We refer to proper first order equations $H(x, u, Du) = 0$ and $u_t + H(x, u, Du) = 0$ as "Hamilton-Jacobi" equations.

Famous second order examples are given by $F(x, r, p, X) = -\text{Trace}\,(X)$ and $F(x, r, p, X) = -\text{Trace}\,(X) - f(x)$ where f is given; the pdes are then Laplace's equation and Poisson's equation:

$$F(D^2 u) = -\sum_{i=1}^{N} u_{x_i x_i} = -\Delta u = 0 \quad \text{and} \quad -\Delta u = f(x)$$

The equations are degenerate elliptic since $X \to \text{Trace}\,(X)$ is monotone increasing on $\mathcal{S}(N)$. We do not rule out the linear case! Incorporating t as an additional variable as above, the heat equation $u_t - \Delta u = 0$ provides another famous example. The convention used here, that Du, $D^2 u$ stand for the spatial gradient and spatial Hessian, will be in force whenever we write "$u_t + F(x, u, Du, D^2 u)$".

Note the preference implied by these examples; we prefer $-\Delta$ to Δ. A reason is that (in various settings), $-\Delta$ has an order preserving inverse. This convention is not uniform; for example, Souganidis [35] does not follow it and reverses the inequality in the definition of degenerate ellipticity.

More generally, the linear equation

$$-\sum_{i,j=1}^{N} a_{i,j}(x) u_{x_i, x_j} + \sum_{i=1}^{N} b_i(x) u_{x_i} + c(x) u - f(x) = 0$$

may be written in the form $F = 0$ by setting

(0.1) $\qquad F(x, r, p, X) = -\text{Trace}\,(A(x)X) + \langle b(x), p \rangle + c(x) r - f(x)$

where $A(x)$ is a symmetric matrix with the elements $a_{i,j}(x)$ and $b(x) = (b_1(x), \ldots, b_N(x))$. This F is degenerate elliptic if $0 \leq A(x)$ and proper if also $0 \leq c(x)$.

In the text we will pose some exercises which are intended to help readers orient themselves (and to replace boring text with pleasant activities). We violate all conventions by doing so even in this introduction. Some exercises are "starred" which means "please do it now" if the fact is not familiar.

Exercise 0.1.* Verify that F given in (0.1) is degenerate elliptic if and only if $A(x)$ is nonnegative.

The second order examples given above are associated with the "maximum principle". Indeed, the calculus of the maximum principle is a fundamental idea in the entire theory.

Exercise 0.2.[*] Show that F is proper if and only if whenever $\varphi, \psi \in C^2$ and $\varphi - \psi$ has a nonnegative maximum (equivalently, $\psi - \varphi$ has a nonpositive minimum) at $\hat{x}$, then

$$F(\hat{x}, \psi(\hat{x}), D\psi(\hat{x}), D^2\psi(\hat{x})) \leq F(\hat{x}, \varphi(\hat{x}), D\varphi(\hat{x}), D^2\varphi(\hat{x})).$$

So far, we have presented nonlinear first order examples and linear second order examples. However, the class of proper equations is very rich. Indeed, if F, G are both proper, then so is $\lambda F + \mu G$ for $0 \leq \lambda, \mu$. More interesting is the following simple fact: if $F_{\alpha, \beta}$ is proper for $\alpha \in \mathcal{A}$, $\beta \in \mathcal{B}$ (some index sets), then so is

$$F = \sup_{\alpha \in \mathcal{A}} \inf_{\beta \in \mathcal{B}} F_{\alpha, \beta}$$

provided only it is finite. This generality is essential to applications of the theory in differential games (see Bardi [2]), while applications in control theory correspond to the case "F_α" in which there is only one index (see Bardi [2] and Soner [34]).

For example, $\max(u_t + |Du|^2 - g(x), -\Delta u - f(x)) = 0$ is a proper equation. The other lecture series will present many examples of scientific significance. We have only attempted here to indicate that that class of proper equations is broad and interesting.

Here we aim at a clear and congenial presentation of the most basic elements of the theory of viscosity solutions of proper equations $F = 0$. These are the notion of a viscosity solution, maximum principle type comparison results for viscosity solutions, and existence results for viscosity solutions via Perron's method. We do not aim at completeness or technical generality, which often distract from ideas.

The text is organized in sections, many of which are quite brief. The descriptions below contain remarks about the logic of the presentation. By the numbers, the topics are:

Section 1: An illustration of the need to be able to consider nondifferentiable functions as solutions of proper fully nonlinear equations is given using first order examples.

Section 2: The notions of viscosity subsolutions, supersolutions and solutions are presented. The convention that the modifier "viscosity" will be dropped thereafter in the text is introduced. It is essential to deal with semicontinuous functions in the theory, and this generality appears here.

Section 3: Striking general existence and uniqueness theorems are presented without proof to indicate the success of viscosity solutions in this arena. The contrast with the examples in Section 1 is dramatic.

Sections 4, 5, 6: A primary test of a notion of generalized solutions is whether or not appropriate uniqueness results can be obtained (when suitable side conditions - boundary conditions, growth conditions, initial conditions, etc. - are satisfied). Actually, one wants a bit more here, that is the sort of *comparison* theorems which follow from the maximum principle. Basic arguments needed in proofs of comparison results for viscosity solutions of first order stationary problems (those without "t") are presented here and typical results are deduced. Section 4 concerns the Dirichlet problem, Section 5 concerns bounded solutions of a problem in IR^N, and Section 6 provides an example of treating unbounded solutions. The second order case is more complex and is not taken up until Section 10. However, nothing is wasted, and all the arguments presented in these sections are invoked in the second order setting.

Section 7: The notions of Section 2 are recast in a form convenient for use in the next section and in the comparison theory in Sections 8 and 9.

Section 8: Two related results, each an important tool, are established. One states roughly that the supremum of a family of subsolutions is again a subsolution, and the other that the limit of a sequence of viscosity subsolutions (supersolutions, solutions) of a converging sequence of equations (meaning the F_n's converge) is a subsolution (respectively, a supersolution, solution) of the limiting equation. We call this last theme "stability" of the notion; it is one of the great tools of the theory in applications. The mathematics involved is elementary with a "point-set" flavor.

Section 9: Existence is proved via Perron's Method using a result of the previous section. The existence theory presupposes "comparison". At this stage, comparison has only been treated in the first order case, and is simply assumed for the second order case. This does not affect either clarity or the basic argument. At this juncture, the most basic ideas have been presented with the exception of comparison for second order equations.

Section 10: The primary difference between the first and second order cases is explained. Then the rather deep result which is used here to bridge the gap, called here "the Theorem on Sums" (an analytical result about semicontinuous functions), is stated without proof. An example is given to show how this tool theorem renders the second order case as easy to treat as the first order case.

Section 11: The Theorem on Sums is proved.

Section 12: In the preceding sections comparison was only demonstrated for various equations of the form $F(x, u, Du, D^2u) = 0$. Here the main additional points needed to treat $u_t + F(x, u, Du, D^2u) = 0$ are sketched.

Regarding notation, we use standard expressions like "$C^2(\Omega)$" (the twice continuously differentiable functions on Ω) and "$|p|$" (the Euclidean length of p) without further comment when it seems reasonable. With some exceptions, we minimize distracting notation.

Regarding the literature, it is too vast to try to summarize in a work like this, which aims at presenting basic ideas and not at technical generality or great precision. We will basically rely on the big brother to this work, the more intense (and reportedly less friendly) [12] for its extensive references, together with those in the other contributions to this volume. (We recommend the current work as preparation for reading [12], especially the topics therein not taken up here.) We do give some references corresponding to the original works initiating the themes treated here. A few more recent papers are cited as appropriate. All references appear at the ends of sections. In addition, we mention the books by Cabré and Caffarelli [7] and Dong [17] for recent expositions of regularity theory of solutions, which is not treated here, as well as the classic text of Gilbarg and Trudinger [23]. Regularity theory is also one of the themes of Evans [20]. The recent book of Barles [3] presents a complete theory of the first order case (which itself fills a book that contains 154 references!). The book of Fleming and Soner [22, Chapters II and V] also nicely covers the basic theory. There are alternative theories for first order equations; see, e.g., [9] and [36]. Of course, MathSciNet now allows one to become nearly current regarding the state of the literature relatively easily, and one can profitably search on any of the leads given above.

A significant limitation of our presentation is that only the Dirichlet boundary condition is discussed at any length, and this in its usual form rather than the generalized version. Other boundary conditions appear in the contributions of Bardi [2] and Soner [34] in an essential way. In addition to the references they give, the reader may refer for example to [12, Section 7] for a discussion in the spirit of this work. Another limitation is that singular equations are not treated at all. Equations with singularities appear in contributions of Evans [20] and Souganidis [35]. See also [12, Section 9]. Finally, only continuous solutions are discussed here, while within applications one meets the discontinuous solutions. The contribution of Bardi [2, Section V] treats this issue, and discontinuous functions appear quickly in the exposition of Souganidis [35].

1. On the Need for Nonsmooth Solutions

The fact is that it is difficult to give examples of solutions (in any sense) of equations $F = 0$ which are not classical solutions unless the equation is pretty "degenerate" (roughly, the monotonicity of $X \to F(x, r, p, X)$ is not strong enough) or "singular" (that is, F may have discontinuities or other types of singularities). (A "classical" solution of an equation $F(x, u(x), Du(x), D^2u(x)) = 0$ is a twice continuously differentiable function which satisfies the equation pointwise; if the equation is first order classical solutions are once continuously differentiable; if the equation has the form $u_t + F(x, u, Du, D^2u) = 0$, then a classical solution will possess the derivatives u_t, Du, D^2u in the classical sense. Similar remarks apply to subsolutions and supersolutions.) The reason is that the regularity theory of sufficiently nondegenerate and nonsingular equations is still unsettled. In particular, it may be that nondegenerate nonsingular equations $F = 0$ with smooth F admit only classical solutions, although some suspect that this is not so.

However, if the equation is first order (so very degenerate), then examples are easy. The next exercise gives a simple problem without classical solutions and for which there are solutions slightly less regular than "classical"; however allowing less regular solutions generates "nonuniqueness".

Exercise 1.1. Put $N = 1$, $\Omega = (-1, 1)$ and $F(x, r, p, X) = |p|^2 - 1$. Verify that there is no classical (here this means $C^1(-1, 1) \cap C([-1, 1])$) solution u of $F(u') = (u')^2 - 1 = 0$ on $(0, 1)$ satisfying the Dirichlet conditions $u(-1) = u(1) = 0$. Verify that $u(x) = 1 - |x|$ and $v(x) = |x| - 1$ are both "strong" solutions: in this case, they are Lipschitz continuous and the equation is satisfied pointwise except at $x = 0$ (so almost everywhere).

Of course, the problem in Exercise 1.1 has a unique solution within our theory, as we will see later (it is $u(x) = 1 - |x|$).

To further establish the desirability of allowing nondifferentiable solutions, we recall the classical method of characteristics as it applies to the Cauchy problem for a Hamilton-Jacobi equation $u_t + H(Du) = 0$:

$$(1.1) \quad \begin{cases} u_t + H(Du) = 0 & \text{for} \quad x \in \mathbb{R}^N, \ t > 0 \\ u(0, x) = \psi(x), & \text{for} \quad x \in \mathbb{R}^N. \end{cases}$$

Suppose that H is smooth and that u is a smooth solution of $u_t + H(Du) = 0$ on $t \geq 0$, $x \in \mathbb{R}^N$. Define $Z(t) \in \mathbb{R}^N$ to be the solution of the initial value problem

$$Z'(t) = \frac{d}{dt} Z(t) = DH(Du(t, Z(t))), \quad Z(0) = \hat{x}$$

over the largest interval for which this solution exists. A computation yields

$$\frac{d}{dt} Du(t, Z(t)) = D\frac{\partial u}{\partial t}(t, Z(t)) + D^2 u(t, Z(t)) Z'(t)$$
$$= D\frac{\partial u}{\partial t}(t, Z(t)) + D^2 u(t, Z(t)) DH(Du(t, Z(t)))$$
$$= 0$$

where the last equation arises from differentiating $u_t + H(Du) = 0$ with respect to x.

Remark 1.1. In calculations such as the above, one has to decide whether the the gradient Dv of a scalar function v is to be a column vector or a row vector. There is no ambiguity about $D^2 v$, for it is to be square and symmetric in any case. In the introduction we wrote the gradient as a row vector, but above interpret it as a column vector. This is consistent with interpreting points of $\mathbb{R}^N$ as column vectors while writing row vectors, and with these sloppy conventions the above is correct.

We conclude that Du is constant on the curve $t \to (t, Z(t))$. It then would follow that $Z(t) = \hat{x} + tDH(D\psi(\hat{x}))$. However, the resulting equation $Du(t, \hat{x} + tDH(D\psi(\hat{x})) \equiv D\psi(\hat{x})$ yields contradictions as soon as we have characteristics crossing, that is $y \neq z$ but $t > 0$ such that $y + tDH(D\psi(y)) = z + tDH(D\psi(z))$. In this case, one says that "shocks form" and there are no smooth solutions u defined for all $t \geq 0$ in general.

Exercise 1.2. (i) Continue the analysis above to find

$$u(t, Z(t)) = \psi(\hat{x}) + t(\langle D\psi(\hat{x}), DH(D\psi(\hat{x}))\rangle - H(D\psi(\hat{x})))$$

where $\langle \cdot, \cdot \rangle$ is the Euclidean inner product.

(ii) If $N = 1$, then shocks will form unless $x \to H'(\psi'(x))$ is monotone.

Under reasonable assumptions, as is shown in elementary courses, analysis by characteristics provides a smooth solution of (1.1) until shocks form. When classical solutions break down, in this area and others, one is led to think of the problem of finding a way to continue past the breakdown with a less regular solution. However, one can also immediately think of the problem of finding solutions in cases where the data does not allow the classical analysis. E.g., what does one do if H and/or ψ above is not smooth? The "breakdown" idea is not central in this view.

Just as in the case of Exercise 1.1, relaxing the regularity requirement for a solution just a tiny bit leads to nonuniqueness for (1.1). One does not expect uniqueness in general for stationary problems, but one does expect uniqueness for initial-value problems.

Exercise 1.3. Consider the equation $u_t + (u_x)^2 = 0$ for $t > 0$, $x \in \mathbb{R}$ coupled with the initial condition $u(0, x) \equiv 0$. Verify that the function

$$v(t, x) \equiv 0 \quad \text{for} \quad 0 < t \le |x|,$$
$$v(t, x) = -t + |x| \quad \text{for} \quad |x| \le t,$$

satisfies the initial condition, is continuous and has all the regularity one desires off the lines $x = 0$, $t = |x|$, and satisfies the equation off these lines. Thus $u \equiv 0$ and v are *distinct* nearly classical - even piecewise linear - solutions of the Cauchy problem.

We have not given second order examples. However, here is a model equation which will be covered under the theory to be described and for which the issue of how smooth solutions are is unsettled. Let $A_i \in \mathcal{S}(N)$, $i = 1, 2, 3$ satisfy $I \le A_i \le 2I$ for $i = 1, 2, 3, 4$ and

$$F(X) = -\max(\text{Trace}\,(A_1 X)\,, \min(\text{Trace}\,(A_2 X)\,, \text{Trace}\,(A_3 X))).$$

This is a uniformly elliptic equation - here this means that there are constants $0 < \lambda < \Lambda$ such that

$$F(X + P) \le F(X) - \lambda \text{Trace}\,(P) \quad \text{and} \quad |F(X) - F(Y)| \le \Lambda \|X - Y\|$$

for $X, Y, P \in \mathcal{S}(N)$, $P \ge 0$. Here $\|X\|$ can be any reasonable matrix norm of X; a good one is the sum of the absolute values of the eigenvalues of X, as it coincides with the trace on nonnegative matrices.

Exercise 1.4. Determine λ, Λ which work above.

It is known that solutions of uniformly elliptic equations typically have Hölder continuous first derivatives, but it is not known if these solutions are necessarily C^2. If the equation is uniformly elliptic and convex in X, regularity is known. See Evans [20], Cabré and Caffarelli [7], Dong [17], the references therein, as well as Trudinger [39] and Świçch [37] for a recent result concerning Sobolev rather than Hölder regularity.

2. The Notion of Viscosity Solutions

As we will see, the theory will require us to deal with semicontinuous functions, there is no escape. Therefore, let us recall the notions of the *upper semicontinuous envelope* u^* and the *lower semicontinuous envelope* u_* of a function $u : \Omega \to \mathbb{R}$:

(2.1)
$$\begin{cases} u^*(x) = \limsup_{r \downarrow 0} \{u(y) : y \in \Omega, |y - x| \leq r\} \\ u_*(x) = \liminf_{r \downarrow 0} \{u(y) : y \in \Omega, |y - x| \leq r\} . \end{cases}$$

Recall that u is upper semicontinuous if $u = u^*$ and lower semicontinuous if $u = u_*$; equivalently, u is upper semicontinuous if $x_k \to x$ implies $u(x) \geq \limsup_{k \to \infty} u(x_k)$, etc. Of course, u^* is upper semicontinuous and u_* is lower semicontinuous.

Exercise 2.1.* In the above definition Ω could be replaced by an arbitrary metric space $\mathcal{O}$ if $|y - x|$ is replaced distance between $x, y \in \mathcal{O}$. Show in this generality that u is upper semicontinuous if and only if $v = -u$ is lower semicontinuous if and only if $\{x \in \mathcal{O} : u(x) \leq r\}$ is closed for each $r \in \mathbb{R}$. Show that a function which is both upper semicontinuous and lower semicontinous is continuous. Show that if $\mathcal{O}$ is compact and u is upper semicontinuous on $\mathcal{O}$, then u has a maximum point $\hat{x}$ such that $u(x) \leq u(\hat{x})$ for $x \in \mathcal{O}$.

Motivation for the following definition is found in Exercise 0.2; see also Exercise 2.4 below. The semicontinuity requirements in the definition are partly explained by the last part of Exercise 2.1 and the fact that we will want to produce the maxima associated with subsolutions, etc., in proofs.

Definition 2.1. *Let F be proper, Ω be open and $u : \Omega \to \mathbb{R}$. Then u is a viscosity subsolution of $F = 0$ in Ω if it is upper semicontinuous and for every $\varphi \in C^2(\Omega)$ and local maximum point $\hat{x} \in \Omega$ of $u - \varphi$, we have $F(\hat{x}, u(\hat{x}), D\varphi(\hat{x}), D^2\varphi(\hat{x})) \leq 0$. Similarly, $u : \Omega \to \mathbb{R}$ is a viscosity supersolution of $F = 0$ in Ω if it is lower semicontinuous and for every $\varphi \in C^2(\Omega)$ and local minimum point $\hat{x} \in \Omega$ of $u - \varphi$, we have $F(\hat{x}, u(\hat{x}), D\varphi(\hat{x}), D^2\varphi(\hat{x})) \geq 0$. Finally, u is a viscosity solution of $F = 0$ in Ω if is both a viscosity subsolution and a viscosity supersolution (hence continuous) of $F = 0$.*

Remark 2.2. Hereafter we use the following conventions: "supersolution", "subsolution" and "solution" mean "viscosity supersolution", "viscosity subsolution" and "viscosity solution" – other notions will carry the modifiers (e.g., classical solutions, etc.). Moreover, the phrases "subsolution of $F = 0$" and "solution of $F \leq 0$" mean the same (and similarly for supersolutions).

Remark 2.3. Explicit subsolutions and supersolutions which are semicontinuous and not continuous will not appear in these lectures. They intervene abstractly in proofs, however, in an essential way.

Exercise 2.2. Reconcile Definition 2.1 with Exercise 0.2 in the following sense: Show that if F is proper, $u \in C^2(\Omega)$ and

$$F(x, u(x), Du(x), D^2u(x)) \leq 0$$

$(F(x, u(x), Du(x), D^2u(x)) \geq 0)$ for $x \in \Omega$, then u is a solution of $F \leq 0$ (respectively $F \geq 0$) in the above sense.

Exercise 2.3. With F as in Exercise 1.1, verify that $u(x) = 1 - |x|$ is a solution of $F = 0$ on (-1,1), but that $u(x) = |x| - 1$ is not. Attempt to show that $u(x) = 1 - |x|$ is the only solution of $F = 0$ in (-1,1) which vanishes at $x = -1, 1$. Verify that $u(x) = |x| - 1$ is a solution of $-(u')^2 + 1 = 0$. In general, verify that if F is proper then u is a solution of $F \leq 0$ if and only if $v = -u$ is a solution of $G \geq 0$ where $G(x, r, p, X) = -F(x, -r, -p, -X)$ and that G is proper. Thus any result about subsolutions provides a dual result about supersolutions.

Exercise 2.4. In general, if Ω is bounded and open in $\mathbb{R}^N$, verify that $u(x) = $ distance$(x, \partial\Omega)$ is a solution of $|Du| = 1$ in Ω.

We mention that the idea of putting derivatives on test functions in this maximum principle context was first used to good effect in Evans [18, 19]. The full definitions above in all their semicontinuous glory, evolved after the uniqueness theory was initiated in [14], [15]. The definition in these works was equivalent to that above, but was formulated differently and all functions were assumed continuous. The paper [16] comments on equivalences and writes proofs more similar to those given today. Ishii's introduction of the Perron method in [24] was a key point in establishing the essential role of semicontinuous functions in the theory. Ishii in fact defines a "solution" to be a function u such that u^* is a subsolution and u_* is a supersolution. See Bardi's lectures [2] in this regard.

3. Statements of Model Existence - Uniqueness Theorems

Recalling the discussion of classical solutions of the Cauchy problem (1.1) and Exercise 1.3, the following results are a striking affirmation that the solutions introduced in Definition 2.1 are appropriate.

For Hamilton-Jacobi equations we have:

Theorem 3.1. *Let $H : \mathbb{R}^N \to \mathbb{R}$ be continuous and $\psi : \mathbb{R}^N \to \mathbb{R}$ be uniformly continuous. Then there is a unique continuous function $u : [0, \infty) \times \mathbb{R}^N \to \mathbb{R}$ with the following properties: u is uniformly continuous in x uniformly in t, u is a solution of $u_t + H(Du) = 0$ in $(0, \infty) \times \mathbb{R}^N$ and u satisfies $u(0, x) = \psi(x)$ for $x \in \mathbb{R}^N$.*

Even more striking is the following even more unequivocal generalization to include second order equations:

Theorem 3.2. *Let $F : \mathbb{R}^N \times S(N) \to \mathbb{R}$ be continuous and degenerate elliptic. Then the statement of Theorem 3.1 remains true with the equation $u_t + H(Du) = 0$ replaced by the equation $u_t + F(Du, D^2u) = 0$.*

The analogue of 3.2 for the stationary problem (i.e., without "t") is

Theorem 3.3. *Let $F : \mathbb{R}^N \times \mathcal{S}(N) \to \mathbb{R}$ be continuous and degenerate elliptic and $f : \mathbb{R}^N \to \mathbb{R}$ be uniformly continuous. Then there is a unique uniformly continuous $u : \mathbb{R}^N \to \mathbb{R}$ which is a solution of $u + F(Du, D^2u) - f(x) = $ in $\mathbb{R}^N$.*

Moreover, the solutions whose unique existence is asserted above are the ones which are demanded by the theories developed in the other lectures in this volume. In Bardi [2] and Soner [34] formulas are given for potential solutions of various problems, in control theoretic and differential games settings, and it is a triumph of the theory that the functions given by the formulas can be shown to be the unique solutions given by the theory.

All of the heavy lifting needed to prove these results is done below. However, some of the details are left for the reader's pleasure. The proof of Theorem 3.1 is indicated at the end of Section 9, the proof of Theorem 3.3 is completed in Exercise 10.3 and the proof of Theorem 3.2 is completed in Exercise 12.1.

4. Comparison for Hamilton-Jacobi Equations: the Dirichlet Problem

The technology of the proof of comparison in the second order case is more complex than in the first order case, so at this first stage we offer some sample first order comparison proofs. As a pedagogical device, we present a sequence of proofs illustrating various technical concerns. We begin with simplest case, that is the Dirichlet problem. The next two sections concern variants. Arguments are the main point, so we do not package the material as "theorems", etc. All of the arguments given are invoked later in the second order case so no time is wasted by passing through the first order case along the way.

Let Ω be a bounded open set in $\mathbb{R}^N$. The Dirichlet problem is:

$$(DP) \qquad H(x, u, Du) = 0 \quad \text{in} \quad \Omega, \quad u = g \quad \text{on} \quad \partial\Omega.$$

Here H is continuous and proper on $\overline{\Omega} \times \mathbb{R} \times \mathbb{R}^N$ and $g \in C(\partial\Omega)$. We say that $u : \overline{\Omega} \to \mathbb{R}$ is a subsolution (supersolution) of (DP) if u is upper semicontinuous (respectively, lower semicontinuous), solves $H \leq 0$ (respectively, $H \geq 0$) in Ω and satisfies $g \leq u$ on $\partial\Omega$ (respectively, $u \geq g$ on $\partial\Omega$).

Exercise 4.1. One does not expect (DP) to have solutions in general. Show that if $N = 1$, $\Omega = (0, 1)$, the Dirichlet problem $u + u' = 1$, $u(0) = u(1) = 0$ does not have solutions (in the sense of Definition 2.1!).

We seek to show that if u is a subsolution of (DP) and v is a supersolution of (DP), then $u \leq v$. We will not succeed without further conditions on H. Indeed, choose Ω to be the unit ball and let $w(x) \in C^1(\overline{\Omega})$ be any function which vanishes on $\partial\Omega$ but does not vanish identically. Then w and $-w$ are distinct classical solutions (and hence viscosity solutions, via Exercise 2.4) of (DP) with $H(x, u, p) = |p|^2 - |Dw|^2$, $g = 0$. We will discover sufficient conditions to guarantee the comparison theorem along the way.

The idea of comparison proofs for viscosity solutions is this: we would like to consider an interior maximum $\hat{x}$ of $u(x) - v(x)$ and use $H(\hat{x}, u(\hat{x}), Du(\hat{x})) \leq 0$, $H(\hat{x}, v(\hat{x}), Dv(\hat{x})) \geq 0$ to conclude that $u(\hat{x}) \leq v(\hat{x})$ or $u \leq v$. A primary difficulty is that u and v need not be differentiable at such a maximum $\hat{x}$. Thus instead one chooses smooth "test functions" $\varphi(x, y)$ for which $u(x) - v(y) - \varphi(x, y)$ has a maximum $(\hat{x}, \hat{y})$. Assuming that $\hat{x}, \hat{y} \in \Omega$, $\hat{x}$ is a maximum of $x \to u(x) - \varphi(x, \hat{y})$ and so, by the definition of subsolution, $H(\hat{x}, u(\hat{x}), D_x\varphi(\hat{x}, \hat{y})) \leq 0$. Similarly, $H(\hat{y}, v(\hat{y}), -D_y\varphi(\hat{x}, \hat{y})) \geq 0$ and then

$$H(\hat{x}, u(\hat{x}), D_x\varphi(\hat{x}, \hat{y})) - H(\hat{y}, v(\hat{y}), -D_y\varphi(\hat{x}, \hat{y})) \leq 0.$$

It remains to conclude that $u \leq v$ by playing with the choice of φ and perhaps making auxiliary estimates.

Pick $\varepsilon > 0$ and small and let us maximize

$$(4.1) \qquad \Phi(x, y) = u(x) - v(y) - \frac{1}{2\varepsilon}|x - y|^2$$

over $\overline{\Omega} \times \overline{\Omega}$. Since Φ is upper semicontinuous a maximum $(\hat{x}_\varepsilon, \hat{y}_\varepsilon)$ exists. The test function $\varphi(x, y) = |x - y|^2/(2\varepsilon)$ is chosen to "penalize" large values of $|x - y|$ when ε is sent to zero. It further has the desirable property that $D_x\varphi = -D_y\varphi$, the utility of which is seen below.

We prepare a useful lemma about penalized maximums of semicontinuous functions for use now and later.

Lemma 4.1. *Suppose $\mathcal{O} \subset \mathbb{R}^N$. Let $w, \Psi : \mathcal{O} \to \mathbb{R}$, $0 \leq \Psi$, and $w, -\Psi$ be upper semicontinous. Let*

$$(4.2) \qquad \mathcal{N} = \{z \in \mathcal{O} : \Psi(z) = 0\} \neq \emptyset,$$

and

$$\sup_{z \in \mathcal{O}}(w(z) - \Psi(z)) < \infty.$$

Let $M_\varepsilon = \sup_{z \in \mathcal{O}}(w(z) - \Psi(z)/\varepsilon)$ for $\varepsilon \leq 1$. If $z_\varepsilon \in \mathcal{O}$ is such that

$$(4.3) \qquad M_\varepsilon - \left(w(z_\varepsilon) - \frac{1}{\varepsilon}\Psi(z_\varepsilon)\right) \to 0$$

then

$$(4.4) \qquad \frac{1}{\varepsilon}\Psi(z_\varepsilon) \to 0.$$

Moreover, and if $\hat{z} \in \mathcal{O}$ is a cluster point of z_ε as $\varepsilon \downarrow 0$, then $\hat{z} \in \mathcal{N}$ and $w(z) \leq w(\hat{z})$ for $z \in \mathcal{N}$.

Proof. M_ε is clearly a decreasing function of $0 < \varepsilon \leq 1$. Since $\sup_\mathcal{N} w \leq M_\varepsilon \leq M_1 < \infty$ where $\mathcal{N}$ is the nonempty set of (4.2), $M_0 = \lim_{\varepsilon \downarrow 0} M_\varepsilon$ exists and is finite. Letting $g(\varepsilon)$ be the left-hand side of (4.3), for $0 < \mu, \varepsilon$ we have

$$M_\mu - \left(\frac{1}{\varepsilon} - \frac{1}{\mu}\right)\Psi(z_\varepsilon) \geq w(z_\varepsilon) - \frac{1}{\mu}\Psi(z_\varepsilon) - \left(\frac{1}{\varepsilon} - \frac{1}{\mu}\right)\Psi(z_\varepsilon) =$$

$$w(z_\varepsilon) - \frac{1}{\varepsilon}\Psi(z_\varepsilon) \geq M_\varepsilon - g(\varepsilon).$$

Taking $\mu = 2\varepsilon$ we conclude from the extreme inequalities that

$$\frac{1}{\varepsilon}\Psi(z_\varepsilon) \leq 2(M_{2\varepsilon} - M_\varepsilon + g(\varepsilon))$$

and the right-hand side tends to zero as $\varepsilon \downarrow 0$.

Assume now that $z_\varepsilon \to \hat{z} \in \mathcal{O}$ along a sequence of ε's tending to zero. Then $0 = \limsup_{\varepsilon \downarrow 0} \Psi(z_\varepsilon) \geq \Psi(\hat{z})$ by lower semicontinuity, and $\hat{z} \in \mathcal{N}$. Moreover, by the upper semicontinuity of w,

$$w(\hat{z}) \geq \lim_{\varepsilon \downarrow 0}\left(w(z_\varepsilon) - \frac{1}{\varepsilon}\Psi(z_\varepsilon)\right) = M_0 \geq \sup_{z \in \mathcal{N}} w(z).$$

$\square$

Since $\overline{\Omega} \times \overline{\Omega}$ is compact, the maximum point $(\hat{x}_\varepsilon, \hat{y}_\varepsilon)$ of Φ of (4.1) has a limit point $\varepsilon \downarrow 0$. It follows from Lemma 4.1 that

(4.5) $$\frac{1}{\varepsilon}|\hat{x}_\varepsilon - \hat{y}_\varepsilon|^2 \to 0$$

and any limit point has the form $(\hat{x}, \hat{x})$. If $\hat{x} \in \partial\Omega$, then $u(\hat{x}) \leq g(\hat{x}) \leq v(\hat{x})$ shows that

$$\limsup_{\varepsilon \downarrow 0} \Phi(\hat{x}_\varepsilon, \hat{y}_\varepsilon) \leq 0.$$

If no such limit $\hat{x} \in \partial\Omega$, then $\hat{x}_\varepsilon, \hat{y}_\varepsilon$ must lie in Ω for small ε. In this case, as explained before, we have

(4.6) $$H\left(\hat{x}_\varepsilon, u(\hat{x}_\varepsilon), \frac{\hat{x}_\varepsilon - \hat{y}_\varepsilon}{\varepsilon}\right) - H\left(\hat{y}_\varepsilon, v(\hat{y}_\varepsilon), \frac{\hat{x}_\varepsilon - \hat{y}_\varepsilon}{\varepsilon}\right) \leq 0.$$

When does this information imply $u \leq v$? For a simple example, let us assume that G has the form $H(x, r, p) = r + G(p) - f(x)$ where f is continuous on $\overline{\Omega}$. Then (4.6) rewrites to

$$u(\hat{x}_\varepsilon) - v(\hat{y}_\varepsilon) \leq f(\hat{x}_\varepsilon) - f(\hat{y}_\varepsilon);$$

in view of (4.5) and the uniform continuity of f the right-hand side tends to zero as $\varepsilon \downarrow 0$ and we conclude again that

$$\limsup_{\varepsilon \downarrow 0} \Phi(\hat{x}_\varepsilon, \hat{y}_\varepsilon) \leq \limsup_{\varepsilon \downarrow 0}(u(\hat{x}_\varepsilon) - v(\hat{y}_\varepsilon)) \leq 0.$$

Since $u(x) - v(x) = \Phi(x,x) \le \Phi(\hat{x}_\varepsilon, \hat{y}_\varepsilon)$, we conclude $u \le v$ in the limit $\varepsilon \downarrow 0$. The case in which H has the form $H(x,r,p) = G(r,p) - f(x)$ and H is strictly increasing in r uniformly in $p \in \mathrm{I\!R}^N$ is essentially the same.

In the above examples, the x dependence is "separated". When it is not, the situation is more subtle and it convenient to use the full force of (4.5).

Exercise 4.2. Establish comparison for (DP) when $H(x,r,p)$ satisfies

$$|H(x,r,p) - H(y,r,p)| \le \omega(|x-y|(1+|p|))$$

for some function satisfying $\omega(0+) = 0$ and H is sufficiently increasing in r. Show that the "sufficiently increasing" (however you formulated it) assumption can be dropped when there is a $c > 0$ such that either u solves $H \le -c$ or v solves $H \ge c$.

It was remarked at the beginning that solutions of (DP) for equations of the form $|Du|^2 = f(x)$ are not necessarily unique. However, they are unique if $f(x) > 0$ in Ω as shown by the next two exercises.

Exercise 4.3.* Let $F(x,r,p)$ be proper, u be a solution of $F \le 0$ (respectively, $F \ge 0$) and consider a change of unknown function according to $u = K(w)$. Here K is continuously differentiable, $K'(r) > 0$ for r in the domain of K and the range of K includes the range of u. Show that w is then a subsolution (respectively, supersolution) of the resulting equation, $G(x, w, Dw) = F(x, K(w), K'(w)Dw) = 0$. Note, however, that G may not be proper. Discuss the second order case.

Exercise 4.4.* Let $f \in C(\overline{\Omega})$, $f(x) > 0$ for $x \in \Omega$. Find a change of unknown in the equation $|Du|^2 - f(x) = 0$ which - with a little massaging - produces a proper equation $H(x, w, Dw) = 0$ for which comparison in the Dirichlet problem holds.

Except for the semicontinuous generality, which plays a small role, comparison results of these forms have been known since [15]. However, the proofs above are certainly clearer than the original ones.

5. Comparison for Hamilton-Jacobi Equations in $\mathrm{I\!R}^N$

The point of this section is to indicate how to handle unbounded domains. The reader may skip ahead now to Section 7 if desired.

We consider the model stationary Hamilton-Jacobi equation on $\mathrm{I\!R}^N$:

$$\text{(SHJE)} \qquad u + H(Du) = f(x) \quad \text{for} \quad x \in \mathrm{I\!R}^N.$$

Means to treat more general equations used in the previous section will also work here, and we focus only the modifications of arguments required by the unbounded domain $\mathrm{I\!R}^N$. Everywhere below, $u, v : \mathrm{I\!R}^N \to \mathrm{I\!R}$, u is a subsolution and v is a supersolution of (SHJE). Moreover, $H, f : \mathrm{I\!R}^N \to \mathrm{I\!R}$, H is continuous and f is uniformly continuous. The goal is again to prove $u \le v$.

We suppose that u, v are bounded on $\mathbb{R}^N$; this is relaxed in the next section. For $0 < \varepsilon, \delta$ define the function

$$\Phi(x,y) = u(x) - v(y) - \frac{1}{2\varepsilon}|x-y|^2 - \frac{\delta}{2}\left(|x|^2 + |y|^2\right)$$

on $\mathbb{R}^N \times \mathbb{R}^N$. The term $\delta(|x|^2 + |y|^2)/2$ is present to guarantee that Φ has a maximum on its unbounded domain and will be removed by sending $\delta \downarrow 0$. Φ is upper semicontinuous and, since $u(x) - v(y)$ is bounded above by assumption, Φ tends to $-\infty$ as $|x|, |y| \to \infty$. Thus Φ has a maximum point $(\hat{x}, \hat{y})$ (it depends on ε, δ; however we no longer indicate this dependence). Proceeding as above we have

(5.1) $\qquad u(\hat{x}) - v(\hat{y}) \le H\left(\dfrac{\hat{x} - \hat{y}}{\varepsilon} - \delta\hat{y}\right) - H\left(\dfrac{\hat{x} - \hat{y}}{\varepsilon} + \delta\hat{x}\right) + f(\hat{x}) - f(\hat{y}).$

From the assumed boundedness of u, v, it follows that

(5.2) $\qquad\qquad\qquad \sup_{x,y}(u(x) - v(y) - \dfrac{1}{2}|x-y|^2) < \infty.$

A slight modification of Lemma 4.1 then yields

(5.3) $\qquad\qquad\qquad \dfrac{|\hat{x} - \hat{y}|^2}{\varepsilon} \le C_{\varepsilon,\delta}$

where

(5.4) $\qquad\qquad\qquad \lim_{\varepsilon \downarrow 0}\lim_{\delta \downarrow 0}\sup C_{\varepsilon,\delta} = 0.$

In addition, for $\varepsilon \le 1$, $u(0) - v(0) \le \Phi(\hat{x}, \hat{y})$ and (5.2) imply

$$\frac{\delta}{2}\left(|\hat{x}|^2 + |\hat{y}|^2\right) \le v(0) - u(0) + u(\hat{x}) - v(\hat{y}) - \frac{1}{2}|\hat{x} - \hat{y}|^2 \le C_1$$

It follows from (5.3) and the above that:

(5.5)
$\qquad$ (i) $\quad |\hat{x} - \hat{y}|/\varepsilon = (|\hat{x} - \hat{y}|^2/\varepsilon)^{1/2}/\sqrt{\varepsilon} \le (C_{\varepsilon,\delta}/\varepsilon)^{1/2},$

$\qquad$ (ii) $\quad |\hat{x} - \hat{y}| \le (C_{\varepsilon,\delta}\varepsilon)^{1/2},$

$\qquad$ (iii) $\quad \delta|\hat{x}| + \delta|\hat{y}| = \sqrt{\delta}((\delta|\hat{x}|^2)^{1/2} + (\delta|\hat{y}|^2)^{1/2}) \le 2\sqrt{\delta}C_1^{1/2}.$

Let ρ_f be the modulus of continuity of f, that is the least nondecreasing function such that

$$|f(x) - f(y)| \le \rho_f(|x-y|);$$

ρ_f is continuous. Likewise, the merely continuous function H is uniformly continuous on compact sets, so there is a least function ρ_H such that

$$|H(p+q) - H(p)| \le \rho_H(R, r) \quad \text{for} \quad |p| \le R, \ |q| \le r.$$

We have as well $\rho_f(0+) = \rho_H(R, 0+) = 0$ for $R > 0$. Returning to (5.1), we may use (i)-(iii) above to conclude (with unseemly precision) that

$$u(\hat{x}) - v(\hat{y}) \leq \rho_H((C_{\varepsilon,\delta}/\varepsilon)^{1/2}, 2\sqrt{\delta}C_1^{1/2}) + \rho_f((\varepsilon C_{\varepsilon,\delta})^{1/2}).$$

Thus

$$\limsup_{\delta \downarrow 0}(u(\hat{x}) - v(\hat{y})) \leq \rho_f((\varepsilon \limsup_{\delta \downarrow 0} C_{\varepsilon,\delta})^{1/2}).$$

Therefore

(5.6)
$$u(x) - v(y) - \frac{|x - y|^2}{2\varepsilon} = \lim_{\delta \downarrow 0} \Phi(x, y) \leq \lim_{\delta \downarrow 0} \Phi(\hat{x}, \hat{y}) \leq$$
$$\limsup_{\delta \downarrow 0}(u(\hat{x}) - v(\hat{y})) \leq \rho_f((\varepsilon \limsup_{\delta \downarrow 0} C_{\varepsilon,\delta})^{1/2}).$$

Putting $x = y$ and letting $\varepsilon \to 0$, we conclude that $u(x) \leq v(x)$ for all x.

Another use of estimates like (5.6) is this: if u itself is a bounded solution of (SHJE), we may take $u = v$ in (5.6) to conclude that

$$u(x) - u(y) \leq \inf_{0 < \varepsilon \leq 1} \left(\frac{|x - y|^2}{2\varepsilon} + \rho_f((\varepsilon \limsup_{\delta \downarrow 0} C_{\varepsilon,\delta})^{1/2}) \right),$$

which provides a modulus of continuity for u determined by f. This method generalizes to allow $H(x, p)$ to depend on x as well; in the current case there is a simpler way to obtain a modulus for a solution u.

Exercise 5.1. Generalize the above to show that if $f, g : \mathbb{R}^N \to \mathbb{R}$ are uniformly continuous and $u, -v$ are upper semicontinuous and bounded, u solves $u + H(Du) \leq f$ and v solves $v + H(Dv) \geq g$, then $u(x) - v(x) \leq \sup_{z \in \mathbb{R}^N}(f(z) - g(z))$. If u is a solution of $u + H(Du) = f$ and $z \in \mathbb{R}^N$, then $v(x) = u(x + z)$ is a solution of $v + H(Dv) = g$ with $g(x) = f(x + z)$. Consequently, $u(x) - u(y) \leq \rho_f(|x - y|)$.

6. Hamilton-Jacobi Equations in $\mathbb{R}^N$: Unbounded Solutions

We treat some technical difficulties "at ∞" caused by allowing unbounded u, v in the problem treated in the Section 5. The devices used adapt to the second order case. The reader may skip ahead to Section 7 at this time without disrupting the flow.

This time, we allow a linear growth at infinity, that is

(LG) $u(x) - v(y) \leq L(|x| + |y| + 1)$ for $x, y \in \mathbb{R}^N$

for some constant L (note that this amounts to bounding u, $-v$ separately from above). We show that then $u \leq v$. A review of the proof of Section 5 shows that bounds on u, v were not in fact needed except in so far as they guaranteed (5.2) (which itself guarantees the existence of the maxima used). If we verify (5.2) (or an close substitute), we will be finished.

Remark 6.1. The linear growth is "critical" in the class of powers of $|x|$. The equation $u - |Du|^\gamma = 0$ with $\gamma > 1$ has the two distinct solutions $u \equiv 0$ and $u = ((\gamma - 1)/\gamma)^{\gamma/(\gamma-1)}|x|^{\gamma/(\gamma-1)}$. Choosing γ large, the growth is as close to linear as we please.

The next exercise is used immediately.

Exercise 6.1.* Let $f : \mathbb{R}^N \to \mathbb{R}$ be uniformly continuous and

$$\rho_f(r) = \sup \{ |f(x) - f(y)| : x, y \in \mathbb{R}^N, \quad |x - y| \le r \}$$

be its modulus of continuity. Show that $\rho_f(r + s) \le \rho_f(r) + \rho_f(s)$ (ρ_f is subadditive) and $\rho_f(r) \le \rho_f(\delta) + (\rho_f(\delta)/\delta)r$ for $0 \le r, s, 0 < \delta$.

Since f is uniformly continuous on $\mathbb{R}^N$, by the above exercise it admits the estimate

(6.1) $$|f(x) - f(y)| - K|x - y| \le K$$

where $K = \rho_f(1)$.

We claim that

$$\sup_{\mathbb{R}^N \times \mathbb{R}^N} (u(x) - v(y) - K|x - y|) < \infty$$

(and then $u(x) - v(y) - K|x - y|^2$ is bounded as well).

In view of (LG), the upper semicontinuous function

$$\Phi(x, y) = u(x) - v(y) - K(1 + |x - y|^2)^{1/2} - \frac{\delta}{2}(|x|^2 + |y|^2)$$

attains its maximum at some point $(\hat{x}, \hat{y})$. (LG) and $u(0) - v(0) \le \Phi(\hat{x}, \hat{y})$ imply that

$$\frac{\delta}{2}(|\hat{x}|^2 + |\hat{y}|^2) \le v(0) - u(0) + u(\hat{x}) - v(\hat{y}) \le C + L(|\hat{x}| + |\hat{y}|)$$

which implies

(6.2) $$\delta(|\hat{x}| + |\hat{y}|) \le C$$

where the constants C are various. Using the equations

(6.3) $$u(\hat{x}) - v(\hat{y}) \le H\left(K \frac{\hat{x} - \hat{y}}{(1 + |\hat{x} - \hat{y}|^2)^{1/2}} - \delta\hat{y} \right) - $$
$$H\left(K \frac{\hat{x} - \hat{y}}{(1 + |\hat{x} - \hat{y}|^2)^{1/2}} + \delta\hat{x} \right) + f(\hat{x}) - f(\hat{y}).$$

The key thing here is that

$$\frac{|\hat{x} - \hat{y}|}{(1 + |\hat{x} - \hat{y}|^2)^{1/2}} \le 1$$

while we also have (6.2). Thus the arguments of H above are bounded independently of δ. Invoking (6.1) as well, we conclude that

$$u(\hat{x}) - v(\hat{y}) \leq C + K|\hat{x} - \hat{y}|.$$

But then

$$u(\hat{x}) - v(\hat{y}) - K(1 + |\hat{x} - \hat{y}|^2)^{1/2} \leq u(\hat{x}) - v(\hat{y}) - K|\hat{x} - \hat{y}| \leq C$$

and finally

$$\Phi(x, y) \leq \Phi(\hat{x}, \hat{y}) \leq u(\hat{x}) - v(\hat{y}) - K(1 + |\hat{x} - \hat{y}|^2)^{1/2} \leq C.$$

Passing to the limit as $\delta \downarrow 0$, we conclude that $u(x) - v(y) - K(1 + |x - y|^2)^{1/2}$ is bounded.

Exercise 6.2. Show that linearly bounded solutions of (SHJE) are uniformly continuous.

See the references given in [12, Section 5D].

7. Definitions Revisited – Semijets

In this section the notion of Definition 2.1 will be recast for later convenience. The definition itself involves extrema of differences $u - \varphi$ and then evaluation of the equation at data from the second order Taylor expansion of φ at these extrema. It is only the information from the expansion of φ which matters, and we now emphasize this.

If $\varphi \in C^1(\Omega)$ and $u - \varphi$ has a local maximum relative to Ω at $\hat{x} \in \Omega$, then

(7.1)
$$\begin{aligned} u(x) \leq &u(\hat{x}) + \varphi(x) - \varphi(\hat{x}) = \\ &u(\hat{x}) + \langle p, x - \hat{x} \rangle + o(|x - \hat{x}|) \quad \text{as} \quad \Omega \ni x \to \hat{x} \end{aligned}$$

and if $\varphi \in C^2(\Omega)$, then

(7.2)
$$\begin{aligned} u(x) \leq &u(\hat{x}) + \varphi(x) - \varphi(\hat{x}) \\ = &u(\hat{x}) + \langle p, x - \hat{x} \rangle + \\ &\frac{1}{2} \langle X(x - \hat{x}), x - \hat{x} \rangle + o(|x - \hat{x}|^2) \quad \text{as} \quad \Omega \ni x \to \hat{x} \end{aligned}$$

where

(7.3)
$$p = D\varphi(\hat{x}) \quad \text{and} \quad X = D^2\varphi(\hat{x}).$$

Conversely, if $p \in \mathbb{R}^N$ and (7.1) holds, then there exists $\varphi \in C^1(\Omega)$ such that $u - \varphi$ has a *strict* maximum at $\hat{x}$ and $D\varphi(\hat{x}) = p$. Here a maximum $\hat{x}$ of $u - \varphi$ is strict if there is a nondecreasing function $h : (0, \infty) \to (0, \infty)$ and $r_0 > 0$ such that

(7.4)
$$u(x) - \varphi(x) \leq u(\hat{x}) - \varphi(\hat{x}) - h(r) \quad \text{for} \quad r \leq |x - \hat{x}| \leq r_0.$$

Let us call h the "strictness" in this situation. The proof goes like this: assume (7.1) and set

$$g(r) = \sup\left\{(u(x) - u(\hat{x}) - \langle p, x - \hat{x}\rangle)^+ : x \in \Omega, \ |x - \hat{x}| \le r\right\}.$$

By (7.1), $g(r) = o(r)$ as $r \downarrow 0$ and is nondecreasing. Choose a continuous nondecreasing majorant $\tilde{g}$ with the same properties; that is, we want $g(r) \le \tilde{g}(r)$, $\tilde{g}(r) = o(r)$ and $\tilde{g}$ is nondecreasing. Now put

$$G(r) = \frac{1}{r}\int_r^{2r} \tilde{g}(s)\,ds$$

and check that

$$\varphi(x) = G(|x - \hat{x}|) + \langle p, x - \hat{x}\rangle + |x - \hat{x}|^4$$

has the desired properties (with $h(r) = r^4$ as the strictness).

Exercise 7.1.* Formulate and prove the corresponding statement for (7.2).

We focus on the second case. The quadratic appearing on the right hand side of (7.2) is defined by the "jet" $(u(\hat{x}), p, X)$ and we write

$$(u(\hat{x}), p, X) \in J^{2,+}u(\hat{x})$$

when (7.2) holds. The quantity $u(\hat{x})$ on the left appears to be redundant, but is incorporated for technical reasons. For any function $u : \Omega \to \mathbb{R}$, $J^{2,+}u$ maps Ω into the set of subsets of $\Omega \times \mathbb{R}^N \times S(N)$ (and the empty set may well be a value).

Exercise 7.2.* Let $u(x) = |x|$ for $x \in \mathbb{R}$. Compute $J^{2,+}u(0)$ and $J^{2,-}u(0)$.

Whenever $J^{2,+}u(x)$ is not empty, it is infinite, since whenever (7.2) holds for p, X it also holds for p, Y whenever $X \le Y$. Likewise, we define $(u(\hat{x}), p, X) \in J^{2,-}u(\hat{x})$ to mean that (7.2) holds with the inequality reversed.

Exercise 7.3. Define the first order analogues $J^{1,+}$, $J^{1,-}$ of the second order semijets. Observe that $(u(x), p, X) \in J^{2,+}u(x)$ implies $(u(x), p) \in J^{1,+}u(x)$, but show by example that the converse fails: $J^{2,+}u(x)$ may be empty while $J^{1,+}u(x)$ is nonempty.

Exercise 7.4. Let Ω be open and $u : \Omega \to \mathbb{R}$ be upper semicontinuous. Show that $\{x \in \Omega : J^{2,+}u(x) \ne \emptyset\}$ is dense in Ω. Show that if $J^{1,+}u(\hat{x}) \cap J^{1,-}u(\hat{x})$ is nonempty, then there is a $p \in \mathbb{R}^N$ such that

$$u(x) = u(\hat{x}) + \langle p, x - \hat{x}\rangle + o(|x - \hat{x}|) \quad \text{for} \quad x \in \Omega;$$

in this case we say that u is differentiable at $\hat{x}$ and $p = Du(\hat{x})$. Conclude that there are continuous functions $u : (0, 1) \to \mathbb{R}$ such that

$$J^{2,+}u(x) \cap J^{2,-}u(x) = \emptyset$$

for every $x \in (0, 1)$.

Exercise 7.5. Let $u : \mathbb{R}^N \to \mathbb{R}$ and $|u(x) - u(y)| \le L|x - y|$ for $x, y \in \mathbb{R}^N$ (i.e., u is Lipschitz continuous with constant L). Show that if $(u(\hat{x}), p) \in J^{1,+}u(\hat{x})$, then $|p| \leq L$. Conversely, if $(u(x), p, X) \in J^{2,+}u(x)$ implies that $|p| \leq L$, show that u is Lipschitz with constant L.

According to Exercise 7.1, an upper semicontinuous function $u : \Omega \to \mathbb{R}$ is a subsolution of a proper equation $G = 0$ if and only if $G(x, u(x), p, X) \leq 0$ for every $x \in \Omega$ and $(u(x), p, X) \in J^{2,+}u(x)$. If G is continuous (or even lower semicontinuous), the relation

$$G(x, u(x), p, X) \leq 0$$

persists under taking limits, and this leads us to define the *closure* $\overline{J}^{2,+}u$ of $J^{2,+}u$. This goes as follows: $(r, p, X) \in \overline{J}^{2,+}u(x)$ if there exists

$$\Omega \ni x_n \to x \quad \text{and} \quad (u(x_n), p_n, X_n) \in J^{2,+}u(x_n)$$

such that

$$(u(x_n), p_n, X_n) \to (r, p, X).$$

We have then that an upper semicontinuous function $u : \Omega \to \mathbb{R}$ is a subsolution of a proper equation $G = 0$ if and only if $G(x, r, p, X) \leq 0$ for every $x \in \Omega$ and $(r, p, X) \in \overline{J}^{2,+}u(x)$. Note that upper semicontinuity of u implies that $r \leq u(x)$, so perhaps $G(x, u(x), p, X) > 0$.

One defines $\overline{J}^{2,-}u$ similarly and then $u : \Omega \to \mathbb{R}$ is a subsolution of a proper equation $G = 0$ if and only if $G(x, r, p, X) \geq 0$ for every $x \in \Omega$ and $(r, p, X) \in \overline{J}^{2,-}u(x)$.

Remark 7.1. The notation in use here differs from that in [12] in that the values of $J^{2,+}$ are taken here to include "$u(x)$" and this was not so in [12]. Nobody much likes this "jet" business and perhaps we should refer to "second order superdifferentials" or some such. There seems to be a law of conservation of pedantic excess in attempts to resolve this issue. It is a bookkeeping question, and when we get to the Theorem on Sums, one needs to do the bookkeeping somehow.

The construction of the test functions used to show that the "jet" formulations are equivalent to the "maximum of $u - \varphi$" formulations appears in Evans [18].

8. Stability of the Notions

We begin by considering two related issues. First, if $\mathcal{F}$ is a collection of solutions of $F \leq 0$, then $(\sup_{u \in \mathcal{F}} u)^*$ (recall (2.1)) is another subsolution. Next, if u_n is a solution of $F_n \leq 0$ for $n = 1, 2, \ldots$, $F_n = 0$, and $u_n \to u$, $F_n \to F$ in a suitable sense, then u is a subsolution of $F = 0$. These facts are linked in that they both rely on the following result.

The reader will notice in the statement below that there is a set $\mathcal{O} \subset \mathbb{R}^N$ which is "locally compact"; for example, both open sets and closed subsets of $\mathbb{R}^N$ contain a compact (relative) neighborhood of each point and so are locally compact, as are various other sets. In fact, the above considerations easily generalize to allow locally compact sets Ω in the definition of the jets, etc. If Ω is locally compact, we take $\varphi \in C^2(\Omega)$ to mean it is the restriction to Ω of a twice continuously differentiable function defined on a neighborhood (in $\mathbb{R}^N$) of Ω. The relation (7.2) needs no modification, for we have already appended "as $\Omega \ni x \to \hat{x}$" to emphasize that Ω may not be open, etc. The jet "operators" $J^{2,+}$, $J^{2,-}$ are written $J_\Omega^{2,+}$, $J_\Omega^{2,-}$ and their closures $\overline{J}_\Omega^{2,+}$ and $\overline{J}_\Omega^{2,-}$ when Ω is not necessarily open. At the moment, we pose this technical generality "because we can" and it doesn't affect the presentation; however, it is essential in various ways in other parts of the theory.

Proposition 8.1. *Let $\mathcal{O} \subset \mathbb{R}^N$ be locally compact, $U : \mathcal{O} \to \mathbb{R}$ be upper semicontinuous, $z \in \mathcal{O}$ and $(U(z), p, X) \in J_\mathcal{O}^{2,+}U(z)$. Suppose also that u_n is a sequence of upper- semicontinuous functions on $\mathcal{O}$ such that*

(8.1)
$$
\begin{aligned}
&\text{(i)} \quad \text{there exists } x_n \in \mathcal{O} \text{ such that } (x_n, u_n(x_n)) \to (z, U(z)) \text{ and} \\
&\text{(ii)} \quad \text{if } z_n \in \mathcal{O} \text{ and } z_n \to x \in \mathcal{O}, \text{ then } \limsup_{n \to \infty} u_n(z_n) \le U(x).
\end{aligned}
$$

Then

(8.2)
$$
\begin{aligned}
&\text{there exists } \hat{x}_n \in \mathcal{O}, \ (u_n(\hat{x}_n), p_n, X_n) \in J_\mathcal{O}^{2,+}u_n(\hat{x}_n) \\
&\text{such that } (\hat{x}_n, u_n(\hat{x}_n), p_n, X_n) \to (z, U(z), p, X).
\end{aligned}
$$

Before proving this result, we use it.

Proposition 8.2. *Let $\mathcal{F}$ be a nonempty collection of solutions of $F \le 0$ on Ω where F is proper and continuous. If $U(x) = \sup_{u \in \mathcal{F}} u(x)$ and U^* is finite on Ω, then U^* is a solution of $F \le 0$ on Ω.*

Proof. Suppose $z \in \Omega$ and $(U^*(z), p, X) \in J^{2,+}U^*(z)$. It is clear that there exists a sequence $u_n \in \mathcal{F}$ and $x_n \in \Omega$ such that $(x_n, u_n(x_n)) \to (z, U^*(z))$ and that then the assumption of Proposition 8.1 are satisfied with U replaced by U^*. Let $(\hat{x}_n, u_n(\hat{x}_n), p_n, X_n) \to (z, U^*(z), p, X)$ be as in Proposition 8.1; by assumption $F(\hat{x}_n, u_n(\hat{x}_n), p_n, X_n) \le 0$ and we conclude $F(z, U^*(z), p, X) \le 0$ in the limit. $\square$

We use Proposition 8.1 again.

For an arbitrary sequence of functions u_n on $\mathcal{O}$ we can form the smallest function U such that if $\mathcal{O} \ni x_n \to x \in \mathcal{O}$, then $\limsup_{n \to \infty} u_n(x_n) \le U(x)$. U is given by

$$
\begin{aligned}
U(x) &= \limsup_{\substack{n \to \infty \\ \mathcal{O} \ni y \to x}} u_n(y) \\
&= \lim_{m \to \infty} \sup \left\{ u_n(y) : n \ge m, \ y \in \mathcal{O}, \ |y - x| \le \frac{1}{m} \right\};
\end{aligned}
$$

we write $U = \limsup^*_{n\to\infty} u_n$. In the opposite sense, we define

$$\liminf_{n\to\infty}{}_* \, u_n = -\limsup^*_{n\to\infty}(-u_n).$$

Note that for any $x \in \mathcal{O}$ there exists sequences $n_j \to \infty$ and $\mathcal{O} \ni x_{n_j} \to x$ such that $u_{n_j}(x_{n_j}) \to U(x)$.

Exercise 8.1. Show $\limsup^*_{n\to\infty} u_n$ is upper semicontinuous and that if $u_n \equiv u$ for all n, then $\limsup^*_{n\to\infty} u_n = u^*$, the upper semicontinuous envelope of u.

The following statement is remarkable, in that it produces a subsolution of a limit problem from an *arbitrary* sequence of subsolutions of approximate problems. No control of derivatives of any kind is assumed.

Theorem 8.3. *For $n = 1, 2, \ldots$, let u_n be a subsolution of a proper equation $F_n = 0$ on Ω. Let $U = \limsup^*_{n\to\infty} u_n$ and F be proper and satisfy*

$$F \leq \liminf_{n\to\infty}{}_* \, F_n.$$

If U is finite, then it is a solution of $F \leq 0$. In particular, if $u_n \to U$, $F_n \to F$ locally uniformly, then U is a solution of $F \leq 0$.

Proof. According to the above discussion, if $(U(z), p, X) \in J^{2,+}U(z)$, then there is a subsequence of the u_n (which we again call u_n) such that the hypotheses of Proposition 8.1 are satisfied. If $(\hat{x}_n, u(x_n), p_n, X_n)$ is as in the proposition, our assumptions imply

$$F(z, U(z), p, X) \leq \liminf_{n\to\infty} F_n(\hat{x}_n, u_n(x_n), p_n, X_n) \leq 0.$$

$\square$

Remark 8.4. Recall Exercise 2.3, whereby results concerning subsolutions automatically imply the corresponding result for supersolutions.

Exercise 8.2. The equation

$$-\Delta_p u = -\sum_{i=1}^{N}(|Du|^{p-2}u_{x_i})_{x_i} = 0$$

is called the "p-Laplace" equation; p is a number and we focus on large p. Carrying out the differentiations, show that this is a degenerate elliptic equation (and then proper as it does not have a "u" dependence). Suppose u_p is a solution of this equation for large p and that $u_p \to u$ uniformly on Ω as $p \to \infty$. Show that u is then a solution of the "infinity Laplace" equation,

$$-\Delta_\infty u = -\sum_{i,j=1}^{N} u_{x_i} u_{x_j} u_{x_i, x_j} = 0.$$

Exercise 8.3. Suppose that $\{u_n\}$ are uniformly bounded and u_n solves

$$u_n + H(Du_n) - \frac{1}{n}\Delta u_n = f(x)$$

on $\mathbb{R}^N$ where f is uniformly continuous. Prove via comparison for $u + H(Du) = f$ that then

$$\limsup_{n \to \infty}{}^* u_n \leq \liminf_{n \to \infty}{}_* u_n.$$

Show this implies that u_n converges locally uniformly to a limit u, and u is the unique bounded solution of $u + H(Du) = f$.

Proof of Proposition 8.1. Without loss of generality we put $z = 0$. By the assumptions and Exercise 7.1, there is an $r > 0$ such that $N_r = \{x \in \mathcal{O} : |x| \leq r\}$ is compact and a twice continuously differentiable function φ defined in a neighborhood of $\mathcal{O}$ such that

$$(8.3) \qquad U(x) - \varphi(x) \leq U(0) - \varphi(0) \quad \text{for} \quad x \in N_r,$$

the maximum 0 is both strict and the only maximum of $U - \varphi$ in N_r, $p = D\varphi(0)$, and $X = D^2\varphi(0)$. By the assumption (8.1) (i), there exists $x_n \in \mathcal{O}$ such that $(x_n, u_n(x_n)) \to (0, U(0))$. Let $\hat{x}_n \in N_r$ be a maximum point of $u_n(x) - \varphi(x)$ over N_r so that

$$(8.4) \qquad u_n(x) - \varphi(x) \leq u_n(\hat{x}_n) - \varphi(\hat{x}_n) \text{ for } x \in N_r.$$

Suppose that (passing to a subsequence if necessary) $\hat{x}_n \to y$ as $n \to \infty$. Putting $x = x_n$ in (8.4) and taking the limit inferior as $n \to \infty$, we find

$$U(0) - \varphi(0) \leq \liminf_{n \to \infty} u_n(\hat{x}_n) - \varphi(y);$$

on the other hand, by (8.1) (ii) $\liminf u_n(\hat{x}_n) \leq U(y)$. Thus $U(0) - \varphi(0) \leq U(y) - \varphi(y)$ and we conclude that $y = 0$ (because 0 is the only maximum) and that the inequality $\liminf u_n(\hat{x}_n) \leq U(0)$ cannot be strict – that is, $(\hat{x}_n, u_n(\hat{x}_n)) \to (0, U(0))$. Since this holds no matter the subsequence, it holds without passing to a subsequence. Finally

$$J^{2,+}u_n(\hat{x}_n) \ni (u_n(\hat{x}_n), D\varphi(\hat{x}_n), D^2\varphi(\hat{x}_n)) \to (U(0), p, X)$$

and we are done. $\qquad\qquad\qquad\qquad\qquad\qquad\qquad\qquad\qquad\qquad\qquad$ $\square$

Remark 8.5. Proposition 8.1 could be restated in the form it is proved; a strict maximum of $U - \varphi$ perturbs to maxima of $u_n - \varphi$ which converge, etc. Which form you prefer is a matter of taste; this author "thinks" in terms of the data at a point while eliminating "jets" in the statement does make it less unattractive.

Proposition 8.2 is due to Ishii [24]. Theorem 8.3 is of great utility in applications and is due to Barles and Perthame [4], [5]. See the comments of [12, Section 6] and Barles [3], for further orientation. In the case of uniform convergence, Evans [18], [19] already employed the essential idea. Sophisticated limit questions are discussed in Souganidis [35] and many references are given. Barles and Souganidis [6] is typical significant contribution reflecting the utility of the ability to take limits easily. Exercise 8.3 reflects the origin of the term "viscosity"; if one can solve regularized equations by adding "artificial viscosity", then passage to the limit is an easy matter. Note that with the technology of this section, one does not even need to estimate the modulus of continuity of the u_n. Exercise 8.2 is light-hearted, and is from Jensen [29], which is quite fascinating. At the moment, a pure viscosity solution proof of the uniqueness theorem in [29] (comparison for the Dirichlet problem for $-\Delta_\infty u = 0$) is not known.

9. Existence Via Perron's Method

We establish existence results for Dirichlet problems via Perron's method. Below (DP) means:

(DP) $$F(x, u, Du, D^2u) = 0 \quad \text{in} \quad \Omega, \quad u = g \quad \text{on} \quad \partial\Omega.$$

Subsolutions, etc., are defined exactly as in Section 4.

F is to be proper and continuous, while g is continuous. We call the following implementation of "Perron's Method" Ishii's Theorem, as it is a good example of the method he introduced into this subject. At this stage, we have not verified the hypotheses of the theorem for second order equations, but will take this up later. The assumptions have been verified in first order cases in Section 4.

Theorem (Ishii). *Let comparison hold for (DP); i.e., if w is a subsolution of (DP) and v is a supersolution of (DP), then $w \leq v$. Suppose also that there is a subsolution $\underline{u}$ and a supersolution $\overline{u}$ of (DP) which satisfy the boundary condition $\underline{u}_*(x) = \overline{u}^*(x) = g(x)$ for $x \in \partial\Omega$. Then*

(9.1) $$W(x) = \sup\{w(x) : \underline{u} \leq w \leq \overline{u} \text{ and } w \text{ is a subsolution of (DP)}\}$$

is a solution of (DP).

The first step in the proof of Ishii's Theorem was given in Proposition 8.2. The second is a simple construction which we now describe. Roughly speaking, it says that if a subsolution of (DP) is not a solution, then it is not a maximal subsolution. Of course, if comparison holds for (DP) and it has a solution, then the solution is the largest subsolution. We have to take care of semicontinuity considerations. Suppose that Ω is open, u is a solution of $F \leq 0$ and that u_* is not a solution of $F \geq 0$; in particular, assume $0 \in \Omega$ and we have

(9.2) $$F(0, u_*(0), D\varphi(0), D^2\varphi(0)) < 0$$

for some $\varphi \in C^2$ such that for some $r_0 > 0$

$$u_*(x) - \varphi(x) > u_*(0) - \varphi(0) + h(r) \quad \text{for} \quad r \leq |x| \leq r_0$$

where $h > 0$ is the strictness of the minimum. Adjusting φ by a constant, we assume $u_*(0) = \varphi(0)$. Then, by continuity,

$$F(x, \varphi(x) + \delta, D\varphi(x), D^2\varphi(x)) < 0$$

for $0 \leq \delta < \delta_0$ and $|x| \leq r_0$ provided δ_0 and r_0 are sufficiently small; that is, $u_\delta = \varphi(x) + \delta$ is a classical solution of $F < 0$ in $|x| \leq r_0$. Since

$$u(x) \geq u_*(x) \geq \varphi(x) + h(r) \quad \text{for} \quad r \leq |x| \leq r_0,$$

if we choose $\delta < (1/2)h(r_0/2)$, then $u(x) > u_\delta(x)$ for $r_0/2 \leq |x| \leq r_0$ and then, by Proposition 8.2, the function

$$U(x) = \begin{cases} \max\{u(x), u_\delta(x)\} \text{ if } |x| < r_0, \\ \qquad\qquad u(x) \text{ otherwise} \end{cases}$$

is a solution of $F \leq 0$ in Ω. The last observation is that in every neighborhood of 0 there are points such that $U(x) > u(x)$; indeed, by definition, there is a sequence $(x_n, u(x_n))$ convergent to $(0, u_*(0))$ and then

$$\lim_{n \to \infty} (U(x_n) - u(x_n)) = u_\delta(0) - u_*(0) = u_*(0) + \delta - u_*(0) > 0.$$

We summarize what this "bump" construction provides in the following lemma, the proof of which consists only of choosing r_0 sufficiently small.

Lemma 9.1. *Let Ω be open, and u be solution of $F \leq 0$ in Ω. If u_* fails to be a supersolution at some point $\hat{x}$, i.e., there exists $(u_*(\hat{x}), p, X) \in J_\Omega^{2,-} u_*(\hat{x})$ for which $F(\hat{x}, u_*(\hat{x}), p, X) < 0$, then for any small $\kappa > 0$ there is a subsolution U_κ of $F \leq 0$ in Ω satisfying*

(9.3)
$$\begin{cases} U_\kappa(x) \geq u(x) \text{ and } \sup_\Omega(U_\kappa - u) > 0, \\ U_\kappa(x) = u(x) \text{ for } x \in \Omega \text{ and } |x - \hat{x}| \geq \kappa. \end{cases}$$

Proof of Ishii's Theorem. With the notation of the theorem observe that $\underline{u}_* \leq W_* \leq W \leq W^* \leq \overline{u}^*$ and, in particular, $W_* = W = W^* = 0$ on $\partial\Omega$. By Proposition 8.2 W^* is a subsolution of (DP) and hence, by comparison, $W^* \leq \overline{u}$. It then follows from the definition of W that $W = W^*$ (so W is a subsolution). If W_* fails to be a supersolution at some point $\hat{x} \in \Omega$, let W_κ be provided by Lemma 9.1. Clearly $\underline{u} \leq W_\kappa$ and $W_\kappa = 0$ on $\partial\Omega$ for sufficiently small κ. By comparison, $W_\kappa \leq \overline{u}$ and since W is the maximal subsolution between $\underline{u}$ and $\overline{u}$, we arrive at the contradiction $W_\kappa \leq W$. Hence W_* is a supersolution of (DP) and then, by comparison for (DP), $W^* = W \leq W_*$, showing that W is continuous and is a solution. $\qquad\square$

Ishii's Theorem leaves open the question of when a subsolution $\underline{u}$ and a supersolution $\overline{u}$ of (DP) which agree with g on $\partial\Omega$ can be found. Some general discussion can be found in [12, Section 4]. Here we only discuss simple illustrative cases to show the power.

First, according to Exercise 4.4, comparison holds for the equation $|Du|^2 - f(x) = 0$ if $f > 0$ on Ω. Put $g = 0$. Assuming that $f > 0$, $H(x, 0) - f(x) \leq 0$, and we have a subsolution, namely $u \equiv 0$. To find a supersolution, we rely on Exercise 2.4. Taking $u = M \text{distance}(x, \partial\Omega)$, $M|Du| = M$ and if $M \geq \sup_\Omega f$, u is a supersolution. We conclude that a unique solution exists. No requirements had to be laid on $\partial\Omega$.

Next we take up the construction of a supersolution for a general class of uniformly elliptic operators. The construction is standard (see e.g. Gilbarg and Trudinger [23]); but the presentation is made consistent with the theme here. The implications, via Ishii's Theorem and comparison results to follow, is a quite general existence and uniqueness theorem. At this juncture, we remark that the applications of the theory to degenerate equations is more significant, but we leave it to the reader to visit [12] or other works in this regard. Nonetheless, even in the uniformly elliptic case, we are outside the realm of classical solutions, regularity is not known for general uniformly elliptic equations.

Define the "trace" norm on $\mathcal{S}(N)$:

$$(9.4) \qquad \|X\| = \sum_{\mu \in \text{eig}(X)} |\mu|$$

where eig(X) is the set of eigenvalues of X counted according to their multiplicity.

Exercise 9.1.[*] Verify that the trace norm is indeed a norm on X.

Recall that F is called *uniformly elliptic* if there exists constants $0 < \lambda \leq \Lambda$ such that

$$(9.5) \qquad \begin{aligned} F(x, r, p, X + Z) &\leq F(x, r, p, X) - \lambda \text{Trace}(Z) \quad \text{and} \\ |F(x, r, p, X) - F(x, r, p, Y)| &\leq \Lambda \|X - Y\| \end{aligned}$$

for $X, Y, Z \in \mathcal{S}(N)$, $Z \geq 0$. For example, $F = -\text{Trace}(X)$ satisfies (9.5) with $\lambda = \Lambda = 1$. The two inequalities in (9.5) say first that $F(x, r, p, X)$ is decreasing with respect to X at an at least linear rate and then that it is Lipschitz continuous in X as well. A bit less intuitive is the condition

$$(9.6) \qquad \begin{aligned} F(x, r, p, X) - \Lambda \text{Trace}(Z) &\leq F(x, r, p, X + Z) \leq \\ F(x, r, p, X) &- \lambda \text{Trace}(Z) \quad \text{for} \quad Z \geq 0, \end{aligned}$$

but it is easy to see the equivalence or (9.5) and (9.6).

We are assuming that F is proper. Rewriting $F = 0$ as

$$F(x, u, Du, D^2u) - F(x, 0, 0, 0) = F(x, 0, 0, 0),$$

we may as well consider the equation $F = f$ and assume that

$$(9.7) \qquad F(x, 0, 0, 0) \equiv 0.$$

Finally, we add a condition of Lipschitz continuity with respect to the gradient:

$$(9.8) \qquad |F(x, u, p, X) - F(x, u, q, X)| \leq \gamma |p - q|$$

for some constant γ.

Exercise 9.2.[*] Show that for $b \in \mathbb{R}^N$, $A \in \mathcal{S}(N)$, $c \in \mathbb{R}$,

$$F(x, r, p, X) = -\text{Trace}\,(AX) + \langle b, p \rangle + cr$$

is proper and satisfies (9.6) if and only if eig(A) $\subset [\lambda, \Lambda]$, $|b| \leq \gamma$ and $c \geq 0$.

The goal is to construct supersolutions of (DP) with $g = 0$ for the equation $F = f$ where $f \in C(\overline{\Omega})$. Concerning the region Ω, it is assumed that there is an r_0 such that every point $x_b \in \partial \Omega$ is on the boundary of a ball of radius r_0 which does not otherwise meet $\overline{\Omega}$. For each x_b let z_b be such that $|z_b - x_b| = r_0 < |x - z_b|$ for $x \in \overline{\Omega}$, $x \neq x_b$. We seek a supersolution of $F = f$ in the form $U(x) = G(r)$ where $r = |x - z_b|$ which will satisfy $U(x_b) = 0$ and $U \geq 0$ in $\overline{\Omega}$.

A computation shows

$$
\begin{aligned}
(G(r)_{x_i})_{x_j} &= \left(G'(r) \frac{(x - z_b)_i}{r} \right)_{x_j} \\
&= \left(G''(r) - \frac{1}{r} G'(r) \right) \frac{(x - z_b)_i (x - z_b)_j}{r^2} + \frac{G'(r)}{r} \delta_{i,j}
\end{aligned}
$$

Thus any vector orthogonal to $x - z_b$ is an eigenvector of $D^2 G$ with $G'(r)/r$ as the eigenvalue, while $x - z_b$ is itself an eigenvector with eigenvalue $G''(r)$. Letting P be the orthogonal projection along $x - z_b$ we have

$$(9.9) \qquad D^2 G(|x - z_b|) = G''(r) P + \frac{1}{r} G'(r)(I - P).$$

Taking

$$(9.10) \qquad G(r) = \frac{1}{r_0^\sigma} - \frac{1}{r^\sigma}$$

for $\sigma > 0$, G is nonnegative, concave and increasing on $r_0 \leq r$. The decomposition (9.9) above then represents $D^2 G$ as an orthogonal sum of a positive matrix and a negative matrix. Using this above in conjunction with F being proper, (9.6), (9.7), and (9.8) we find

$$
\begin{aligned}
F(x, G, DG, D^2 G) &\geq F(x, 0, DG, D^2 G) \\
&\geq F(x, 0, 0, , D^2 G) - \gamma |DG| \\
&\geq F(x, 0, 0, G''(r) P) - \frac{\Lambda}{r} G'(r) \text{Trace}\,(I - P) - \gamma G'(r) \\
&\geq F(x, 0, 0, 0) - \lambda G''(r) - \frac{\Lambda(N - 1)}{r} G'(r) - \gamma G'(r) \\
&= -\lambda G''(r) - \frac{\Lambda(N - 1)}{r} G'(r) - \gamma G'(r).
\end{aligned}
$$

Using (9.10), we find

$$(9.11) \qquad F(x, G, DG, D^2G) \geq \frac{\sigma}{r^{\sigma+2}}((\sigma+1)\lambda - \Lambda(N-1) - \gamma r).$$

Taking σ sufficiently large, we can guarantee that $F(x, G, DG, D^2G) \geq \kappa > 0$ on Ω and then if $M = \sup_\Omega f^+/\kappa$, we have

$$F(x, MG, MDG, MD^2G) \geq f$$

on ω.

We do not yet have our supersolution, since MG does not satisfy the boundary condition $MG = 0$ on $\partial\Omega$. However

$$\inf_{x_b \in \partial\Omega} MG(|x - z_b|)$$

does. Moreover, it is continuous since the family $\{MG(|x - z_b| : x_b \in \partial\Omega\}$ is equicontinuous; it is then also a supersolution by Proposition 8.2.

Exercise 9.3. Construct a supersolution for a general continuous boundary function g on $\partial\Omega$. (Hint: $g(x_b) + \varepsilon + MG(|x - z_b|)$.)

Exercise 9.4. Verify that if each of $F_{A,B}$ satisfies (9.6) with fixed constants λ, Λ, then so does $\sup_A \inf_B F_{A,B}$. Observe that the supersolution just constructed depends only on structure conditions and the implications of this.

As a last example, we take up the stationary Hamilton-Jacobi equation of Section 6 for which we established comparison (if you skipped that section, you can either return or skip this example). First of all, it is clear from the proof given for the Dirichlet problem that Perron's method applies to $u + H(Du) = f$ in $\mathbb{R}^N$ where f is uniformly continuous. Coupled with the comparison proved in Section 6, to prove the unique existence of a uniformly continuous solution, we need only produce linearly bounded sub and supersolutions. We seek a supersolution in the form $u(x) = A(1 + |x|^2)^{1/2} + B$ for some constants A and B. Since f is uniformly continuous, $f(x) \leq K|x| + K$ for some K and it suffices to have

$$A(1 + |x|^2)^{1/2} + B + H\left(A\frac{x}{(1 + |x|^2)^{1/2}}\right) \geq K|x| + K \quad \text{for} \quad x \in \mathbb{R}^N.$$

We may take $A = K$ and then B large enough to guarantee this inequality (and subsolutions are obtained in a similar way).

Ishii [24] introduced Perron's method into this arena. The construction carried out for uniformly elliptic equations can be modified (incorporating an angular variable) to the situation in which there are exterior cones rather than balls at each point of $\partial\Omega$, see Miller [33]. The existence theory in the uniformly elliptic case is completed in Exercise 10.2. It is interesting that the Dirichlet problem for the equation $F(u, Du, D^2u) = f(x)$ where f is merely in $L^N(\Omega)$ can be shown to have unique "viscosity solutions" as well. Of course, this requires adapting the notions and other machinery. See Caffarelli, Crandall, Kocan, and Święch [8].

10. The Uniqueness Machinery for Second Order Equations

Let us begin by indicating why uniqueness for second equations cannot be treated by simple extensions of the arguments for the first order case.

We consider a Dirichlet problem

$$\text{(DP)} \qquad \begin{cases} u + F(Du, D^2u) = f(x) & \text{in} \quad \Omega \\ u(x) = \psi(x) & \text{on} \quad \partial\Omega \end{cases}$$

and seek to prove a comparison result for subsolutions and supersolutions. For a change, we package a comparison result as a formal theorem.

Theorem 10.1. *Let Ω be a bounded open subset of $\mathbb{R}^N$, $f \in C(\overline{\Omega})$, and $F(p, X)$ be continuous and degenerate elliptic. Let, $u, v : \overline{\Omega} \to \mathbb{R}$ be upper semicontinuous and lower semicontinuous respectively, u be a solution of $u + F(Du, D^2u) \le f$, v be a solution of $v + F(Dv, D^2v) \ge f$ in Ω, and $u \le v$ on $\partial\Omega$. Then $u \le v$ in Ω.*

The strategy of the first order proof already given suggests that we consider a maximum of $u(x) - v(y) - |x - y|^2/(2\varepsilon)$ over $\overline{\Omega} \times \overline{\Omega}$ and let $\varepsilon \downarrow 0$. Following the corresponding proof in Section 3, we may assume $(\hat{x}, \hat{y})$ lies in $\Omega \times \Omega$ if ε is small, and apply the definitions to find that

$$u(\hat{x}) + F\left(\frac{\hat{x} - \hat{y}}{\varepsilon}, \frac{1}{\varepsilon}I\right) \le f(\hat{x}), \quad v(\hat{y}) + F\left(\frac{\hat{x} - \hat{y}}{\varepsilon}, -\frac{1}{\varepsilon}I\right) \ge f(\hat{y}),$$

which turns out to be useless since $I \ge -I$. We need refined information about this maximum $(\hat{x}, \hat{y})$, which turns out to be a substantial and interesting issue. The information we need corresponds to the fact that if $\Phi(x, y) = u(x) - v(y) - |x - y|^2/(2\varepsilon)$ has a maximum at $(\hat{x}, \hat{y})$ and u, v are twice differentiable, then the full Hessian of Φ in (x, y) is nonpositive, or

$$\begin{pmatrix} D^2u(\hat{x}) & 0 \\ 0 & -D^2v(\hat{y}) \end{pmatrix} \le \frac{1}{\varepsilon}\begin{pmatrix} I & -I \\ -I & I \end{pmatrix};$$

the failed attempt above did reflect the full second order test for a maximum in the doubled variables (x, y). (The notation "I" is used here and later to denote the identity in any dimension.) Since the matrix on the right annihilates $\begin{pmatrix} \xi \\ \xi \end{pmatrix}$ for $\xi \in \mathbb{R}^N$, the above inequality implies that $D^2u(\hat{x}) \le D^2v(\hat{y})$, which is the sort of thing we need. To get there, we require some preparations.

The basic fact concerns semijets of differences (equivalently, sums), as hinted above. We call it the "Theorem on Sums".

Theorem on Sums. *Let $\mathcal{O}$ be a locally compact subset of $\mathbb{R}^N$. Let $u, -v : \mathcal{O} \to \mathbb{R}$ be upper semicontinuous and φ be twice continuously differentiable in a neighborhood of $\mathcal{O} \times \mathcal{O}$. Set*

$$w(x, y) = u(x) - v(y) \quad \text{for} \quad x, y \in \mathcal{O}$$

and suppose $(\hat{x}, \hat{y}) \in \mathcal{O} \times \mathcal{O}$ is a local maximum of $w(x, y) - \varphi(x, y)$ relative to $\mathcal{O} \times \mathcal{O}$. Then for each $\kappa > 0$ with $\kappa D^2 \varphi(\hat{x}, \hat{y}) < I$ there exists $X, Y \in \mathcal{S}(N)$ such that

$$(u(\hat{x}), D_x\varphi(\hat{x}, \hat{y}), X) \in \overline{J}_{\mathcal{O}}^{2,+} u(\hat{x}), \quad (v(\hat{y}), -D_y\varphi(\hat{x}, \hat{y}), Y) \in \overline{J}_{\mathcal{O}}^{2,-} v(\hat{y})$$

and the block diagonal matrix with entries $X, -Y$ satisfies

(10.1) $$-\frac{1}{\kappa} I \le \begin{pmatrix} X & 0 \\ 0 & -Y \end{pmatrix} \le (I - \kappa D^2\varphi(\hat{x}, \hat{y}))^{-1} D^2\varphi(\hat{x}, \hat{y}).$$

We use the Theorem on Sums to prove Theorem 10.1.

Proof of Theorem 10.1. Set

$$\Phi(x, y) = u(x) - v(y) - \frac{1}{2\varepsilon} |x - y|^2$$

and consider a maximum $(\hat{x}, \hat{y})$ over $\overline{\Omega} \times \overline{\Omega}$.

Let

$$A = \frac{1}{2\varepsilon} D^2 |x - y|^2 \Big|_{x = \hat{x}, y = \hat{y}} = \frac{1}{\varepsilon} \begin{pmatrix} I & -I \\ -I & I \end{pmatrix}$$

and note that $A \le (2/\varepsilon) I$; moreover, for $2\kappa/\varepsilon < 1$

$$(I - \kappa A)^{-1} A = \frac{1}{\varepsilon - 2\kappa} \begin{pmatrix} I & -I \\ -I & I \end{pmatrix}.$$

By Theorem on Sums, there exists $X, Y \in \mathcal{S}(N)$ such that

$$\left(u(\hat{x}), \frac{\hat{x} - \hat{y}}{\varepsilon}, X\right) \in \overline{J}^{2,+} u(\hat{x}), \quad \left(v(\hat{y}), \frac{\hat{x} - \hat{y}}{\varepsilon}, Y\right) \in \overline{J}^{2,-} v(\hat{y})$$

(10.2) $$-\frac{1}{\kappa} I \le \begin{pmatrix} X & 0 \\ 0 & -Y \end{pmatrix} \le \frac{1}{\varepsilon - 2\kappa} \begin{pmatrix} I & -I \\ -I & I \end{pmatrix}.$$

Since the right hand side annihilates $\begin{pmatrix} \xi \\ \xi \end{pmatrix}$ for $\xi \in \mathbb{R}^N$, we conclude that $X \le Y$ and then $F((\hat{x} - \hat{y})/\varepsilon, Y) \le F((\hat{x} - \hat{y})/\varepsilon, X)$ since F is degenerate elliptic. Moreover,

$$u(\hat{x}) + F\left(\frac{\hat{x} - \hat{y}}{\varepsilon}, X\right) \le f(\hat{x}), \quad v(\hat{y}) + F(\frac{\hat{x} - \hat{y}}{\varepsilon}, Y) \ge f(\hat{y})$$

so

$$u(\hat{x}) - v(\hat{y}) \le f(\hat{x}) - f(\hat{y})$$

and we conclude (using Lemma 4.1) that $u(\hat{x}) - v(\hat{y}) \to 0$ as $\varepsilon \downarrow 0$. Since

$$u(x) - v(x) = \Phi(x, x) \le \Phi(\hat{x}, \hat{y}) \le u(\hat{x}) - v(\hat{y})$$

we conclude. $\square$

Note that the choice $\kappa = \varepsilon/3$ in (10.2) yields

$$(10.3) \qquad -\frac{3}{\varepsilon} I \le \begin{pmatrix} X & 0 \\ 0 & -Y \end{pmatrix} \le \frac{3}{\varepsilon} \begin{pmatrix} I & -I \\ -I & I \end{pmatrix}.$$

Exercise 10.1. Extend the above comparison proof to the equation

$$F(x, u, Du, D^2 u) = 0$$

in place of $u + F(Du, D^2 u) = f(x)$ under the conditions that there exists $\gamma > 0$ such that

$$(10.4) \qquad \gamma(r - s) \le F(x, r, p, X) - F(x, s, p, X)$$

for $r \ge s$, $(x, p, X) \in \overline{\Omega} \times \mathbb{R}^N \times \mathcal{S}(N)$ and there is a function $\omega \colon [0, \infty] \to [0, \infty]$ which satisfies $\omega(0+) = 0$ such that

$$(10.5) \quad F\left(y, r, \frac{x-y}{\varepsilon}, Y\right) - F\left(x, r, \frac{x-y}{\varepsilon}, X\right) \le \omega(|x - y|(1 + |x - y|/\varepsilon)).$$

whenever $x, y \in \Omega$, $r \in \mathbb{R}$, $X, Y \in \mathcal{S}(N)$ and (10.3) holds.

See [12, Section 3] regarding verifying the assumption (10.5).

Exercise 10.2.* Show, as in Exercise 4.2 that the "strictly increasing in r" assumption of the previous exercise can be dropped under the assumption that for $c > 0$ such that either u solves $F \le -c$ or v solves $F \ge c$. Show that if F satisfies (9.6) and v solves $F \ge 0$, then there are arbitrarily small radial perturbations ψ such that $v + \psi$ solves $F \ge c > 0$.

Remark 10.2. At this point, via Perron's Method, the supersolutions constructed in Section 9, and the preceding exercise, we have uniquely solved (for example) the Dirichlet problem for $F(u, Du, D^2 u) = f(x)$.

Exercise 10.3. Adapt the above comparison proof to handle linear growth of the sub and supersolutions in the problem $u + F(Du, D^2 u) = f(x)$ on $\mathbb{R}^N$. Use Perron's method to complete the proof of Theorem 3.3. (Hint: try supersolutions of the form employed for the first order case at the end of Section 9.)

The second order uniqueness theory has a long history with important contributions by R. Jensen ([27] contains the first proof for second order equations without convexity conditions and introduced new ideas), H. Ishii, P. L. Lions and P. E. Souganidis (e.g., [31], [30], [25], [26]). See the comments ending [12, Section 3]. The result called here the "Theorem on Sums", which makes life so easy, is a mild refinement of the result called the "maximum principle for semi-continuous functions" in Crandall and Ishii [11] and was preceded by Crandall [10]. Note also "Ishii's Lemma" in [22, Chapter V]. The proof below is slicker too.

11. Proof of the Theorem on Sums

In this section we sketch the proof of Theorem on Sums. By now, you know that it can be used to good effect. It was stated before in the form used in the theory, but for notational simplicity we change the statement a trifle, so it is, at last, about sums.

Theorem on Sums. *Let $\mathcal{O}$ be a locally compact subset of $\mathbb{R}^N$. Let $u, v : \mathcal{O} \to \mathbb{R}$ and φ be twice continuously differentiable in a neighborhood of $\mathcal{O} \times \mathcal{O}$. Set*

$$w(x, y) = u(x) + v(y) \quad \text{for} \quad x, y \in \mathcal{O}$$

and suppose $(\hat{x}, \hat{y}) \in \mathcal{O} \times \mathcal{O}$ is a local maximum of $w(x, y) - \varphi(x, y)$ relative to $\mathcal{O} \times \mathcal{O}$. Then for each $\kappa > 0$ with $\kappa D^2\varphi(\hat{x}, \hat{y}) < I$, there exists $X, Y \in S(N)$ such that

$$(u(\hat{x}), D_x\varphi(\hat{x}, \hat{y}), X) \in \overline{J}_{\mathcal{O}}^{2,+} u(\hat{x}), \quad (v(\hat{y}), D_y\varphi(\hat{x}, \hat{y}), Y) \in \overline{J}_{\mathcal{O}}^{2,+} v(\hat{y})$$

and the block diagonal matrix with entries X, Y satisfies

$$(11.1) \qquad -\frac{1}{\kappa} I \leq \begin{pmatrix} X & 0 \\ 0 & Y \end{pmatrix} \leq (I - \kappa D^2\varphi(\hat{x}, \hat{y}))^{-1} D^2\varphi(\hat{x}, \hat{y}).$$

[1] Consider the function $u(x) = x \sin x$, $u(0) = 0$ on $\mathbb{R}$. Show that $J^{2,+}u(0)$ is empty while $\overline{J}^{2,+}u(0)$ is quite large.

In order to prove this result we will need to "regularize" merely semicontinuous functions. The method we will use is called "sup convolution". This operation is introduced next, the relevant properties are established, and these results are then used in the proof.

Let $\mathcal{O}$ be a closed subset of $\mathbb{R}^M$, $\kappa > 0$, $\psi : \mathcal{O} \to \mathbb{R}$ and $\zeta \in \mathbb{R}^M$, and put

$$(11.2) \qquad \hat{\psi}(\zeta) = \sup_{z \in \mathcal{O}} \left(\psi(z) - \frac{1}{2\kappa} |z - \zeta|^2 \right).$$

Obviously the supremum is assumed and finite if $\psi \not\equiv -\infty$ and

$$(11.3) \qquad \limsup_{z \in \mathcal{O}, |z| \to \infty} \frac{\psi(z)}{|z|^2} < \frac{1}{2\kappa},$$

which we always assume, often without comment. If we extend ψ to all of $\mathbb{R}^M$ by

$$\psi(z) = -\infty \quad \text{for} \quad z \notin \mathcal{O}$$

the formula becomes

$$(11.4) \qquad \hat{\psi}(\zeta) = \max_{z \in \mathbb{R}^M} (\psi(z) - \frac{1}{2\kappa} |z - \zeta|^2).$$

We make this extension without further comment, and treat upper semicontinous $\psi : \mathbb{R}^M \to [-\infty, \infty)$ where we now allow $-\infty$ as a value, but insist that $\psi \not\equiv -\infty$.

Obviously $\hat{\psi}$ depends on κ which is not indicated in our notation; κ will be obvious from the context. We note some obvious properties of sup convolution:

(11.5)
$$\begin{cases} \text{(i) If } \psi \leq \varphi, \text{ then } \hat{\psi} \leq \hat{\varphi}, \\ \text{(ii) } \psi \leq \hat{\psi}. \\ \text{(iii) } \hat{\psi}(\zeta) + (1/2\kappa)|\zeta|^2 \text{ is convex.} \end{cases}$$

Property (i) needs no comment. Property (ii) is seen by putting $z = \zeta$ in the defining supremum. Property (iii) holds because $\psi(z) - (1/2\kappa)|z-\zeta|^2 + (1/2\kappa)|\zeta|^2$ is affine in ζ, and the supremum of a family convex functions is convex.

The property (iii) is called "semiconvexity"; $\hat{\psi}$ is semiconvex (with constant $1/(2\kappa)$). Less obvious properties include that sup convolution is an approximation of the identity, that is
$$\lim_{\kappa \downarrow 0} \hat{\psi}(\zeta) = \psi(\zeta)$$

for $\zeta \in \mathbb{R}^N$ when the left-hand side of (11.3) is finite.

Exercise 11.1. Prove this claim.

Hence sup convolution provides an approximation of ψ which is pretty regular, that is semiconvex. The function $\hat{\psi}$ enjoys all the regularity of convex functions.

We later use the following special case:

Exercise 11.2.* If $B \in \mathcal{S}(M)$ and $2\kappa B < I$, prove that if $\psi(z) = \langle Bz, z \rangle$, then
$$\hat{\psi}(\zeta) = \langle B(I - 2\kappa B)^{-1}\zeta, \zeta \rangle .$$

Finally, we note the "magic" properties of sup convolution – it respects $J^{2,+}$ in the following sense:

Theorem On Magic Properties. *Let* $\psi : \mathbb{R}^N \to \mathbb{R}$ *be upper semicontinuous and satisfy (11.3). If* $\hat{\zeta} \in \mathbb{R}^M$ *and*
$$(\hat{\psi}(\hat{\zeta}), p, X) \in J^{2,+}\hat{\psi}(\hat{\zeta})$$

then for
$$\hat{z} = \hat{\zeta} + \kappa p$$

and for every real $M \times M$ *matrix* T,
$$(\psi(\hat{\zeta} + \kappa p), p, \frac{1}{\kappa}(I - T^*)(I - T) + T^* X T) \in J^{2,+}\psi(\hat{z})$$

where T^* *is the adjoint of* T. *Moreover,* $\hat{z}$ *is the unique point for which*

(11.6)
$$\hat{\psi}(\hat{\zeta}) = \psi(\hat{z}) - \frac{1}{2\kappa}|\hat{z} - \hat{\zeta}|^2.$$

In particular, choosing $T = I$,

$$(\psi(\hat{\zeta} + \kappa p), p, X) \in J^{2,+}\psi(\hat{z}).$$

Proof. We may assume that there exists $\varphi \in C^2(\mathbb{R}^M)$ such that

$$\hat{\psi}(\zeta) - \varphi(\zeta) \le \hat{\psi}(\hat{\zeta}) - \varphi(\hat{\zeta})$$

for $\zeta \in \mathbb{R}^M$ and $D\varphi(\hat{\zeta}) = p$, $D^2\varphi(\hat{\zeta}) = X$. From the definition of $\hat{\psi}$ this implies there exists $\hat{z}$ such that

$$\psi(z) - \frac{1}{2\kappa}|z - \zeta|^2 - \varphi(\zeta) \le \psi(\hat{z}) - \frac{1}{2\kappa}|\hat{z} - \hat{\zeta}|^2 - \varphi(\hat{\zeta})$$

for all $z \in \mathbb{R}^M$. Putting first $z = \hat{z}$, we find that $|\hat{z} - \zeta|^2/(2\kappa) + \varphi(\zeta)$ has a minimum at $\hat{\zeta}$, and thus by the first and second order test for a maximum,

(11.7) $$\hat{z} = \hat{\zeta} + \kappa D\varphi(\hat{\zeta}) = \hat{\zeta} + \kappa p, \quad D^2\varphi(\hat{\zeta}) = X \ge -\frac{1}{\kappa}I.$$

Next, leave z free but put $\zeta = T(z - \hat{z}) + \hat{\zeta}$. This leads to

$$\psi(z) - \Phi(z) \le \psi(\hat{z}) - \Phi(\hat{z})$$

where

$$\Phi(z) = \frac{1}{2\kappa}|(I - T)z + T\hat{z} - \hat{\zeta}|^2 + \varphi(Tz - T\hat{z} + \hat{\zeta}).$$

We conclude that

$$(\psi(\hat{z}), D\Phi(\hat{z}), D^2\Phi(\hat{z})) \in J^{2,+}\psi(\hat{z}).$$

Finally, by (11.7) and direct computation

$$D\Phi(\hat{z}) = \frac{1}{\kappa}(\hat{z} - \hat{\zeta}) = p,$$

$$D^2\Phi(\hat{z})$$
$$= \frac{1}{\kappa}(I - T^*)(I - T) + T^*D^2\varphi(\hat{\zeta})T = \frac{1}{\kappa}(I - T^*)(I - T) + T^*XT.$$

$\square$

Remark 11.1. It is natural to ask what is the smallest matrix Y which can be written in the form

$$Y = \frac{1}{\kappa}(I - T^*)(I - T) + T^*XT,$$

where we recall that $X \ge -1/\kappa$. There is no optimal choice if $-1/\kappa$ is an eigenvalue of X – this can be seen from the scalar case: putting $X = -1/\kappa, T = t$ leads to $Y = (1 - 2t)/\kappa$, which is not bounded below. If $X > -1/\kappa$, then the answer is (exercise)

$$T = (I + \kappa X)^{-1}.$$

which yields, with a little algebra,

$$(11.8) \qquad \frac{1}{\kappa}(I - (I + \kappa X)^{-1})^2 + (I + \kappa X)^{-2}X = (I + \kappa X)^{-1}X.$$

Exercise 11.3. Compute $\hat{\psi}$ if $\psi(0) = 1$ and $\psi(x) = 0$ for $x \neq 0$.

Exercise 11.4.* Show that if u is a subsolution of a proper equation

$$F(u, Du, D^2u) = 0,$$

on $\mathbb{R}^N$, then $\hat{u}$ is as well. Using (11.6), show

$$|\hat{z} - \hat{\zeta}|^2 \leq 2\kappa(\psi(\hat{z}) - \psi(\hat{\zeta})).$$

Conclude that if u is a bounded subsolution of an equation $F(u, Du, D^2u) = f(x)$ where f is uniformly continuous, then $\hat{u}$ is a solution of $F(\hat{u}, D\hat{u}, D^2\hat{u}) \leq f(x) + \delta_\kappa$ for some constant $\delta_\kappa \to 0$ as $\kappa \downarrow 0$. Discuss the general case, $F(x, Du, D^2u) \leq 0$.

It is always an option to use the method (direct use of the approximations in the equations) as indicated by Exercise **Exercise 11.5.*** in place of the Theorem on Sums while working on problems in this area.

We will also employ two nontrivial facts about semiconvex functions. The first assertion is a classical result of Aleksandrov:

Theorem (Aleksandrov). *Let* $g : \mathbb{R}^M \to \mathbb{R}$ *be semiconvex. Then* g *is twice differentiable almost everywhere on* $\mathbb{R}^N$.

Here g is "twice differentiable" at $\hat{z}$ means that

$$g(z) = g(\hat{z}) + \langle p, z - \hat{z} \rangle + \frac{1}{2}\langle X(z - \hat{z}), z - \hat{z} \rangle + o(|z - \hat{z}|^2)$$

for some $p \in \mathbb{R}^M$, $X \in \mathcal{S}(M)$, and then we say $Dg(\hat{z}) = p$, $D^2g(\hat{z}) = X$. We will not prove this classical result.

The next result we will need concerning semiconvex functions follows. In the statement, $B_r(z)$ is the ball of radius r about z. We will prove this result after completing the proof of the Theorem on Sums. It is a variant of an argument of Aleksandrov.

Lemma 11.2. *Let* $\varphi : \mathbb{R}^M \to \mathbb{R}$ *be semiconvex and* $\hat{x}$ *be a strict local maximum point of* φ. *For* $p \in \mathbb{R}^N$, *set* $\varphi_p(x) = \varphi(x) + \langle p, x \rangle$. *Then for* $0 < r$ *sufficiently small and all* $\delta > 0$,

$$\{y \in B_r(\hat{x}) : \exists p \in B_\delta(0) \text{ such that } \varphi_p(x) \leq \varphi_p(y) \text{ for } x \in B_r(\hat{x})\}$$

has positive measure.

Proof of the Theorem on Sums. In the proof, we may as well assume that $\mathcal{O} = \mathbb{R}^N$. Indeed, if not, we first restrict u, v to compact neighborhoods N_1, N_2 of $\hat{x}, \hat{y}$ in $\mathcal{O}$ and then extend the restrictions to $\mathbb{R}^N$ by $u(x) = v(y) = -\infty$ for $x \notin N_1$

and $y \notin N_2$. One then checks that $\bar{J}_O^{2,+} u(\hat{x}) = \bar{J}^{2,+} u(\hat{x})$ (provided $-\infty < u(\hat{x})$), etc. We may assume that φ is C^2 in a neighborhood of $N_1 \times N_2$, and then on $\mathbb{R}^N$ by modification off $N_1 \times N_2$. It is clear that $(\hat{x}, \hat{y})$ is still a local maximum of $w - \varphi$ relative to $\mathbb{R}^N \times \mathbb{R}^N$.

Further we may as well assume that

$$\hat{x} = \hat{y} = 0, \; u(0) = v(0) = 0$$

and that

$$\varphi(x, y) = \frac{1}{2} \left\langle A \begin{pmatrix} x \\ y \end{pmatrix}, \begin{pmatrix} x \\ y \end{pmatrix} \right\rangle$$

for some $A \in S(2N)$ is a pure quadratic, and 0 is a global maximum of $w - \varphi$. Indeed, a translation puts $\hat{x}, \hat{y}$ at the origin and then by replacing $\varphi(x, y), u(x), v(y)$ by $\varphi(x, y) - (\varphi(0) + \langle D_x \varphi(0), x \rangle + \langle D_y \varphi(0), y \rangle)$ and $u(x) - (u(0) + \langle D_x \varphi(0), x \rangle)$, $v(y) - (v(0) + \langle D_y \varphi(0), y \rangle)$ we reduce to the situation $\hat{x} = \hat{y} = D_x \varphi(0) = D_y \varphi(0) = 0$, and $\varphi(0) = u(0) = v(0) = 0$. Then, since

$$\varphi(x, y) = \frac{1}{2} \left\langle A \begin{pmatrix} x \\ y \end{pmatrix}, \begin{pmatrix} x \\ y \end{pmatrix} \right\rangle + o(|x|^2 + |y|^2)$$

where $A = D^2 \varphi(0)$, and $w(x, y) - \varphi(x, y) \le w(0) - \varphi(0) = 0$ for small x, y, if $\eta > 0$, we will have

(11.9) $$w(x, y) - \frac{1}{2} \left\langle (A + \eta I) \begin{pmatrix} x \\ y \end{pmatrix}, \begin{pmatrix} x \\ y \end{pmatrix} \right\rangle < 0$$

for small $x, y \ne 0$. Globality of the (strict) maximum at 0 may be achieved by localizing further as above. If the result holds in this case, we then pass to the limit as $\eta \downarrow 0$ to obtain the general result.

From

$$w(x, y) = u(x) + v(y) \le \frac{1}{2} \left\langle A \begin{pmatrix} x \\ y \end{pmatrix}, \begin{pmatrix} x \\ y \end{pmatrix} \right\rangle$$

and Exercise 11.2 we have

(11.10) $$\hat{w}(\zeta) = \hat{u}(\xi) + \hat{v}(\eta) \le \frac{1}{2} \langle A(I - \kappa A)^{-1} \zeta, \zeta \rangle$$

where we are writing

$$\zeta = \begin{pmatrix} \xi \\ \eta \end{pmatrix}.$$

Since $u \le \hat{u}$, etc., $0 = u(0) \le \hat{u}(0)$; similarly $0 \le \hat{v}(0)$. On the other hand, by (11.10) we have $\hat{u}(0) + \hat{v}(0) \le 0$, and then $\hat{u}(0) = \hat{v}(0) = 0$ and

$$\hat{u}(\xi) + \hat{v}(\eta) - \frac{1}{2} \langle A(I - \kappa A)^{-1} \zeta, \zeta \rangle$$

has a maximum at the origin, which is in fact strict (or increase A a little if you don't want to check this). By Jensen's Lemma and Aleksandrov's Theorem, for each $0 < \delta$ we have $p_\delta, q_\delta \in B_\delta(0)$ such that

$$\hat{u}(\xi) + \hat{v}(\eta) - \frac{1}{2}\left\langle A(I - \kappa A)^{-1}\zeta, \zeta \right\rangle + \langle p_\delta, \xi \rangle + \langle q_\delta, \eta \rangle$$

has a maximum $(\hat{\xi}_\delta, \hat{\eta}_\delta) \in B_\delta(0) \times B_\delta(0)$ and $\hat{u}$, $\hat{v}$ are twice differentiable at $\hat{\zeta}_\delta, \hat{\eta}_\delta$. We now apply the magic properties, and let $\delta \downarrow 0$ to reach the conclusion. The second order test at a maximum and semiconvexity imply

$$-\frac{1}{\kappa}I \le \begin{pmatrix} D^2\hat{u}(\hat{\xi}_\delta) & 0 \\ 0 & D^2\hat{v}(\hat{\eta}_\delta) \end{pmatrix} \le (1 - \kappa A)^{-1}A;$$

therefore we can assume

$$X_\delta = D^2\hat{u}(\hat{\xi}_\delta) \to X, \quad Y_\delta = D^2\hat{v}(\hat{\eta}_\delta) \to Y$$

along a sequence $\delta \downarrow 0$ and

$$-\frac{1}{\kappa}I \le \begin{pmatrix} X & 0 \\ 0 & Y \end{pmatrix} \le (1 - \kappa A)^{-1}A.$$

Let P project $\mathbb{R}^N \times \mathbb{R}^N$ on its first coordinates. By the magic properties,

$$(u(\hat{\xi}_\delta + \kappa(p_\delta + PA\hat{\zeta}_\delta)), p_\delta + PA\hat{\zeta}_\delta, X_\delta) \in J^{2,+}u(\hat{\zeta}_\delta + \kappa(p_\delta + PA\hat{\zeta}_\delta)).$$

By the definitions and magic properties,

$$u(\hat{\xi}_\delta + \kappa(p_\delta + PA\hat{\zeta}_\delta)) - \frac{\kappa}{2}|p_\delta + PA\hat{\zeta}_\delta|^2 = \hat{u}(\hat{\xi}_\delta);$$

since $\hat{u}$ is continuous, we conclude that

$$u(\hat{\xi}_\delta + \kappa(p_\delta + PA\hat{\zeta}_\delta)) \to \hat{u}(0) = u(0) = 0$$

and then

$$(u(0), 0, X) \in \overline{J}^{2,+}u(0)$$

follows upon letting $\delta \downarrow 0$. The analogous comments hold for v. $\qquad\square$

Proof of Jensen's Lemma. We assume that r is so small that φ has $\hat{x}$ as a unique maximum point in $B(\hat{x}, r)$ and assume for the moment that φ is C^2. It follows from this that if δ is sufficiently small and $p \in B_\delta(0)$, then every maximum of φ_p with respect to $B(\hat{x}, r)$ lies in the interior of $B(\hat{x}, r)$. Since $D\varphi + p = 0$ holds at maximum points of φ_p, $D\varphi(K) \supset B_\delta(0)$. Let $\lambda \ge 0$ and $\varphi(x) + (\lambda/2)|x|^2$ be convex; we then have $-\lambda I \le D^2\varphi$; moreover, on K, $D^2\varphi \le 0$ and then

$$-\lambda I \le D^2\varphi(x) \le 0 \quad \text{for} \quad x \in K.$$

In particular, $|\det D^2\varphi(x)| \le \lambda^N$ for $x \in K$. Thus

$$\mathrm{meas}(B_\delta(0)) \le \mathrm{meas}(D\varphi(K)) \le \int_K |\det D^2\varphi(x)|dx \le \mathrm{meas}(K)|\lambda|^n$$

and we have a lower bound on the measure of K depending only on λ.

In the general case, in which φ need not be smooth, we approximate it via mollification with smooth functions φ_κ which have the same semiconvexity constant λ and which converge uniformly to φ on $B(\hat{x}, r)$. The corresponding sets K_κ obey the above estimates for small κ and

$$K \supset \cap_{n=1}^\infty \cup_{m=n}^\infty K_{\frac{1}{m}}$$

is evident. The result now follows. $\qquad\square$

Exercise 11.6. Extend the Theorem on Sums to an arbitrary number of summands.

Concerning the "magic properties", perhaps they are not so magic if viewed through the lens explained by Evans [20] under the subheading "Jensen's regularizations"; in this regard see also Lasry and Lions [32] and Jensen, Lions and Souganidis [30]. The precise formulation and proof of the magic properties given above is from Crandall, Kocan, Soravia, and Święch [13]. It is interesting that the Theorem on Sums does not refer to regularizations; the information is stored as a general fact about semicontinuous functions. Of course, as noted before, using regularizations themselves in pdes is an important tool. Aleksandrov's theorem is from [1]; see also [21]. Jensen's lemma is proved in [27].

12. Briefly Parabolic

We briefly indicate how to extend the results of the preceding sections to problems involving the "parabolic" equation

(PE) $$u_t + F(t, x, u, Du, D^2 u) = 0$$

where now u is to be a function of (t, x) and Du, $D^2 u$ mean $D_x u(t, x)$ and $D_x^2 u(t, x)$. We do this by discussing comparison for the Cauchy–Dirichlet problem on a bounded domain; it will then be clear how to modify other proofs as well. Let $\mathcal{O}$ be a locally compact subset of $\mathbb{R}^N$, $T > 0$, and $\mathcal{O}_T = (0, T) \times \mathcal{O}$. We denote by $P_{\mathcal{O}}^{2,+}$, $P_{\mathcal{O}}^{2,-}$ the "parabolic" variants of the semijets $J_{\mathcal{O}}^{2,+}$, $J_{\mathcal{O}}^{2,-}$. For example, if $u : \mathcal{O}_T \to \mathbb{R}$, $P_{\mathcal{O}}^{2,+} u$ is defined by $(u(s, z), a, p, X) \in \mathbb{R} \times \mathbb{R}^N \times \mathcal{S}(N)$ lies in $P_{\mathcal{O}}^{2,+} u(s, z)$ if $(s, z) \in \mathcal{O}_T$ and

(12.1)
$$u(t, x) \le u(s, z) + a(t - s) + \langle p, x - z \rangle + \frac{1}{2} \langle X(x - z), x - z \rangle$$
$$+ o(|t - s| + |x - z|^2) \text{ as } \mathcal{O}_T \ni (t, x) \to (s, z);$$

similarly, $P_{\mathcal{O}}^{2,-} u = -P_{\mathcal{O}}^{2,+}(-u)$. The corresponding definitions of $\overline{P}_{\mathcal{O}}^{2,+}$, $\overline{P}_{\mathcal{O}}^{2,-}$ are then clear. This definition reflects that (PE) is first order in t.

The notions of a subsolution, etc., of (PE) on an open set are contained in the previous discussion. As in Section 7, they may be reformulated, and with a little work one sees that we have: a subsolution of (PE) on $\mathcal{O}_T$ is an upper semicontinuous function $u : \mathcal{O}_T \to \mathbb{R}$ such that

(12.2)
$$a + F(t, x, r, p, X) \le 0 \text{ for } (t, x) \in \mathcal{O}_T$$

whenever $(r, a, p, X) \in P_{\mathcal{O}}^{2,+} u(t, x)$; likewise, a supersolution is a lower semicontinuous function v such that

(12.3) $$a + F(t, x, r, p, X) \geq 0 \text{ for } (t, x) \in \mathcal{O}_T$$

whenever $(r, a, p, X) \in \overline{P}_{\mathcal{O}}^{2,-} v(t, x)$.

We show how to treat the Cauchy–Dirichlet problem for (PE), exhibiting the considerations which do not occur in the stationary case. Consider the problem

(12.4)
$$\begin{cases} \text{(E)} & u_t + F(t, x, u, Du, D^2 u) = 0 \text{ in } (0, T) \times \Omega \\ \text{(BC)} & u(t, x) = 0 \text{ for } 0 \leq t < T \text{ and } x \in \partial\Omega \\ \text{(IC)} & u(0, x) = \psi(x) \text{ for } x \in \overline{\Omega}. \end{cases}$$

where $\Omega \subset \mathbb{R}^N$ is open and $T > 0$ and $\psi \in C(\overline{\Omega})$ are given. By a subsolution of (12.4) on $[0, T) \times \overline{\Omega}$ we mean an upper semicontinuous function $u : [0, T) \times \overline{\Omega} \to \mathbb{R}$ such that u is a subsolution of (PE) in $(0, T) \times \Omega$, $u(t, x) \leq 0$ for $0 \leq t < T$ and $x \in \partial\Omega$ and $u(0, x) \leq \psi(x)$ for $x \in \overline{\Omega}$ and so on for supersolutions and solutions.

Theorem 12.1. *Let* $\Omega \subset \mathbb{R}^N$ *be open and bounded. Let* $F \in C([0, T] \times \overline{\Omega} \times \mathbb{R} \times \mathbb{R}^N \times \mathcal{S}(N))$ *be continuous, proper and satisfy (10.5) for each fixed* $t \in [0, T)$, *with the same function* ω. *If* u *is a subsolution of (12.4) and* v *is a supersolution of (12.4), then* $u \leq v$ *on* $[0, T) \times \Omega$.

To continue, we require the parabolic analogue of The Theorem on Sums. It takes the form:

Theorem 12.2. *Let* $\mathcal{O}$ *be a locally compact subset of* $\mathbb{R}^N$ *and* $u, v : (0, T) \times \mathcal{O} \to \mathbb{R}$ *be upper semicontinuous. Let* φ *be defined on an open neighborhood of* $(0, T) \times \mathcal{O} \times \mathcal{O}$ *and such that* $(t, x, y) \to \varphi(t, x, y)$ *is once continuously differentiable in* t *and twice continuously differentiable in* (x, y). *Let* $(\hat{t}, \hat{x}, \hat{y}) \in (0, T) \times \mathcal{O} \times \mathcal{O}$ *and*

$$w(t, x, y) \equiv u(t, x) + v(t, y) - \varphi(t, x, y) \leq w(\hat{t}, \hat{x}, \hat{y})$$

for $0 < t < T$ *and* $x, y \in \mathcal{O}$. *Assume, moreover, that there is an* $r > 0$ *such that for every* $M > 0$ *there is a* C *such that*

(12.5) $$b \leq C \quad \text{whenever} \quad (u(t, x), b, q, X) \in P_{\mathcal{O}}^{2,+} u(t, x)$$

and

$$|x - \hat{x}| + |t - \hat{t}| \leq r \quad \text{and} \quad |u(t, x)| + |q| + |X| \leq M.$$

Assume the same condition on v *(with* $\hat{x}$ *replaced by* $\hat{y}$ *just above). Then for each* $\kappa > 0$ *with* $\kappa D^2 \varphi(\hat{t}, \hat{x}, \hat{y}) < I$ *there are* $X, Y \in \mathcal{S}(N)$, $b_1, b_2 \in \mathbb{R}$ *such that*

$$(u(\hat{t}, \hat{x}), b_1, D_x \varphi(\hat{t}, \hat{x}, \hat{y}), X) \in \overline{P}_{\mathcal{O}}^{2,+} u(\hat{t}, \hat{x}),$$
$$(v(\hat{t}, \hat{y}), b_2, D_y \varphi(\hat{t}, \hat{x}, \hat{y}), Y) \in \overline{P}_{\mathcal{O}}^{2,+} v(\hat{t}, \hat{y})$$

and

$$-\frac{1}{\kappa}I \le \begin{pmatrix} X & 0 \\ 0 & Y \end{pmatrix} \le (I - \kappa D^2\varphi(\hat{t},\hat{x},\hat{y}))^{-1} D^2\varphi(\hat{t},\hat{x},\hat{y})$$

and

$$b_1 + b_2 = \varphi_t(\hat{t},\hat{x},\hat{y}).$$

Observe that the condition (12.5) is guaranteed if u is a subsolution of a parabolic equation (and likewise for v).

Proof of Theorem 12.1. First observe that (decreasing T if necessary), we can assume that u is bounded above and v is bounded below. Next, let c, $\varepsilon > 0$, and notice that $\tilde{u} = u - ct - \varepsilon/(T-t)$ is also a subsolution of (12.4) and in fact satisfies (PE) with a strict inequality:

$$\tilde{u}_t + F(t,x,\tilde{u},D\tilde{u},D^2\tilde{u}) \le -c - \frac{\varepsilon}{(T-t)^2}.$$

Since $u \le v$ follows from $\tilde{u} \le v$ in the limit $c,\varepsilon \downarrow 0$, it will simply suffice to prove the comparison under the additional assumptions

$$(12.6) \qquad \begin{cases} (i) \ \ u_t + F(t,x,u,Du,D^2u) \le -c < 0 \text{ and} \\ (ii) \ \lim_{t\uparrow T} u(t,x) = -\infty \text{ uniformly on } \overline{\Omega}. \end{cases}$$

Let $(\hat{t},\hat{x},\hat{y})$ be a maximum point of $u(t,x) - v(t,y) - |x-y|^2/(2\varepsilon)$ over $[0,T) \times \overline{\Omega} \times \overline{\Omega}$ where $\alpha > 0$; such a maximum exists in view of the assumed bound above on u, $-v$, the compactness of $\overline{\Omega}$ and (12.6) (ii). By Lemma 4.1,

$$(12.7) \qquad \frac{|\hat{x} - \hat{y}|^2}{\varepsilon} \to 0$$

as $\varepsilon \downarrow 0$. Let $(\tilde{t},\tilde{x},\tilde{x})$ be a limit point of $(\hat{t},\hat{x},\hat{y})$ as $\varepsilon \downarrow 0$. If

$$(\tilde{t},\tilde{x},\tilde{x}) \in \{0\} \times \overline{\Omega} \cup [0,T) \times \partial\Omega,$$

upper semicontinuity and the side conditions imply that

$$\liminf_{\varepsilon\downarrow 0}\left(u(\hat{t},\hat{x}) - v(\hat{t},\hat{y}) - \frac{|\hat{x}-\hat{y}|^2}{2\varepsilon}\right) \le u(\tilde{t},\tilde{x}) - v(\tilde{t},\tilde{x}) \le 0.$$

Hence

$$(12.8) \qquad u(t,x) - v(t,x) \le \lim_{\varepsilon\downarrow 0}\left(u(\hat{t},\hat{x}) - v(\hat{t},\hat{y}) - \frac{|\hat{x}-\hat{y}|^2}{2\varepsilon}\right) = 0$$

and we are done.

If $\tilde{t} > 0$ and $\tilde{x} \notin \partial\Omega$, then we may assume that $\hat{x},\hat{y} \in \Omega$ and use Theorem 12.2 at $(\hat{t},\hat{x},\hat{y})$ to learn that there are numbers a, b and X, $Y \in \mathcal{S}(N)$ such that

$$(u(\hat{t},\hat{x}), a, \alpha(\hat{x}-\hat{y}), X) \in \overline{P}_{\mathcal{O}}^{2,+} u(\hat{t},\hat{x}),$$

and

$$(v(\hat{t}, \hat{y}), b, \alpha(\hat{x} - \hat{y}), Y) \in \overline{P}_{\mathcal{O}}^{2,-} v(\hat{t}, \hat{y})$$

such that

(12.9) $a - b = 0$ and $-\dfrac{3}{\varepsilon}\begin{pmatrix} I & 0 \\ 0 & I \end{pmatrix} \leq \begin{pmatrix} X & 0 \\ 0 & -Y \end{pmatrix} \leq \dfrac{3}{\varepsilon}\begin{pmatrix} I & -I \\ -I & I \end{pmatrix}.$

We may assume that $v(\hat{t}, \hat{x}) < u(\hat{t}, \hat{x})$ for otherwise we have (12.8). Then the relations

$$a + F(\hat{t}, \hat{x}, u(\hat{t}, \hat{x}), (\hat{x} - \hat{y})/\varepsilon, X) \leq -c, \quad b + F(\hat{t}, \hat{y}, v(\hat{t}, \hat{y}), (\hat{x} - \hat{y})/\varepsilon, Y) \geq 0,$$

and (12.9) imply

$$c \leq F\left(\hat{t}, \hat{y}, v(\hat{t}, \hat{y}), \frac{\hat{x} - \hat{y}}{\varepsilon}, Y\right) - F\left(\hat{t}, \hat{x}, u(\hat{t}, \hat{x}), \frac{\hat{x} - \hat{y}}{\varepsilon}, X\right) \leq$$
$$\omega\left(\frac{|\hat{x} - \hat{y}|^2}{\varepsilon} + |\hat{x} - \hat{y}|\right)$$

which leads to a contradiction via (12.7).

Exercise 12.1.* Work out the full proof of Theorem 3.2. The primary steps are to adapt the above proof to the pure initial-value problem via the devices used in Sections 5 and 6, to note that Perron's Method applies, and to produce subsolutions and supersolutions. All is routine, except perhaps the last step. Here is a brute force way to go about this. First, via comparison and stability, it suffices to discuss Lipschitz continuous ψ (as uniformly continuous functions on $\mathbb{R}^N$ are precisely those which are uniform limits of Lipschitz continuous functions). If ψ has Lipschitz constant L, then for every $\varepsilon > 0$, $z \in \mathbb{R}^N$, $\psi \leq \psi_{z,\varepsilon}$ where

$$\psi_{z,\varepsilon}(x) = \psi(z) + L(|x - z|^2 + \varepsilon)^{1/2}.$$

Show that there is an A_ε such that $u_{\varepsilon,z} = A_\varepsilon t + \psi_{z,\varepsilon}$ is a supersolution for the initial value problem, that $\inf_{\varepsilon,z} u_{\varepsilon,z}$ is continuous on $[0, \infty) \times \mathbb{R}^N$, and agrees with ψ at $t = 0$.

Regarding the parabolic theorem on sums, see [11] and note again "Ishii's Lemma" in [22, Chapter V].

REFERENCES

[1] A. D. Aleksandrov, *Almost everywhere existence of the second differential of a convex function and some properties of convex functions*, Leningrad University Annals (Mathematical Series) **37** (1939), 3–35 (in Russian).

[2] M. Bardi, *Some applications of viscosity solutions to optimal control and differential games*, this volume.

[3] G. Barles, **Solutions de viscosité des équations de Hamilton-Jacobi**, Springer-Verlag, New York, 1994.

[4] G. Barles and B. Perthame, *Discontinuous solutions of deterministic optimal stopping–time problems*, Modèl. Math. et Anal. Num. **21** (1987), 557–579.

[5] ──────, *Exit time problems in optimal control and the vanishing viscosity method*, SIAM J. Control Optim. **26** (1988), 1133–1148.

[6] G. Barles and P. E. Souganidis, *Convergence of approximation schemes for fully nonlinear second order equations*, Asymp. Anal. **4** (1989), 271–283.

[7] X. Cabré and L. Caffarelli, **Fully Nonlinear Elliptic Equations**, Amer. Math. Society, Providence, 1995.

[8] L. Caffarelli , M. Crandall, M. Kocan and A. Święch, *On viscosity solutions of fully nonlinear equations with measurable ingredients*, Comm. Pure Appl. Math. **49** (1996), 365–397.

[9] F. H. Clarke , Y. S. Ledyaev, R. J. Stern, and P. R. Wolenksi, *Qualitative properties of trajectories of control systems - a survey*, Journal of Dynamical and Control Systems **1** (1995), 1–48.

[10] M. G. Crandall, *Quadratic forms, semidifferentials and viscosity solutions of fully nonlinear elliptic equations*, Ann. I.H.P. Anal. Non. Lin. **6** (1989), 419–435.

[11] M. G. Crandall and H. Ishii, *The maximum principle for semicontinuous functions*, Diff. and Int. Equations **3** (1990), 1001–1014.

[12] M. G. Crandall , H. Ishii and P. L. Lions, *User's Guide to viscosity solutions of second order partial differential equations*, Bull. Amer. Math. Soc. (N.S.) **27** (1992), 1–67.

[13] M. G. Crandall, M. Kocan, P. Soravia and A. Święch: *On the equivalence of various weak notions of solutions of elliptic pdes with measurable ingredients*, in **Progress in elliptic and parabolic partial differential equations**, Alvino et. al. eds, Pitman Research Notes 350, Addison Wesley Longman, 1996, p 136-162.

[14] M. G. Crandall and P. L. Lions, *Condition d'unicité pour les solutions generalisées des équations de Hamilton-Jacobi du premier ordre*, C. R. Acad. Sci. Paris **292** (1981), 183–186.

[15] ──────, *Viscosity solutions of Hamilton-Jacobi equations*, Trans. Amer. Math. Soc. **277** (1983), 1–42.

[16] M G. Crandall , P. L. Lions and L. C. Evans, *Some properties of viscosity solutions of Hamilton-Jacobi equations*, Trans. Amer. Math. Soc. **282** (1984), 487–502.

[17] G. Dong, **Nonlinear Partial Differential Equations of Second Order**, Translations of Mathematical Monographs 95, American Mathematical Society, Providence, 1994.

[18] L. C. Evans, *A convergence theorem for solutions of nonlinear second order elliptic equations*, Indiana Univ. J. **27** (1978), 875–887.

[19] _____, *On solving certain nonlinear differential equations by accretive operator methods*, Israel J. Math. **36** (1980), 225–247.

[20] _____, *Regularity for fully nonlinear elliptic equations and motion by mean curvature*, this volume.

[21] L. C. Evans and R. Gariepy, **Measure Theory and Fine Properties of Functions**, Studies in Advanced Mathematics, CRC Press, Boca Raton, 1992.

[22] W. H. Fleming and H. Mete Soner, **Controlled Markov Processes and Viscosity Solutions**, Applications of Mathematics 25, Springer-Verlag, New York, 1993.

[23] D. Gilbarg and N. S. Trudinger, **Elliptic Partial Differential Equations of Second order, 2nd Edition**, Springer-Verlag, New York, 1983.

[24] H. Ishii, *Perron's method for Hamilton-Jacobi equations*, Duke Math. J. **55** (1987), 369–384.

[25] _____, *On uniqueness and existence of viscosity solutions of fully nonlinear second order elliptic PDE's*, Comm. Pure Appl. Math. **42** (1989), 14–45.

[26] H. Ishii and P. L. Lions, *Viscosity solutions of fully nonlinear second order elliptic partial differential equations*, J. Diff. Equa. **83** (1990), 26–78.

[27] R. Jensen, *The maximum principle for viscosity solutions of fully nonlinear second order partial differential equations*, Arch. Rat. Mech. Anal. **101** (1988), 1–27.

[28] _____, *Uniqueness criteria for viscosity solutions of fully nonlinear elliptic partial differential equations*, Indiana U. Math. J. **38** (1989), 629–667.

[29] _____, *Uniqueness of Lipschitz extensions - Minimizing the sup norm of the gradient*, Arch. Rat. Mech. Anal **123** (1993), 51–74.

[30] R. Jensen , P. L. Lions and P. E. Souganidis, *A uniqueness result for viscosity solutions of second order fully nonlinear partial differential equations*, Proc. AMS **102** (1988), 975–978.

[31] P. L. Lions, *Optimal control of diffusion processes and Hamilton-Jacobi-Bellman equations. Part 1: The dynamic programming principle and applications and Part 2: Viscosity solutions and uniqueness*, Comm. P. D. E. **8** (1983), 1101–1174 and 1229-1276.

[32] J. M. Lasry and P. L. Lions, *A remark on regularization in Hilbert spaces*, Israel J. Math **55** (1986), 257–266.

[33] K. Miller, *Barriers on cones for uniformly elliptic equations*, Ann. di Mat. Pura Appl. **LXXVI** (1967), 93–106.

[34] M. Soner, *Controlled Markov processes, viscosity solutions and applications to mathematical finance*, this volume.

[35] P. E. Souganidis, *Front Propagation: Theory and applications*, this volume.

[36] A. Subbotin, **Solutions of First-order PDEs. The Dynamical Optimization Perspective**, Birkhauser, Boston, 1995.

[37] A. Święch, $W^{1,p}$-*interior estimates for solutions of fully nonlinear, uniformly elliptic equations*, preprint.

[38] N. S. Trudinger, *Comparison principles and pointwise estimates for viscosity solutions of second order elliptic equations*, Rev. Mat. Iberoamericana **4** (1988), 453–468.

[39] _____, *Hölder gradient estimates for fully nonlinear elliptic equations*, Proc. Roy. Soc. Edinburgh, Sect. A **108** (1988), 57–65.

Some Applications of Viscosity Solutions to Optimal Control and Differential Games

Martino Bardi

Dipartimento di Matematica Pura ed Applicata

Università di Padova

Via Belzoni, 7. I-35131 Padova - Italy

e-mail: bardi@math.unipd.it

Contents

Introduction

When the theory of viscosity solutions for first order Hamilton-Jacobi (HJ) equations was initiated in the early 80s by M. Crandall and P.L. Lions [CL1], one of its main motivations was the study of the HJ-Bellman equations arising in the Dynamic Programming approach to deterministic optimal control problems, as Lions stressed in the introductory Chapter 1 of his book [L1]. Indeed, many of the early papers on viscosity solutions were devoted to applications in this field, e.g. [L1], [E], [CD], [CDE], [CDI], and in the closely related field of differential games [ES], while Lions was setting the bases for the applications to stochastic optimal control [L2].

Later the theory of viscosity solutions was extended to a unified treatment of 1st and 2nd order degenerate elliptic-parabolic equations and the number and diversity of applications increased considerably, as the reader can see from the other articles in this volume and the survey paper [CIL], but optimal control remains one of the largest and most active fields of application, whose literature includes hundreds of papers by dozens of authors.

In these notes I describe the classical method of Dynamic Programming for a choice of five deterministic optimal control problems (Section 1.A) and a differential game (Sect. 3), and give the characterization of their value function as the unique viscosity solution of an appropriate boundary value problem for the HJB equation (Sections 1.B and 5), or for the HJ-Isaacs equation in the case of differential games (Sect. 3). Moreover I illustrate the use of these equations for computing optimal multivalued feedbacks once the value function is known (Sect. 2), and their use for computing simultaneously approximations of the value function and almost optimal feedbacks (Sections 4 and 5). Sections 1-4 deal with problems with continuous value function, so one needs only the "standard" theory of continuous viscosity solutions as settled in [CIL]; Section 5, instead, focuses on a class of problems with discontinuous value function, namely, time-optimal control without controllability assumptions. I test on these problems three different extensions of the theory which allow the solutions to be discontinuous. Some of the results in the last section are very recent and work for HJ-Isaacs equations of 1st and 2nd order as well.

I give an extensive list of references for each of the topics I deal with; though the bibliography contains more than 100 titles, it is not exhaustive. All the material in these notes is taken from my forthcoming book with Capuzzo Dolcetta [BCD].

There are some important topics in optimal control where viscosity solutions apply successfully and which I do *not* treat in these notes, in particular

- asymptotic problems, such as large deviations, ergodic problems, penalization, systems with fast components..., see e.g. [FS], [Ba2], [BCD], [CDL], [BBCD];
- stochastic control, see e.g. [FS], [L2], [L4], [S2], [FSou];
- optimal control of infinite dimensional systems, see e.g. [CL2], [CL3], [CDP], [I4], [L4], [CGS], [T];

and of course the reader should also consult the large number of references therein. I also do not treat here the basic theory of viscosity solutions, and instead I refer the reader to the classical article of Crandall, Evans and Lions [CEL] or to [Ba2], [BCD] for 1st order HJ equations, and to [CIL], [FS] or [C] for the general theory including 2nd order equations.

I taught this material in a 6 hours course at the C.I.M.E. Session "Viscosity solutions and applications" in Montecatini, June 1995. I would like to thank the scientific organizers of the Session, Italo Capuzzo Dolcetta and Pierre-Louis Lions, for their kind invitation to give this course. I am also grateful to Silvia Faggian for useful comments on the first draft of this paper.

1. The Dynamic Programming Method and Hamilton-Jacobi-Bellman Equations

A. The classical Dynamic Programming method.

We are given a *controlled dynamical system*

$$(S) \qquad \begin{cases} y'(t) = f(y(t), a(t)), & t > 0, \\ y(0) = x \end{cases}$$

where $a(\cdot) \in \mathcal{A} := \{a : [0, +\infty[\to A \quad \text{measurable}\}$, A is a given compact metric space, $x \in \mathbb{R}^N$,

$$f : \mathbb{R}^N \times A \to \mathbb{R}^N \quad \text{continuous}$$

$$|f(x, a) - f(y, a)| \leq L|x - y| \quad \forall x, y \in \mathbb{R}^N, a \in A.$$

We call $y(t)$ and $a(\cdot)$, respectively, the *state* and the *open loop control* of the system. Notation: $y_x(\cdot, a) = y_x(\cdot)$ is the solution of (S).

We are also given a *cost functional*, depending on the initial state $x \in \mathbb{R}^N$ and the control $a(\cdot) \in \mathcal{A}$, to be chosen among the following:

- *Infinite Horizon:*

$$J_\infty(x, a) := \int_0^\infty \ell(y_x(t), a(t)) e^{-t} dt$$

for a given function $\ell : \mathbb{R}^N \times A \to \mathbb{R}$, continuous and bounded, and such that

$$|\ell(x, a) - \ell(y, a)| \leq \omega_\ell(|x - y|) \quad \forall x, y \in \mathbb{R}^N, a \in A,$$

where ω_ℓ is a modulus [a modulus is a nondegreasing function $\omega : \mathbb{R}^+ \to \mathbb{R}^+$, continuous in 0, $\omega(0) = 0$];

- *Finite Horizon* (Mayer problem): given $g \in C(\mathbb{R}^N)$, $t > 0$,

$$J(x, t, a) := g(y_x(t, a));$$

- *Minimum Time*: given $\mathcal{T} \subseteq \mathbb{R}^N$ closed target,

$$t_x(a) := \begin{cases} \min\{s : y_x(s, a) \in \mathcal{T}\} & \text{if } \{s : y_x(s, a) \in \mathcal{T}\} \neq \emptyset \\ +\infty & \text{otherwise}; \end{cases}$$

- *Discounted Minimum Time*:

$$J(x, a) := \begin{cases} \displaystyle\int_0^{t_x(a)} e^{-s} ds = 1 - e^{-t_x(a)}, & \text{if } t_x(a) < +\infty \\ 1 & \text{otherwise}. \end{cases}$$

We want to minimize these functionals either for $a(\cdot) \in \mathcal{A}$ (so the only constraints are on the control a), or for

$$a(\cdot) \in \mathcal{A}_x := \{a \in \mathcal{A} : y_x(t, a) \in \bar{\Omega}, \forall t > 0\}$$

where $\Omega \subseteq \mathbb{R}^N$ is a given open set: this is the problem with *state constraints*.

If the infimum is attained at some control $a^*(\cdot)$, then $a^*(\cdot)$ is called an optimal control for the initial position x (and the horizon t, in the finite horizon problem).

The Dynamic Programming method begins with introducing the *value function* of the problem:

- *Infinite Horizon*: $V_\infty(x) := \displaystyle\inf_{a(\cdot) \in \mathcal{A}} J_\infty(x, a);$
- *Infinite Horizon with State Constraints*: $V_c(x) := \displaystyle\inf_{a(\cdot) \in \mathcal{A}_x} J_\infty(x, a);$
- *Finite Horizon*: $v(x, t) := \displaystyle\inf_{a(\cdot) \in \mathcal{A}} J(x, t, a);$
- *Minimum Time*: $T(x) := \displaystyle\inf_{a(\cdot) \in \mathcal{A}} t_x(a);$
- *Discounted Minimum Time*: $V(x) := \displaystyle\inf_{a(\cdot) \in \mathcal{A}} J(x, a) = 1 - e^{-T(x)}.$

Next, one writes an equation satisfied by the value, which expresses the intuitive remark that the minimum cost is achieved if one behaves as follows:

(a) let the system evolve for a small time s choosing an arbitrary control $a(\cdot)$ on the interval $[0, s]$;

(b) pay the corresponding cost;

(c) pay what remains to pay after time s with best possible controls;

(d) minimize the sum of these two costs over all possible controls on the interval $[0, s]$.

The functional equation for the value is called *Dynamic Programming Principle* (briefly, DPP).

Proposition 1.1. *For all* $s > 0$

Infinite Horizon : $V_\infty(x) = \inf\limits_{a(\cdot)\in\mathcal{A}} \left\{ \int_0^s \ell(y_x(t), a(t))e^{-t}dt + V_\infty(y_x(s, a))e^{-s} \right\}$;

Finite Horizon : $v(x, t) = \inf\limits_{a(\cdot)\in\mathcal{A}} v(y_x(s, a), t - s)$, *if* $s \le t$;

Minimum Time : $T(x) = \inf\limits_{a(\cdot)\in\mathcal{A}} \{s + T(y_x(s, a))\}$, *if* $s \le T(x) < +\infty$;

Discounted Minimum Time :

$$V(x) = \inf_{a(\cdot)\in\mathcal{A}} \left\{ \int_0^s e^{-t}dt + V(y_x(s, a))e^{-s} \right\}, \quad \textit{if} \quad s \le T(x). \qquad \square$$

The elementary proof of the DPP is based only on the semigroup property of the solutions of (S) :

$$y_x(s + t, a) = y_{y_x(s,a)}(t, a(\cdot + s)),$$

and the following properties of admissible controls:

1. $a(\cdot) \in \mathcal{A}$, $t > 0 \Rightarrow a(\cdot + t) \in \mathcal{A}$;

2. $a_1(\cdot), a_2(\cdot) \in \mathcal{A}$, $t > 0$ and $a(s) := \begin{cases} a_1(s), & \text{if } s \le t \\ a_2(s), & \text{if } s > t \end{cases} \Rightarrow a(\cdot) \in \mathcal{A}.$

Proof of DPP for the Minimum Time . For all $a(\cdot) \in \mathcal{A}$

$$t_x(a) = s + t_{y_x(s,a)}(a(\cdot + s)) \ge s + T(y_x(s, a)),$$

so

$$T(x) = \inf_{a(\cdot)\in\mathcal{A}} t_x(a) \ge \inf_{a(\cdot)\in\mathcal{A}} \{s + T(y_x(s, a))\}.$$

To prove the opposite inequality fix any $a(\cdot) \in \mathcal{A}$, set $z := y_x(s, a)$ and assume for simplicity there exists $a_1(\cdot) \in \mathcal{A}$ such that $T(z) = t_z(a_1)$. Define

$$\bar{a}(t) := \begin{cases} a(t) & \text{if } t \le s \\ a_1(t - s) & \text{if } t > s. \end{cases}$$

Then

$$T(x) \le t_x(\bar{a}) = s + t_z(a_1) = s + T(y_x(s, a)).$$

Since $a(\cdot)$ is arbitrary we obtain the desired inequality. ∎

Here is another useful form of the DPP, which is closer to Bellman's original Principle of Optimality.

Proposition 1.2. *For all $a(\cdot) \in \mathcal{A}$ the following function is nondecreasing:*

Infinite Horizon $: s \mapsto \displaystyle\int_0^s \ell(y_x(t), a(t)) e^{-t} dt + V_\infty(y_x(s,a)) e^{-s}, \ s \in [0, +\infty[;$

Finite Horizon $: s \mapsto v(y_x(s,a), t-s), \ s \in [0, t];$

Minimum Time $: s \mapsto s + T(y_x(s,a)), \ s \in [0, t_x(a)], \ if \ T(x) < +\infty;$

Discounted Minimum Time $: s \mapsto \displaystyle\int_0^s e^{-t} dt + V(y_x(s,a)) e^{-s}, \ s \in [0, t_x(a)].$

Moreover this function is constant if and only if the control $a(\cdot)$ is optimal for the initial position x (and the horizon t in the finite horizon problem).

Proof for the Minimum Time. For any $a(\cdot) \in \mathcal{A}$, the DPP with initial position $y_x(s, a)$ gives

$$T(y_x(s,a)) \leq \varepsilon + T(y_x(s+\varepsilon, a))$$

for $\varepsilon > 0$ small enough, and this proves the first statement.

If $h(s) := s + T(y_x(s,a))$ is constant, then $h(s) \equiv h(0) = T(x)$. Therefore $0 \leq T(x) < +\infty$ implies $t_x(a) < +\infty$ and $h(t_x(a)) = t_x(a)$ because $T \equiv 0$ on the target $\mathcal{T}$. Then $T(x) = t_x(a)$.

Finally, if $a(\cdot) \in \mathcal{A}$ is optimal for x, then

$$h(0) = T(x) = t_x(a),$$

while the proof of the DPP gives

$$t_x(a) \geq h(s),$$

thus $h(0) = h(s)$ because h is nondecreasing. ∎

Next we derive the Hamilton-Jacobi-Bellman equation (briefly, HJB), which is an infinitesimal version of the DPP.

Proposition 1.3. *Assume the value function is C^1 in a neighbourhood of x (of (x,t) for the finite horizon problem). Then*

Infinite Horizon $: V_\infty(x) + \max_{a \in A}\{-f(x,a) \cdot DV_\infty(x) - \ell(x,a)\} = 0;$

Finite Horizon $: \dfrac{\partial v}{\partial t}(x,t) + H(x, D_x v(x,t)) = 0, \quad t > 0,$

Minimum Time $: H(x, DT(x)) = 1, \quad x \notin \mathcal{T}, \ T(x) < +\infty,$

Discounted Minimum Time $: V(x) + H(x, DV(x)) = 1, \quad x \notin \mathcal{T},$

where

$$H(x,p) := \max_{a \in A}\{-f(x,a) \cdot p\}.$$

Proof for the Minimum Time. We first prove the inequality "$\leq$". Fix a constant control $\bar{a}(t) \equiv a_0$ and set $y(t) = y_x(t, \bar{a})$. The DPP gives

$$T(x) - T(y(s)) \leq s \quad \text{for} \quad 0 \leq s < T(x),$$

so we divide by $s > 0$ and let $s \to 0$ to get

$$-DT(x) \cdot y'(0) \leq 1.$$

Since $y'(0) = f(x, a_0)$ and $a_0 \in A$ is arbitrary, we obtain

$$\max_{a \in A}\{-f(x,a) \cdot DT(x)\} \leq 1.$$

Next we prove the inequality "$\geq$". For $\varepsilon, s > 0$ small, by the DPP there is $\bar{a} \in \mathcal{A}$ such that

$$T(x) \geq s + T(y(s)) - \varepsilon s,$$

where $y(s) := y_x(s, \bar{a})$. Then

$$\begin{aligned}
1 - \varepsilon &\leq \frac{T(x) - T(y(s))}{s} \\
&= -\frac{1}{s}\int_0^s \frac{d}{ds}T(y(s))ds \\
&= -\frac{1}{s}\int_0^s DT(y(s)) \cdot y'(s)ds \\
&= -\frac{1}{s}\int_0^s DT(x) \cdot f(x, \bar{a}(s))ds + o(1), \quad \text{as } s \to 0, \\
&\leq \max_{a \in A}\{-DT(x) \cdot f(x,a)\} + o(1).
\end{aligned}$$

By letting s and ε go to 0 we get $1 \leq H(x, DT(x))$. ∎

Each of these PDEs can be coupled with a natural boundary condition, so we have a boundary (or initial) value problem (briefly, BVP) whose candidate solution is the value function:

Infinite Horizon :

$(\infty HP) \qquad u + \max_{a \in A}\{-f(x,a) \cdot Du(x) - \ell(x,a)\} = 0 \quad \text{in } \mathbb{R}^N;$

Infinite Horizon with State Constraints :

$(\infty HPC) \qquad \begin{cases} u + \max_{a \in A}\{-f(x,a) \cdot Du(x) - \ell(x,a)\} = 0 \quad \text{in } \Omega, \\ u + \max_{a \in A}\{-f(x,a) \cdot Du(x) - \ell(x,a)\} \geq 0 \quad \text{on } \partial\Omega; \end{cases}$

Finite Horizon:

$$(CP) \qquad \begin{cases} \dfrac{\partial u}{\partial t} + H(x, D_x u) = 0 & \text{in } \mathbb{R}^N \times]0, +\infty[, \\[2mm] u(x, 0) = g(x) & \text{in } \mathbb{R}^N \times \{0\}; \end{cases}$$

Minimum Time:

$$(FBP) \qquad \begin{cases} H(x, Du) = 1 & \text{in } \Omega \setminus \mathcal{T}, \\ u = 0 & \text{on } \partial \mathcal{T}, \\ u(x) \to +\infty & \text{as } x \to \partial \Omega, \end{cases}$$

where $\Omega \supseteq \mathcal{T}$ is an open set to be determined;

Discounted Minimum Time:

$$(DP) \qquad \begin{cases} u + H(x, Du) = 1 & \text{in } \mathbb{R}^N \setminus \mathcal{T}, \\ u = 0 & \text{on } \partial \mathcal{T}. \end{cases}$$

The BVPs for the finite horizon and the discounted minimum time are standard: they are, respectively, a Cauchy problem and a Dirichlet problem (note, however, that the open set $\mathbb{R}^N \setminus \mathcal{T}$ is unbounded if, for instance, the target $\mathcal{T}$ is compact). The PDE for the infinite horizon is set in the whole space; since V_∞ is bounded, boundedness of the solution can be considered as a boundary condition "at infinity". The boundary condition for the problem with state constraints is unusual and it was first pointed out by Soner [S1]; it is easy to check by the argument of Proposition 1.3 that the value function V_c satisfies it at every point $x \in \partial \Omega$ such that V_c is C^1 in a neighbourhood of x. The BVP for the minimum time problem is a free boundary problem (see [BS1]); one expects that $\Omega = \mathcal{R} := \{x : T(x) < +\infty\}$, and it is not hard to prove that $\mathcal{R}$ is open and $T(x) \to +\infty$ as $x \to \partial \mathcal{R}$, under a controllability assumption on the system near $\mathcal{T}$.

The classical theory of Dynamic Programming continues by assuming one has a solution of the BVP for the HJB equations (or, sometimes, just a subsolution) and using it to give sufficient conditions of optimality. This is usually called a *Verification Theorem*. Next we give an example of such a result for the discounted minimum time problem. For this problem we define a *classical verification function* as a bounded function $u \in C(\mathbb{R}^N) \cap C^1(\mathbb{R}^N \setminus \mathcal{T})$ such that

$$(1.1) \qquad \begin{cases} u + H(x, Du) \leq 1 & \text{in } \mathbb{R}^N \setminus \mathcal{T}, \\ u \leq 0 & \text{on } \partial \mathcal{T}. \end{cases}$$

Theorem 1.4. *Assume u is a classical verification function for the discounted minimum time problem, $x \notin \mathcal{T}$, $a^*(\cdot) \in \mathcal{A}$.*

(i) If $u(x) \geq J(x, a^*)$ then $a^*(\cdot)$ is optimal for x.

(ii) If

(1.2)
$$\begin{cases} u(y^*(t)) - f(y^*(t), a^*(t)) \cdot Du(y^*(t)) = 1 & \text{for a.e. } t \le t_x(a^*), \\ u = 0 & \text{on } \partial \mathcal{T}, \end{cases}$$

where $y^*(\cdot) := y_x(\cdot, a^*)$, then $a^*(\cdot)$ is optimal for x.

Proof. (i) We want to prove that $u(x) \le V(x)$, where V is the value function, so $V(x) = J(x, a^*)$ and $a^*(\cdot)$ is optimal for x. We take any $a(\cdot) \in \mathcal{A}$, $y(\cdot) = y_x(\cdot, a)$, and use (1.1) to get

$$\frac{d}{dt}\left[-e^{-t}u(y(t))\right] = e^{-t}\left[u(y(t)) - Du(y(t)) \cdot f(y(t), a(t))\right] \le e^{-t}$$

for a.e. $t \le t_x(a)$. Then we integrate both sides and obtain

$$u(x) - e^{-t}u(y(t)) \le 1 - e^{-t}.$$

Now we let $t \to t_x(a)$ and use the boundedness of u if $t_x(a) = +\infty$, and $u \le 0$ on $\partial \mathcal{T}$ if $t_x(a) < +\infty$, to get

$$J(x, a) = 1 - e^{-t_x(a)} \ge u(x).$$

Since a is arbitrary we conclude that $V(x) \ge u(x)$.

(ii) Define

(1.3)
$$h(s) := \int_0^s e^{-t}dt + u(y^*(s))e^{-s}.$$

Claim: h is nonincreasing (actually it is constant). Then, since $u = 0$ on $\partial \mathcal{T}$,

$$u(x) = h(0) \ge h(t_x(a^*)) = 1 - e^{-t_x(a^*)} = J(x, a^*),$$

and we conclude by part (i). Since h is C^1 the claim is proved by the following computation

$$h'(s) = e^{-s}\left[1 - u(y^*(s)) + Du(y^*(s)) \cdot f(y^*(s), a^*(s))\right] = 0$$

where the last equality follows from (1.2). ∎

Remark 1.5. If the verification function u is a solution of the HJB equation

$$u + H(x, Du) = 1 \quad \text{in } \mathbb{R}^N \setminus \mathcal{T},$$

then the sufficient condition of optimality (1.2) is equivalent to

$$-f(y^*(s), a^*(s)) \cdot Du(y^*(s)) = \max_{a \in A}\{-f(y^*(s), a) \cdot Du(y^*(s))\}$$
$$= H(y^*(s), Du(y^*(s)))$$

for a.e. $0 < s < t_x(a^*)$. □

Remark 1.6. If we take the value function V itself as a verification function (but this is allowed only if V is smooth...) then the sufficient condition of optimality (1.2) is also necessary. In fact, by the DPP (Proposition 1.2), if a^* is optimal for x, then the function h defined by (1.3) with $u = V$ is constant. Then

$$0 = h'(s) = e^{-s} \left[1 - V(y^*(s)) + DV(y^*(s)) \cdot f(y^*(s), a^*(s)) \right],$$

so (1.2) holds with $u = V$. □

The final step of the classical Dynamic Programming method is the attempt to construct an optimal control in feedback form from the knowledge of the value function. Let us illustrate this step on the discounted minimum time problem, even if we will be forced to make some very irrealistic assumptions. We suppose the value function V is smooth and consider the following subset of A

$$S(z) := \operatorname*{argmin}_{a \in A} \, f(z, a) \cdot DV(z)$$
$$= \{ a \in A \, : \, H(z, DV(z)) = -f(z, a) \cdot DV(z) \}.$$

This is the set of controls which allow the steepest descent of the value function V in the direction of the corresponding vector f. Let us call *admissible feedback* a map $\Phi : \mathbb{R}^N \to A$ such that for all $x \in \mathbb{R}^N$ there is a unique solution of

$$\begin{cases} y' = f(y, \Phi(y)), & t > 0, \\ y(0) = x, \end{cases}$$

and say that it is optimal for x if $\Phi(y(\cdot)) \in \mathcal{A}$ is an optimal open loop control for x. By Theorem 1.4 and Remark 1.6 a control $a^*(\cdot) \in \mathcal{A}$ is optimal for x if and only if

$$a^*(t) \in S(y_x(t, a^*)) \quad \text{for a.e. } t > 0.$$

Therefore an admissible feedback such that

$$\Phi(z) \in S(z) \quad \text{for all } z \in \mathbb{R}^N,$$

is optimal for all initial points $x \in \mathbb{R}^N$.

This method works for problems involving linear systems and quadratic costs. In this case the value function is a quadratic form which can be computed by solving an equation much simpler than the HJB equation (the Riccati equation), $S(z)$ is a singleton for all z, and its only element $\Phi(z)$ is a smooth function of z, so it defines an admissible optimal feedback. However in most problems one meets the following difficulties:

(a) the value function V is not smooth;

(b) even in subsets where V is smooth $S(z)$ is not a singleton;

(c) there are no admissible feedbacks such that $\Phi(z) \in S(z)$ for all z.

B. Viscosity solutions of HJB equations.

In the following we will see how viscosity solutions help to overcome some of the difficulties of the classical Dynamic Programming method. The first result is the rigorous derivation of the HJB equation when the value function is merely continuous. This is originally due to P. L. Lions [L1], the proof we give here follows [E] and [B].

For the sake of completeness we recall the definition of viscosity solution.

Definition 1.7. *Given $E \subseteq \mathbb{R}^M$ we denote by $USC(E)$ (resp. $LSC(E)$) the set of upper (resp. lower) semicontinuous functions $E \to \mathbb{R}$. Let $F : E \times \mathbb{R} \times \mathbb{R}^M \to \mathbb{R}$ be continuous. A function $u \in USC(E)$ (resp. $LSC(E)$) is a viscosity subsolution (resp. supersolution) of*

$$(1.4) \qquad F(x, u, Du) = 0 \quad in \ E,$$

—or a solution of $F(x, u, Du) \leq 0$ (resp. ≥ 0) in E — if, for any $\phi \in C^1(E)$ and any local maximum (resp. minimum) point x of $u - \phi$,

$$F(x, u(x), D\phi(x)) \leq 0 \quad (resp. \ \geq 0).$$

A viscosity solution of (1.4) is a subsolution which is also a supersolution.

Theorem 1.8. *Assume the value function is continuous. Then it is a viscosity solution of (∞HP) for the infinite horizon problem, of (∞HPC) for the same problem with state constraints, of (CP) for the finite horizon problem, of (FBP) with $\Omega = \mathcal{R}$ for the minimum time problem, of (DP) for the discounted minimum time problem.*

Proof for the Minimum Time. In the proof of Proposition 1.3 we showed that $T \in C^1$, $T(x) - T(y_x(s)) \leq s$ for s small, for all trajectories $y_x(\cdot)$ of the system, implies $H(x, DT(x)) \leq 1$. If we only know $T \in C(\overline{\mathbb{R}^N \setminus \mathcal{T}})$ and $\phi \in C^1$ is such that $T - \phi$ has a local maximum at x, then

$$\phi(x) - \phi(y_x(s)) \leq T(x) - T(y_x(s)) \leq s$$

for s small. Therefore we can replace T with ϕ in the previous proof and get

$$H(x, D\phi(x)) \leq 1,$$

so T is a viscosity subsolution.

Similarly, in the proof of Proposition 1.3 we showed that $T \in C^1$, $T(x) - T(y_x(s)) \geq s(1-\varepsilon)$ for s, ε small, for some trajectories $y_x(\cdot)$ of the system, implies $H(x, DT(x)) \geq 1$. Now, if $\phi \in C^1$ is such that $T - \phi$ has a local minimum at x, then

$$\phi(x) - \phi(y_x(s)) \geq T(x) - T(y_x(s)) \geq s(1-\varepsilon)$$

for s, ε small. Then again the previous proof gives

$$H(x, D\phi(x)) \geq 1,$$

so T is a viscosity supersolution as well. ∎

Next we characterize the value function as the unique viscosity solution of the appropriate BVP. This was first done, for different problems, by Lions [L1] and Capuzzo Dolcetta and Evans [CDE]. We begin with a rather general statement and then make several comments on it.

Theorem 1.9. *Assume the value function is continuous and bounded, and Ω is a smooth domain in the infinite horizon problem with state constraints. Then the value function is the minimal supersolution and the maximal subsolution of, respectively, (∞HP), (CP), (DP) among bounded functions, and of (∞HPC), among bounded and continuous functions. In particular it is the unique bounded and continuous viscosity solution.* □

The meaning of super and subsolution of a BVP is the obvious one. A supersolution of (∞HPC) is a viscosity solution of

$$u(x) + \max_{a \in A}\{-f(x,a) \cdot Du(x) - \ell(x,a)\} \geq 0 \quad \text{in } \overline{\Omega},$$

and a subsolution solves

$$u(x) + \max_{a \in A}\{-f(x,a) \cdot Du(x) - \ell(x,a)\} \leq 0 \quad \text{in } \Omega.$$

A supersolution of (CP) (resp. (DP)) is a viscosity supersolution of the PDE such that

$$u(x,0) \geq g(x), \quad x \in \mathbb{R}^N,$$

(resp.

$$u(x) \geq 0, \quad x \in \partial \mathcal{T}),$$

and subsolutions are defined in the obvious symmetric way.

The proof of this Theorem for (∞HP), (CP), (DP), is an immediate consequence of the comparison principles for Dirichlet and Cauchy problems for Hamilton-Jacobi equations, see e.g. [CL1], [CEL], [Ba2], [C], [BCD]. The smoothness assumptions on the Hamiltonian requested in the comparison principles are satisfied because

$$|H(x,p) - H(y,q)| \le K(1+|x|)|p-q| + |q|L|x-y|, \quad \forall x,y,p,q, \tag{1.5}$$

for a suitable constant K, and the Hamiltonian in (∞HP) satisfies a similar estimate with the additional term $\omega_\ell(|x-y|)$.

The comparison principle for (∞HPC) needs some non trivial modifications in the proof and the assumption that $\partial\Omega$ is locally the graph of Lipschitz functions with a uniform bound on their Lipschitz constant, and Ω lies on one side of it. We refer the reader to [S1], [BS2] or [BCD] for the proof of this result; see also [CDL].

The boundedness assumption on the value function in Theorem 1.9 is not restrictive for the infinite horizon and the discounted minimum time problems because

$$\sup |V_\infty| \le \sup |\ell|,$$

and

$$0 \le V \le 1,$$

as it is easy to check.

In the finite horizon problem v is bounded if and only if g is bounded, and this may be a restrictive assumption. This assumption can be dropped and the comparison and uniqueness assertions of Theorem 1.9 remain true in $C(\mathbb{R}^N \times [0,+\infty[)$, but the proof is different and relies on the property of the cone of dependence for solutions of (CP). We refer the reader to [I1] and [BCD].

The free boundary problem (FBP) for the minimum time is not covered by Theorem 1.9 because its solution is unbounded and the set Ω is unknown. Here is the uniqueness result in this case.

Theorem 1.10. *Assume the value function T is continuous. Then the pair $(T,\mathcal{R})$ is the unique solution of (FBP) among pairs (u,Ω) with u continuous and bounded below and $\Omega \supseteq \mathcal{T}$ open set.* □

The proof of this result is based on control theoretic methods and we do not know a purely PDE proof of it. Comparison results can also be proved. We refer the reader to [BS1], [BS2], [Sor1], [BCD], [EJ].

The last comments on Theorem 1.9 concern the assumption of continuity of the value function. For the infinite and finite horizon problems this assumption is not restrictive. In fact it is not hard to prove the following

Proposition 1.11. *$V_\infty \in C(\mathbb{R}^N)$ and $v \in C(\mathbb{R}^N \times [0,\infty[)$.* □

The situation is different in control problems with state constraints or involving an exit time from a domain, where the value may be discontinuous in general. Next result gives some simple sufficient conditions for continuity in the infinite horizon problem with state constraints and in the minimum time problem. The continuity of V in the discounted minimum time problem is obviously equivalent to the continuity of T.

Theorem 1.12. *Assume Ω is a smooth domain and $\mathcal{T}$ is the closure of a smooth domain, and let $n(\cdot)$ be the exterior normal to Ω or $\mathcal{T}$. If*

$$(1.6) \qquad \min_{a \in A} f(x, a) \cdot n(x) < 0$$

for all $x \in \partial\Omega$ and all $x \in \partial\mathcal{T}$, then V_c and T are continuous. □

For the proof we refer to [S1], [BF] or [BCD]. There are many deeper results on this problem based on geometric control theory, see e.g. [BiS], [M], [BCD] and the references therein. For the minimum time problem the continuity of T is equivalent to the *small time controllability* of the system around the target. This important property is not satisfied in several examples, such as the Zermelo Navigation Problem which is one of the most classical problems in optimal control theory, see [Ca], [BSt] and the references therein. This is a motivation to the extensions of the theory of viscosity solutions to allow discontinuous solutions: they will be described in Section 5.

As a first application of Theorem 1.9 to the optimal control problems we started from, we extend part (i) of the Verification Theorem 1.4 to become a necessary and sufficient condition of optimality. This is possible because the theory of viscosity solutions allows to consider nonsmooth verification functions. For simplicity we limit ourselves to the discounted minimum time problem.

Definition 1.13. *A verification function for the discounted minimum time problem is a bounded viscosity solution $u \in USC(\mathbb{R}^N)$ of*

$$u + H(x, Du) \le 0 \quad in \ \mathbb{R}^N \setminus \mathcal{T}$$

such that

$$u \le 0 \quad on \ \partial\mathcal{T}.$$

Corollary 1.14. *Assume the value function V of the discounted minimum time problem is continuous. Then a control $a^*(\cdot)$ is optimal for $x \in \mathbb{R}^N \setminus \mathcal{T}$ if and only if there exists a verification function u such that $u(x) \ge J(x, a^*)$.*

Proof. If $a^*(\cdot)$ is optimal then V is a verification function such that $V(x) = J(x, a^*)$. Viceversa, by Theorem 1.9 $u \le V$, so $J(x, a^*) \le V(x)$ and $a^*(\cdot)$ is optimal for x. ∎

The continuity assumption on the value V can be dropped if one uses some results on discontinuous viscosity solutions that we present in Section 5, see [BSt]. In the next section we extend part *(ii)* of the Verification Theorem 1.4 to nonsmooth verification functions. For some previous results on this subject using different tools of nonsmooth analysis we refer the reader to Clarke's books [Cl1], [Cl2] and the references therein.

To end this section we remark that many other optimal control problems have been studied in the framework of viscosity solutions, e.g. problems with switching costs [CDE], optimal control of reflected processes [L3], systems with impulses [Ba2], [BJM], [MR1], problems with unbounded controls [Ba0], [D], [MR2], [Sor7], the problem of minimizing the maximum of a running cost over a given time interval [BI], systems with constraints on the control which depend on the current state [Bo].

2. Necessary and sufficient conditions of optimality, optimal multivalued feedbacks.

In this section we choose the discounted minimum time as the model problem. Let us recall what we know from the classical theory when the value $V \in C^1(\Omega)$, $\Omega := \mathbb{R}^N \setminus T$. Let $x \in \Omega$ be fixed, $a(\cdot) \in \mathcal{A}$, $y(\cdot) = y_x(\cdot, a)$. Then the following statements are equivalent:

(i) $a(\cdot)$ is optimal for x;

(ii) $V(y(s)) - y'(s) \cdot DV(y(s)) = 1$, for a.e. $s > 0$;

(iii) $a(s) \in \underset{a \in A}{\operatorname{argmax}} \ \{-f(y(s), a) \cdot DV(y(s))\}$, for a.e. $s > 0$.

Next theorem is a version of this result for nonsmooth V by means of the sub and the superdifferential D^+ and D^- of the theory of viscosity solutions. Here is their definition:

$$D^+V(z) := \{p \in \mathbb{R}^N : \ V(y) - V(z) \le p \cdot (y - z) + o(|y - z|) \ as \ y \to z\}$$

$$D^-V(z) := \{p \in \mathbb{R}^N : \ V(y) - V(z) \ge p \cdot (y - z) + o(|y - z|) \ as \ y \to z\}.$$

We recall that a function $u \in USC(\Omega)$ (resp. $LSC(\Omega)$) is a subsolution (resp. supersolution) of (1.4) if and only if $F(x, u, p) \le 0$ for all $p \in D^+u(x)$ (resp. ≥ 0 for all $p \in D^-u(x)$).

We denote

$$D^{\pm}V(z) := D^{+}V(z) \cup D^{-}V(z)$$

and

$$D^{*}V(z) := \{p = \lim_{n} DV(z_n) : z_n \to z\}.$$

Theorem 2.1. (a) *Assume V is locally Lipschitz and $D^{\pm}V(y(s)) \neq \emptyset$ for a.e. $s > 0$. Then the following statements are equivalent:*

(i) $a(\cdot)$ *is optimal for x;*

(ii)$_V$ *for a.e. $s > 0$ and $\forall p \in D^{\pm}V(y(s))$*

$$(2.1) \qquad\qquad V(y(s)) - y'(s) \cdot p = 1;$$

(ii)$'_V$ *for a.e. $s > 0$ $\exists p \in D^{\pm}V(y(s))$ such that (2.1) holds.*

(b) *Assume V is locally Lipschitz and $D^{+}V(y(s)) \supseteq D^{*}V(y(s))$ for a.e. s. Then the following statements are equivalent:*

(i) $a(\cdot)$ *is optimal for x;*

(iii)$_V$ $a(s) \in \operatorname*{argmax}_{a \in A} \{-f(y(s), a) \cdot p\}$, *for a.e. s and $\forall p \in D^{\pm}V(y(s))$.* $\square$

This result is taken from [BCD], and some of its statements can be found also in [CF] and [Z2].

Remark 2.2. If V is semiconcave in Ω, i.e. for every convex and compact set $K \subseteq \Omega$ there is a constant C_K such that

$$V(x+h) - 2V(x) + V(x-h) \leq C_K |h|^2$$

for any h small, then $D^{+}V = co D^{*}V \neq \emptyset$ everywhere and V is locally Lipschitz, so Theorem 2.1 applies. The semiconcavity of generalized solutions of Hamilton-Jacobi equations has been extensively studied, see e.g. [K], [L1], [I1] and the references therein. Cannarsa and Sinestrari [CS] have recently proved that the minimal time function T, and therefore V, has this regularity under the assumptions of Theorem 1.12 if in addition $\partial f / \partial x$ is locally Lipschitz in x uniformly with respect to the control $a \in A$. $\square$

Note that (iii)$_V$ is a form of Maximum Principle and it says that

$$-f(y(s), a(s)) \cdot p = H(y(s), p).$$

An immediate application of this result is the synthesis of an optimal multivalued feedback. Here is an example of non-existence of single-valued optimal admissible feedbacks.

EXAMPLE. Consider $y' = a \in [-1,1]$, $\mathcal{T} = \mathbb{R}\backslash] - 1, 1[$. Then $T(x) = 1 - |x|$, $V(x) = 1 - e^{|x|-1}$. The obvious candidate optimal feedback is $\Phi(y) = 1$ if $y > 0$, $\Phi(y) = -1$ if $y < 0$. No matter how you define $\Phi(0)$, there are two solutions of

$$\begin{cases} y' = \Phi(y), & t > 0 \\ y(0) = 0 \end{cases}$$

namely $y(t) = t$ and $y(t) = -t$ (both optimal), so Φ is not admissible. $\qquad\square$

Since we have to give up the uniqueness of trajectories generated by an optimal feedback, we are justified to consider multivalued instead of single-valued feedbacks. We define the multivalued feedback

$$S_V : \mathbb{R}^N \to \mathcal{P}(A)$$

$$S_V(z) := \bigcap_{p \in D^\pm V(z)} \operatorname*{argmax}_{a \in A} \{-f(z,a) \cdot p\}.$$

By Theorem 2.1, if V is semiconcave, then a control $a(\cdot) \in \mathcal{A}$ is optimal for x if and only if

$$a(s) \in S_V(y_x(s,a)) \quad \text{for a.e. } s > 0.$$

Therefore we have the following

Corollary 2.3. *Assume that V is semiconcave and for all initial points x there exists an optimal open loop control in $\mathcal{A}$ (e.g. $f(z, A)$ is convex for all z). Then the differential inclusion*

$$\begin{cases} y' \in f(y, S_V(y)), & a.e. \ s > 0 \\ y(0) = x \end{cases}$$

has at least one solution for all $x \in \mathbb{R}^N$, and all such solutions are optimal for x.

$\qquad\square$

We summarize the conclusions of the Corollary by saying that S_V is a *fully optimal feedback*.

Now we go back to Theorem 2.1 and its proof. We split it into several parts and begin with a sufficient condition of optimality.

Proposition 2.4. *If V is locally Lipschitz and for a.e. s there exists either $p \in D^\pm V(y(s))$ such that (2.1) holds, or $p \in D^+V(y(s)) \cap D^*V(y(s))$ such that $(iii)_V$ holds, then $a(\cdot)$ is optimal for x.*

Proof. By the DPP Proposition 1.2 it is enough to prove that

$$(2.2) \qquad\qquad h(s) := \int_0^s e^{-t}dt + V(y(s))e^{-s}$$

is nonincreasing. Since V is locally Lipschitz, h is differentiable a.e. and

$$(2.3) \qquad h'(s) = e^{-s} \left(1 - V(y(s)) + \lim_{\varepsilon \to 0} \frac{V(y(s+\varepsilon)) - V(y(s))}{\varepsilon} \right).$$

For all $p \in D^+ V(y(s))$

$$(2.4) \qquad V(y(s+\varepsilon)) - V(y(s)) \leq p \cdot (y(s+\varepsilon) - y(s)) + o(\varepsilon) \quad \text{as } \varepsilon \to 0,$$

so we divide by $\varepsilon > 0$ and obtain that the limit in (2.3) is $\leq p \cdot y'(s)$ for a.e. s. By (2.1) we get $h'(s) \leq 0$.

If $p \in D^- V(y(s))$ the proof is similar: the inequality (2.4) is reversed, so we divide by $\varepsilon < 0$ and get the same conclusion.

Now let $p \in D^+ V(y(s)) \cap D^* V(y(s))$ and take $x_n \to y(s)$ such that $DV(x_n) \to p$. We pass to the limit in

$$V(x_n) + H(x_n, DV(x_n)) = 1$$

to get

$$V(y(s)) + H(y(s), p) = 1.$$

We rewrite $(iii)_V$, as

$$-f(y(s), a(s)) \cdot p = H(y(s), p) \quad \text{for a.e. } s,$$

to obtain (2.1) for a.e. s, and we conclude that $h' \leq 0$ by the first part of the proof. ∎

Remark 2.5. In the first part of Proposition 2.4, V can be replaced by any locally Lipschitz verification function u, see Definition 1.13 : if for a.e. s there exists $p \in D^{\pm} u(y(s))$ such that $u(y(s)) - y'(s) \cdot p = 1$, then $a(\cdot)$ is optimal for x. This is a nonsmooth extension of the classical Verification Theorem 1.4 (ii). The proof is easily obtained by the arguments of the proofs of Theorem 1.4 (ii), Proposition 2.4, and Corollary 1.14. □

Next we prove some necessary conditions of optimality. Note that in this result we do not need the Lipschitz continuity of V.

Proposition 2.6. *If $a(\cdot)$ is optimal for x, then (2.1) and $(iii)_V$ hold for all $p \in D^+ V(y(s))$, for a.e. s.*

Proof. If $p \in D^+ V(y(s))$ and s is a time where $\lim_{\varepsilon \to 0} (V(y(s+\varepsilon)) - V(y(s)))/\varepsilon$ exists, then we divide (2.4) by $\varepsilon < 0$ and let $\varepsilon \to 0$ to get that this limit is $\geq p \cdot y'(s)$. Since h defined by (2.2) is constant by Proposition 1.2, from (2.3) we get

$$0 = e^s h'(s) \geq 1 - V(y(s)) + p \cdot y'(s) \quad \text{for a.e. } s.$$

Then

$$V(y(s)) - p \cdot f(y(s), a(s)) \geq 1 \quad \text{for a.e. } s$$

while

$$V(y) - p \cdot f(y, a) \leq 1 \quad \forall y \notin T, \, \forall a \in A$$

by definition of subsolution. Thus we get all the conclusions. ∎

For the complete proof of Theorem 2.1 we need the following result of independent interest, which was first pointed out by Mirica [Mi1] [Mi2].

Proposition 2.7. $V(x) + H(x, p) = 1$ for all $x \in \Omega, p \in D^- V(x)$. □

This says that V is a viscosity supersolution of

(2.5) $$V + H(x, DV) = 1 \quad \text{in } \Omega,$$

and

$$-V - H(x, DV) = -1 \quad \text{in } \Omega,$$

so we say that V is a *bilateral supersolution* of the HJB equation (2.5). The proof is similar to that of Theorem 1.8 if one employs the following Backward Dynamic Programming Principle, first observed by Soravia [Sor2].

Lemma 2.8. For all $x \in \mathbb{R}^N$ and $s > 0$

$$V(x) \geq \sup_{a \in \mathcal{A}} \left\{ V(z(s)) e^s - \int_0^s e^t dt \right\},$$

where $z(\cdot)$ *solves*

$$\begin{cases} z' = -f(z, a), & s > 0, \\ z(0) = x. \end{cases}$$

□

This Lemma is not hard to prove by means of the DPP Proposition 1.1. Proposition 2.7 can also be obtained from the following general property of viscosity solutions to first order equations with convex Hamiltonian.

Theorem 2.9. *Assume* $\lambda \geq 0$ *and* $F : \Omega \times \mathbb{R}^N \to \mathbb{R}$ *is continuous,* $F(x, \cdot)$ *is convex for all* x *and*

$$F(x, p) - F(y, p) \leq \omega(|x - y|(1 + |p|))$$

for all x, y, p, *where* ω *is a modulus. Then* $u \in C(\Omega)$ *is a viscosity solution of*

$$\lambda u + F(x, Du) = 0 \quad \text{in } \Omega$$

if and only if

$$\lambda u(x) + F(x,p) = 0 \quad \forall x \in \Omega, \ p \in D^- u(x). \qquad \square$$

This result was first proved by Barron and Jensen [BJ2] for evolutive equations, and then by Barles [Ba1] for stationary problems by a simpler method, see also [BCD].

Now the proof of Theorem 2.1 can be completed by arguments very similar to those in the proofs of Proposition 2.4 and 2.6. We leave this to the reader:

EXERCISE. Use Proposition 2.7 to prove that

if $a(\cdot)$ is optimal for x, then (2.1) and $(iii)_V$ hold for all $p \in D^- V(y(s))$ and a.e. s;

if $(iii)_V$ holds for all $p \in D^- V(y(s))$ and a.e. s and V is locally Lipschitz, then $a(\cdot)$ is optimal for x. $\qquad \square$

A similar verification theorem which avoids some of the regularity assumptions on V in Theorem 2.1 can be obtained by considering the (lower) Dini directional derivatives of V

$$\frac{\partial^- V(z)}{\partial q} := \liminf_{t \to 0+} \frac{V(z+tq) - V(z)}{t}$$

instead of the sub and the superdifferentials. Let us first rewrite the classical verification theorem for smooth V by using directional derivatives. The following statements are equivalent:

(j) $a(\cdot) \in \mathcal{A}$ is optimal for x;

(jj) $V(y(s)) - \dfrac{\partial V(y(s))}{\partial y'(s)} = 1$ for a.e. $s > 0$,

(jjj) $f(y(s), a(s)) \in \underset{q \in co\overline{f}(y(s),A)}{argmax} \left\{ -\dfrac{\partial V(y(s))}{\partial q} \right\}$, for a.e. $s > 0$.

The easy proof of this result is left to the reader. Here is its version for nonsmooth V.

Theorem 2.10. *Assume V is locally Lipschitz. Then the following statements are equivalent:*

(j) $a(\cdot) \in \mathcal{A}$ *is optimal for x;*

$(jj)_D$ $V(y(s)) - \dfrac{\partial^- V(y(s))}{\partial y'(s)} = 1$ *for a.e. $s > 0$,*

$(jjj)_D$ $f(y(s), a(s)) \in \underset{q \in co\overline{f}(y(s),A)}{argmax} \left\{ -\dfrac{\partial^- V(y(s))}{\partial q} \right\}$, *for a.e. $s > 0$.* $\qquad \square$

The proof of this result is based again on the DPP Proposition 1.2 and it is obtained by differentiating the function h defined by (2.2). For the proof of

Theorem 2.10 and some related results we refer to [Fr1], [Be1], [RV] for the finite horizon problem, and to [BCD] for the infinite horizon problem.

An immediate consequence of Theorem 2.10, analogous to Corollary 2.3, is that the multivalued feedback

$$S_D(z) := \left\{ a \in A \ : \ V(z) - \frac{\partial^- V(z)}{\partial f(z, a)} = 1 \right\}$$

is fully optimal if for every initial point there exists an optimal open loop control and V is locally Lipschitz continuous (and not necessarily semiconcave as in Corollary 2.3).

We refer to [Sua], [Fr1], [CF], [Z1], [Z2] and [BCD] for the connection between the conditions of optimality of this section and the classical Pontryagin Maximum Principle, and to [Mi3] [Mi4] for verification theorems and construction of optimal multivalued feedbacks by various tools of nonsmooth analysis.

3. Differential Games.

We are given a dynamical system controlled by *two players*

(S_2)
$$\begin{cases} y'(t) = f(y(t), a(t), b(t)), & t > 0 \\ y(0) = x \end{cases}$$

where $a(\cdot) \in \mathcal{A}$ and $b(\cdot) \in \mathcal{B} := \{b : [0, +\infty[\to B \text{ measurable}\}$, B is a given compact metric space, $x \in \mathbb{R}^N$,

$$f : \mathbb{R}^N \times A \times B \to \mathbb{R}^N \quad \text{is continuous}$$

and

$$|f(x, a, b) - f(y, a, b)| \leq L|x - y| \quad \forall x, y \in \mathbb{R}^N, \ a \in A, \ b \in B.$$

We denote by $y_x(\cdot, a, b)$ the solution of (S_2). We are also given a functional J which the first player wants to minimize by choosing $a(\cdot)$ and the second player wants to maximize by choosing $b(\cdot)$. This is called a *two-person zero-sum differential game* (the sum of the costs the two players have to pay is zero). As a model problem we consider the discounted minimum time problem, that is,

$$J(x, a, b) := \int_0^{t_x(a, b)} e^{-s} ds$$

for $x \in \mathbb{R}^N$, $a(\cdot) \in \mathcal{A}$, $b(\cdot) \in \mathcal{B}$, where t_x is the first time the system reaches a given closed target $\mathcal{T} \subseteq \mathbb{R}^N$, i.e.

$$t_x(a, b) = \begin{cases} \min\{t \ : \ y_x(t, a, b) \in \mathcal{T}\}, & \text{if } \{t \ : \ y_x(t, a, b) \in \mathcal{T}\} \neq \emptyset, \\ +\infty & \text{otherwise.} \end{cases}$$

Here are two classical problems which can be modeled in this framework.

Pursuit-evasion games. In these games each player controls an object, and the first wants to approach the second as soon as possible, the second wants to be approached as late as possible. Here the state variable splits $y = (y_A, y_B) \in \mathbb{R}^M \times \mathbb{R}^M$, the differential equation in (S_2) is decoupled

$$y'_A = f_A(y_A, a), \quad y'_B = f_B(y_B, b),$$

and the target is, for instance,

$$\mathcal{T} := \{(y_A, y_B) : |y_{A,i} - y_{B,i}| \leq \varepsilon \quad \forall 1 \leq i \leq k\}$$

for some $\varepsilon \geq 0$ and integer $k \leq M$ (here $y_{A,i}$ denotes the i-th component of y_A). See [Is] and [Le] for these games.

Control under lack of information. Suppose we really have only one controller $a(\cdot)$ of the system, but the behaviour of the system is affected by some unkown disturbance $b(\cdot)$. This is the typical situation modeled by stochastic control theory. However in some cases this theory is not appropriate because either we do not know any statistics of $b(\cdot)$ or it is not safe to minimize the expected value of the cost functional: in these circumstances we may want to minimize in the case of the worst possible behaviour of the disturbance. Then it is appropriate to model $b(\cdot)$ as a second player wishing to maximize our cost. This is strongly related to topics such as H_∞, robust and risk-sensitive control, see [BJ1], [BaBe], [FME], [J], [KK], [ME], [Sor3], [Sor4], [Sor8].

In order to define the value function of the game we have to make some assumptions on the information available to each player at a given instant of time. There are several possibilities and we list here some of them.

The static game. Suppose that at time $t = 0$ one player chooses his whole response to the whole future behaviour of the other player, then the other player makes his choice knowing this response. Then we have the static lower and upper value of the game, respectively

$$v_s(x) := \sup_{b(\cdot) \in \mathcal{B}} \inf_{a(\cdot) \in \mathcal{A}} J(x, a, b),$$

$$u_s(x) := \inf_{a(\cdot) \in \mathcal{A}} \sup_{b(\cdot) \in \mathcal{B}} J(x, a, b).$$

Here the decision process is not dynamical, and this is not realistic in most examples.

The game with feedback controls. Suppose that at each time t both players take their decision knowing only the current position of the state variable $y(t)$. This is quite a realistic information pattern, which is typical in automatic control. One can model it by considering feedback controls $\Phi : \mathbb{R}^N \to A$, $\Psi : \mathbb{R}^N \to B$, and trying to solve

(3.1)
$$\begin{cases} y' = f(y, \Phi(y), \Psi(y)), & t > 0, \\ y(0) = x \end{cases}$$

The existence and uniqueness of the trajectory is ensured under the assumption that f, Φ and Ψ are Lipschitz continuous. However this regularity of the feedbacks is not realistic if one is studying optimization problems because the typical examples of optimal feedbacks are discontinuous. Moreover it is not clear whether a Dynamic Programming Principle holds for Lipschitz feedbacks.

The games with nonanticipating strategies. Suppose that at each time t the first player chooses the current value of the control, knowing the control chosen by the opponent in the past and up to time t, i.e. $b(\cdot)_{|[0,t]}$. In other words, the first player chooses among the *nonanticipating strategies*, namely

$$\Gamma := \{\alpha : \mathcal{B} \to \mathcal{A} : \forall t > 0, \ b(\cdot)_{|[0,t]} = \hat{b}(\cdot)_{|[0,t]} \text{ implies } \alpha[b](s) = \alpha[\hat{b}](s), \ \forall s \leq t\}.$$

This leads to the Varayia, Roxin, Elliot and Kalton definition of *lower value* of the game

$$V(x) := \inf_{\alpha \in \Gamma} \sup_{b(\cdot) \in \mathcal{B}} J(x, \alpha[b], b).$$

Of course one can reverse the roles of the players in this information pattern and define the *upper value* of the game

$$U(x) := \sup_{\beta \in \Delta} \inf_{a(\cdot) \in \mathcal{A}} J(x, a, \beta[a]),$$

where Δ denotes the set of non anticipating strategies for the second player.

EXERCISE 3.1. Prove that $v_s \leq V \leq u_s$ and $v_s \leq U \leq u_s$. □

The names "lower" and "upper" value are motivated by the inequality

(3.2)
$$V(x) \leq U(x) \quad \text{for all } x,$$

which, however, is not obvious at the first glance. We will prove it later in a rather indirect way, by means of the Hamilton-Jacobi equations for V and U obtained by the Dynamic Programming method. The intuitive reason why (3.2) holds is that the player choosing strategies has the informational advantage of knowing at each time the choice the other player is making at the same time. The reader can easily convince himself of this advantage by doing the following exercise, where $V(x) < U(x)$ for all $x \notin \mathcal{T}$.

EXERCISE 3.2. Let $N = 1$, $f(x, a, b) = (a - b)^2$, $A = B = [-1, 1]$, $T = [0, +\infty[$. Show that $V(x) = 1 - e^x$, $U(x) = 1$ for all $x < 0$. □

The information pattern just described is not realistic in many examples, such as pursuit-evasion games, because of the advantage given to the player choosing strategies. However it is reasonable to believe that the outcome of any more fair game is between $V(x)$ and $U(x)$. This justifies the following

Definition 3.1. *If* $V(x) = U(x)$ *we say that the game with initial position* x *has a value, and we call* $V(x)$ *the value of the game.*

Two more information patterns will be described later, one at the end of this section and the other in next section.

Next we develop the Dynamic Programming method for games with nonanticipating strategies.

Theorem 3.2 (Dynamic Programming Principle). *For all* $x \notin T$ *and small* $s > 0$

$$V(x) = \inf_{\alpha \in \Gamma} \sup_{b(\cdot) \in \mathcal{B}} \left\{ \int_0^s e^{-t} dt + V(y_x(s, \alpha[b], b)) e^{-s} \right\},$$

$$U(x) = \sup_{\beta \in \Delta} \inf_{a(\cdot) \in \mathcal{A}} \left\{ \int_0^s e^{-t} dt + U(y_x(s, a, \beta[a])) e^{-s} \right\}.$$

□

See [ES] or [BCD] for the proof. The infinitesimal versions of the DPP are the lower and upper Hamilton-Jacobi-Isaacs equations (briefly, HJI) which involve the following Hamiltonians

$$H(x, p) := \min_{b \in B} \max_{a \in A} \{ -f(x, a, b) \cdot p \},$$

$$\tilde{H}(x, p) := \max_{a \in A} \min_{b \in B} \{ -f(x, a, b) \cdot p \}.$$

Theorem 3.3. ([EI], [BS0]) *and continuous viscosity solution of*

(3.3)
$$\begin{cases} V + H(x, DV) = 1 & \text{in } \mathbb{R}^N \setminus T, \\ V = 0 & \text{on } T. \end{cases}$$

If the upper value function U *is continuous, then it is the unique bounded and continuous viscosity solution of*

(3.4)
$$\begin{cases} U + \tilde{H}(x, DU) = 1 & \text{in } \mathbb{R}^N \setminus T, \\ U = 0 & \text{on } T. \end{cases}$$

□

We refer to [ES] or [BCD] for the proof that each value satisfies the corresponding HJI equation. We just remark that in the proofs that V is a subsolution of the lower HJI equation and U is a supersolution of the upper HJI equation we really use the information pattern where the player using strategies knows the other player's choice at the current time. The proofs of uniqueness follow immediately from the general comparison principle for the Dirichlet problem, see e.g. [CL1], [CEL], [Ba2], [C], [BCD]. It is enough to check that the Hamiltonians H and $\tilde{H}$ satisfy the same local Lipschitz estimate (1.5) as for problems with a single player. As in Theorem 1.9, the comparison principle implies that the lower value V (resp. the upper value U) is the minimal supersolution and the maximal subsolution of the boundary value problem (3.3) (resp. (3.4)) among bounded functions. From this, one can derive verification theorems similar to Corollary 1.14, see [BCD]. Another consequence is the announced proof of the inequality (3.2).

Corollary 3.4. *If V and U are continuous, then $V \leq U$.*

Proof. Since $H \geq \tilde{H}$, V is a subsolution of (3.4). Then the comparison principle gives $V \leq U$ because U is the solution of (3.4). ∎

Finally we give a sufficient condition for the existence of the value of the game.

Corollary 3.5. *If V and U are continuous and*

$$(3.5) \qquad H(x,p) = \tilde{H}(x,p) \quad for \ all \ x, p \in \mathbb{R}^N,$$

then $V = U$ and the game has a value for all initial points. □

This is an obvious consequence of Theorem 3.3 because the lower and the upper HJI equations are the same. The equality (3.5) is called *Isaacs' condition* in the western literature and *solvability of the small game* in the russian literature. The meaning of the last name is easily explained: (3.5) is equivalent to the existence of a saddle for the static (two-person, zero-sum) game over the sets A and B whose payoff is $-f(x,a,b) \cdot p$, for all choices of x and p. The simplest case where this condition is satisfied is that of systems with the separation property

$$f(x,a,b) = f_1(x,a) + f_2(x,b).$$

Note that pursuit-evasion games have this property.

The games with feedback strategies. Suppose that at each time t the first player knows the whole trajectory of the state variable $y(\cdot)$ in the interval $[0,t]$ and takes his decision accordingly. Note that this information pattern is intermediate between feedback controls and nonanticipating strategies. The mathematical description is the following. Let $\mathcal{Y}$ be the set of all trajectories of the system

$$\mathcal{Y} := \{y : [0,+\infty[\ \to \mathbb{R}^N \ : \ \exists x \in \mathbb{R}^N, a(\cdot) \in \mathcal{A}, b(\cdot) \in \mathcal{B}$$

$$\text{such that } y(\cdot) = y_x(\cdot,a,b)\}.$$

Definition 3.6. A feedback strategy *for the first player is a map* $\zeta : \mathcal{Y} \rightarrow \mathcal{A}$ *which is* nonanticipating, *i.e.*

$$y_{|[0,t]} = \hat{y}_{|[0,t]} \quad implies \quad \zeta[y](s) = \zeta[\hat{y}](s) \ for \ all \ s \leq t,$$

and playable, *i.e. for all* $x \in \mathbb{R}^N$, $b(\cdot) \in \mathcal{B}$ *and* $T > 0$ *there exists a unique solution of*

(3.6)
$$\begin{cases} q'(t) = f(q(t), \zeta[q](t), b(t)), & 0 < t \leq T \\ q(0) = x. \end{cases}$$

A feedback strategy for the second player is a nonanticipating *and* playable *map* $\xi : \mathcal{Y} \rightarrow \mathcal{B}$.

We denote with $\mathcal{F}$ and $\mathcal{G}$, respectively, the sets of feedback strategies for the first and the second player. We also denote with $y_x(\cdot; \zeta, b)$ the solution of (3.6), with $y_x(\cdot; a, \xi)$ the trajectory of the system corresponding to $a(\cdot) \in \mathcal{A}$ and $\xi \in \mathcal{G}$, and with $J(x; \zeta, b)$ and $J(x; a, \xi)$ the corresponding payoffs in the discounted minimum time problem, that is, for instance,

$$J(x; \zeta, b) := 1 - exp(-\inf\{t : y_x(t; \zeta, b) \in \mathcal{T}\}).$$

Now we can define the A-feedback value

$$v_A(x) := \inf_{\zeta \in \mathcal{F}} \sup_{b(\cdot) \in \mathcal{B}} J(x; \zeta, b)$$

the B-feedback value

$$v_B(x) := \sup_{\xi \in \mathcal{G}} \inf_{a(\cdot) \in \mathcal{A}} J(x; a, \xi),$$

and when they coincide we call $v_A(x) = v_B(x)$ the *feedback value* of the game. These definitions are special cases of those given by Soravia [Sor4] for games with partial information.

EXERCISE 3.3. Prove that $V \leq v_A$ and $v_B \leq U$. □

The Dynamic Programming Principle still holds for the games with feedback strategies:

Theorem 3.7. (DPP). *For all* $x \notin \mathcal{T}$ *and small* $s > 0$

$$v_A(x) = \inf_{\zeta \in \mathcal{F}} \sup_{b(\cdot) \in \mathcal{B}} \left\{ \int_0^s e^{-t} dt + v_A(y_x(s; \zeta, b)) e^{-s} \right\},$$

$$v_B(x) = \sup_{\xi \in \mathcal{G}} \inf_{a(\cdot) \in \mathcal{A}} \left\{ \int_0^s e^{-t} dt + v_B(y_x(s; a, \xi)) e^{-s} \right\}.$$

□

For the proof we refer to [BCD]. Unfortunately we cannot prove without further assumptions that v_A or v_B satisfies the lower or upper HJI equation. However it is not hard to deduce from the DPP the following

Corollary 3.8. *Assume v_A and v_B are continuous. Then they are supersolutions of (3.3) and subsolutions of (3.4).*   □

This partial result is enough to prove the existence of the feedback value under the Isaacs' condition (3.5) by means of the comparison principle for the Dirichlet problem (3.3). In fact we obtain immediately the following

Theorem 3.9. *If v_A and v_B are continuous then*

$$V \leq v_A \leq U, \qquad V \leq v_B \leq U.$$

In particular, if $H = \tilde{H}$, then $v_A = v_B = V = U$.   □

For more details and examples on the games with feedback strategies and for some remarks on optimal strategies and saddles we refer the reader to Chapter VIII of the book [BCD]. We refer also to the books of Krasovski and Subbotin [KS], [Su2] and [KK] for the theory of *positional differential games* where the game with feedback controls (3.1) is modelled by using constant feedbacks over short time intervals and then letting the length of these intervals go to 0. Under the Isaacs' condition (3.5), the value function defined in this theory is the unique *minimax solution* of the BVP (3.3) (and (3.4)), a notion introduced independently by Subbotin and then proved to be equivalent to viscosity solutions, see [Su2] [LS] and the references therein. Therefore the value function of Krasovski and Subbotin coincides with the value of Definition 3.1 and with the feedback value by Theorems 3.3 and 3.9, provided it is continuous. In the next section we will recall one more notion of value, due to Fleming [F] in the early 60s, and prove its equivalence with the other values under the Isaacs' condition. Also the value functions defined by Friedman and Berkovitz have been studied within the theory of viscosity solutions, see [Sor0], [Be2] and the references therein.

Finally let us comment on the continuity of the value functions we have assumed in Theorems 3.3 and 3.9. This assumption is satisfied in general for problems where the state variables are unrestricted, such as the Infinite Horizon and the Finite Horizon problems in Section 1, but it does *not* hold in general for the discounted minimum time problem we are discussing. A sufficient condition of continuity can be given as in Theorem 1.12 by replacing (1.6) with

$$\min_{a \in A} \max_{b \in B} f(x, a, b) \cdot n(x) < 0,$$

see [BS0]. However this is a strong assumption and there are indeed many classical problems where the value function is discontinuous, see [Is], [Le]. This is our main motivation to weaken further the notion of viscosity solution so that discontinuous solutions are allowed: this will be the subject of Section 5.

The reader interested in a viscosity solution treatment of other two-person zero-sum differential games is referred to [ES] for the finite horizon problem, [BCD] for the infinite horizon problem, [BS1], [BS2], [Sor1] for the undiscounted minimum time problem, [L3] for games with reflected processes, [A] and [Ko] for problems with state constraints, [BJM] and [Y2] for systems with impulses, [Y1] for problems with switching costs, [Br] for games with maximum cost, [ME], [Sor3], [Sor8], [R] for problems where at least one player can use unbounded controls, [FSou] for stochastic games. The stability of dynamical systems controlled by two competing players has also been studied with these methods by Soravia [Sor1] [Sor6].

4. Approximation of viscosity solutions and construction of almost optimal feedbacks.

In this section we come back to the problem of the synthesis of an optimal feedback, which is the most important step of the Dynamic Programming method from the point of view of applications. In Section 2 we showed how to construct an optimal multivalued feedback from the knowledge of the value function, provided the value function is regular enough, at least Lipschitz continuous. In practice, however, the value function is very seldom known explicitely, and there is no hope to compute it exactly for general nonlinear systems of realistic dimension.

In this section we show how the theory of viscosity solutions can be used to prove the convergence of approximation schemes in very general situations. The method is illustrated in part B on a simple semidiscrete scheme based on Dynamic Programming for discrete-time systems. An important feature of this scheme is the simultaneous computation of the value function and of an almost optimal feedback. Therefore it can be used to solve optimal control problems numerically, with no need of the theory and the assumptions of Section 2. Two important technical tools of independent interest will be introduced in part A of this section: the interpretation of Dirichlet boundary conditions in an appropriate generalized sense and certain relaxed semi-limits often called *weak limits in the viscosity sense*. They are due, respectively, to Ishii [I3] and Barles and Perthame [BP1] [BP2], and were first applied to the convergence of discrete approximations by the author and Falcone [BF1] and the author and Soravia [BS3]. For the many other applications of these methods we refer the reader to [FS], [Ba2], [BCD] and their bibliographies.

A. Boundary conditions and weak limits in the viscosity sense.

The Dirichlet boundary value problem with boundary conditions in viscosity sense is the following

$$(4.1) \qquad \begin{cases} F(x, u, Du) = 0 & \text{in } \Omega, \\ u = g \quad \text{or} \quad F(x, u, Du) = 0 & \text{on } \partial\Omega, \end{cases}$$

where $\Omega \subseteq \mathbb{R}^N$ is open and $F : \overline{\Omega} \times \mathbb{R} \times \mathbb{R}^N \to \mathbb{R}$ is continuous. The definition of solution for this boundary value problem is the following.

Definition 4.1. *A function $u \in USC(\overline{\Omega})$ (resp. $LSC(\overline{\Omega})$) is a viscosity subsolution (resp. supersolution) of (4.1) if it is a subsolution (resp. supersolution) of $F(x, u, Du) = 0$ in Ω and, for any $\phi \in C^1(\overline{\Omega})$ and any local maximum (resp. minimum) point $x \in \partial\Omega$ of $u - \phi$,*

$$(4.2) \qquad \min\{u(x) - g(x), F(x, u(x), D\phi(x))\} \le 0,$$

(resp.

$$\max\{u(x) - g(x), F(x, u(x), D\phi(x))\} \ge 0).$$

The main property of the boundary value problem (4.1) is the stability with respect to certain weak limit operations which we define next.

Definition 4.2. *The upper weak limit of the sequence $u_n : \overline{\Omega} \to \mathbb{R}$ is*

$$\overline{u}(x) := \limsup_{n \to \infty}{}^* u_n(x)$$
$$:= \lim_{j \to \infty} \sup\{u_n(y) \,:\, n \ge j,\; y \in \overline{\Omega},\; |y - x| \le 1/j\},$$

and the lower weak limit is

$$\underline{u}(x) := \liminf_{n \to \infty}{}_* u_n(x)$$
$$:= \lim_{j \to \infty} \inf\{u_n(y) \,:\, n \ge j,\; y \in \overline{\Omega},\; |y - x| \le 1/j\}.$$

Here is the precise statement of the stability property of (4.1).

Proposition 4.3. *If u_n are subsolutions (resp. supersolutions) of (4.1), then the upper weak limit $\overline{u}$ (resp. the lower weak limit $\underline{u}$) is a subsolution (resp. supersolution) of (4.1).* $\qquad\square$

The boundary value problem (4.1) is indeed stable also with respect to perturbations of the equation in (4.1), even singular perturbations such as the regularization via vanishing viscosity. Next result is an example which justifies the name given to the boundary condition.

Proposition 4.4. *Let u_n be a classical solution of the Dirichlet problem*

$$(4.3) \qquad \begin{cases} -\dfrac{1}{n}\Delta u_n + F_n(x, u_n, Du_n) = 0 & in \ \Omega, \\[2mm] u_n = g & on \ \partial\Omega, \end{cases}$$

with $g \in C(\partial\Omega)$, and assume F_n converge uniformly to F on compact sets and for some C

$$(4.4) \qquad \sup_{\overline{\Omega}} |u_n| \le C, \quad for \ all \ n.$$

Then the upper weak limit $\bar{u}$ of u_n is a subsolution of (4.1) and the lower weak limit $\underline{u}$ is a supersolution. $\qquad\qquad\square$

The proofs of Proposition 4.3 and 4.4 are easy consequences of the definition of viscosity solution and the following elementary lemma, first observed in this context by Barles and Perthame [BP1].

Lemma 4.5. *Let $\varphi \in C^1(\overline{\Omega})$ and $\bar{u} = \limsup^{*}_{n\to\infty} u_n$, with u_n upper semicontinuous. If $\bar{u} - \varphi$ has a strict maximum at $x \in \overline{\Omega}$, then there is a subsequence $\{n_k\}$ such that $u_{n_k} - \varphi$ has a local maximum at x_{n_k} with*

$$(4.5) \qquad \lim_k x_{n_k} = x, \qquad \lim_k u_{n_k}(x_{n_k}) = \bar{u}(x).$$ $\qquad\square$

The proof of this lemma is left to the reader or can be found in [Ba2], [BCD], [CIL], [C]. Of course a corresponding statement can be given for the lower weak limit.

Proof of Proposition 4.4. By (4.4) both weak limits are finite everywhere. Let $\phi \in C^1(\overline{\Omega})$ and $x \in \overline{\Omega}$ be a maximum point of $\bar{u} - \phi$. We consider the case $x \in \partial\Omega$, since the other is similar and easier. By adding a quadratic term to ϕ we can assume without loss of generality that x is a strict maximum for $\bar{u} - \phi$. We consider the subsequence of local maxima of $u_n - \phi$ given by Lemma 4.5 and relabel it $\{x_n\}$.

Case (a). If a subsequence of $\{x_n\}$ lies in $\partial\Omega$, by the boundary condition in (4.3) we have

$$u_n(x_n) = g(x_n)$$

so by (4.5)

$$\bar{u}(x) = g(x)$$

and (4.2) is satisfied by $u = \bar{u}$.

Case (b). If case (a) does not occur there is a subsequence of $\{x_n\}$ lying in Ω. Then by standard calculus

$$D(u_n - \phi)(x_n) = 0, \qquad \Delta(u_n - \phi)(x_n) \leq 0.$$

From the PDE in (4.3) we get

$$-\frac{1}{n}\Delta\phi(x_n) + F_n(x_n, u_n(x_n), D\phi(x_n)) \leq 0$$

and letting $n \to \infty$

$$F(x, \bar{u}(x), D\phi(x)) \leq 0$$

which gives (4.2) with $u = \bar{u}$ also in this case.

The proof for $\underline{u}$ is similar. ∎

The argument of the previous proof can be easily used to demonstrate the following statement which includes both Proposition 4.3 and 4.4 as special cases.

EXERCISE. Let u_n be viscosity subsolutions of

$$\begin{cases} F_n(x, u, Du, D^2u) = 0 & \text{in } \Omega, \\ u = g \text{ or } F_n(x, u, Du, D^2u) = 0 & \text{on } \partial\Omega, \end{cases}$$

where $F_n : \overline{\Omega} \times \mathbb{R} \times \mathbb{R}^N \times S(N) \to \mathbb{R}$ and $S(N)$ is the set of real symmetric $N \times N$ matrices. The definition of viscosity subsolution for second order equations is analogous to Definitions 1.7 and 4.1, and can be found e.g. in [CIL], [C]. Assume $g \in C(\partial\Omega)$, $F_n \to F$ uniformly on compact sets, F continuous, and (4.4). Then $\bar{u} = \limsup^*_{n\to\infty} u_n$ is a subsolution of

$$\begin{cases} F(x, u, Du, D^2u) = 0 & \text{in } \Omega, \\ u = g \text{ or } F(x, u, Du, D^2u) = 0 & \text{on } \partial\Omega. \end{cases} \qquad \square$$

The stability properties of (4.1) with respect to weak limits are particularly useful when coupled with the following comparison theorem for the boundary value problem (4.1).

Theorem 4.6. *Let $u_1 \in USC(\overline{\Omega})$ and $u_2 \in LSC(\overline{\Omega})$ be, respectively, a sub and a supersolution of*

$$u + H(x, Du) = 0 \quad in \quad \Omega$$

where H satisfies (1.5) and Ω is a bounded open set with Lipschitz boundary. If u_1 is continuous at each point of $\partial\Omega$ and u_2 satisfies

$$u_2 \geq u_1 \quad or \quad u_2 + H(x, Du_2) \geq 0 \quad on \quad \partial\Omega$$

in the viscosity sense, then $u_1 \leq u_2$ in $\overline{\Omega}$. The same conclusion holds if u_2 is continuous at points of $\partial\Omega$ and

$$u_1 \leq u_2 \quad or \quad u_1 + H(x, Du_1) \leq 0 \quad on \quad \partial\Omega \qquad \qquad \square$$

The proof of this result can be found in [I3] and [BCD], an extension to unbounded Ω is in [BS2]. It is easy to combine Proposition 4.4 and Theorem 4.6 to get the following convergence result for the vanishing viscosity problem (4.3).

Corollary 4.7. *Under the hypotheses of Proposition 4.4, assume in addition* $F(x, r, p) = r + H(x, p)$ *with* H *satisfying* (1.5), *and* Ω *bounded with Lipschitz boundary. If there exists a solution* $v \in C(\overline{\Omega})$ *of*

$$(4.6) \qquad \begin{cases} v + H(x, Dv) = 0 & in \ \Omega, \\ v = g & on \ \partial\Omega, \end{cases}$$

then $u_n \to v$ *uniformly.*

Proof. By Proposition 4.4, $\underline{u} = \liminf_* u_n$ is a supersolution of (4.1), thus (4.6) and the first statement of Theorem 4.6 give $v \leq \underline{u}$. Simlilarly, the last statement of Theorem 4.6 gives $\bar{u} \leq v$. Then $\underline{u} = \bar{u} = v$, and it is not hard to show that this implies the uniform convergence of u_n to v. ∎

We call the reader's attention on the very weak compactness assumptions of this convergence result: we have passed to the limit in a singular perturbation problem for a fully nonlinear equation with just an L^∞ uniform a priori estimate on the approximating u_n (and indeed local L^∞ estimates are enough).

B. Convergence of semidiscrete approximations.

We consider the Euler scheme of step $h > 0$ for the dynamical system controlled by two players (S_2), namely

$$(DS) \qquad \begin{cases} y_{n+1} = y_n + hf(y_n, a_n, b_n), \\ y_0 = x, \end{cases}$$

where $a. := \{a_n\} \in A^{\mathbf{N}}$, $b. := \{b_n\} \in B^{\mathbf{N}}$. We regard (DS) as a discrete time dynamical system and we consider the natural counterpart for this system of the discounted minimum time differential game described in Section 3. The payoff of the discrete time game is

$$J_h(x, a., b.) := 1 - e^{-h n_h(x, a., b.)}$$

where

$$n_h(x, a., b.) := \min\{n \in \mathbf{N} : y_n \in \mathcal{T}, \ y. \ solution \ of \ (DS)\}$$

and $\mathcal{T}$ is the target. Note that one expects that $hn_h(x, a_., b_.)$ approximates $t_x(a, b)$ for h small, where $a(t) := a_{[t/h]}$, $b(t) := b_{[t/h]}$ ([s] denotes the integer part of s). We consider the *minorant game* where at time nh the first player chooses a_n knowing the choice b_n the other player is making. The value of this game is

$$v_h(x) := \inf_{\alpha \in \Lambda} \sup_{b_.} J_h(x, \alpha[b_.], b_.),$$

where Λ is the set of nonanticipating discrete strategies for the first player

$$\Lambda := \{\alpha : B^{\mathbf{N}} \to A^{\mathbf{N}} \mid b_j = \hat{b}_j \ for \ j \leq n \ implies \ \alpha[b_.]_j = \alpha[\hat{b}_.]_j \ for \ j \leq n\}.$$

The *majorant game* is defined by giving the informational advantage to the second player instead of the first, and its value is

$$u_h(x) := \sup_{\beta \in \Theta} \inf_{a_.} J_h(x, a_., \beta[a_.])$$

where Θ is the set of nonanticipating discrete strategies for the second player. The *Dynamic Programming Principle* for the minorant game is the following.

Proposition 4.8. *For a function* $u : \mathbf{R}^N \to \mathbf{R}$ *define*

$$Su(x) := \sup_{b \in B} \inf_{a \in A} e^{-h} u(x + hf(x, a, b)) + 1 - e^{-h}.$$

Then

(4.7) $$v_h(x) = Sv_h(x) \ for \ all \ x \notin \mathcal{T}. \qquad \square$$

Equation (4.7) plays the role of Isaacs' equation for the discrete time minorant game. Of course a similar equation holds for the value u_h of the majorant game. Next result characterizes v_h as the unique solution of (4.7) in the set

$$B_0(\mathbf{R}^N) = \{u : \mathbf{R}^N \to \mathbf{R} \ bounded \ and \ null \ on \ \mathcal{T}\}.$$

Lemma 4.9. *The map* $S : B_0(\mathbf{R}^N) \to B_0(\mathbf{R}^N)$ *is a contraction and preserves order. In particular* v_h *is the unique fixed point of* S *in* $B_0(\mathbf{R}^N)$. $\qquad \square$

The final result on discrete time games says that one can construct an optimal feedback strategy for the first player and an optimal feedback control for the second player from the knowledge of the value v_h by solving for all $y \in \mathbf{R}^N$ the finite dimensional problem

$$\max_{b \in B} \min_{a \in A} v_h(y + hf(y, a, b)).$$

Proposition 4.10. *Let* $F : \mathbf{R}^N \times B \to A$ *be such that*

$$F(y, b) \in \operatorname*{argmin}_{a \in A} v_h (y + hf(y, a, b)).$$

For any sequence $b_. \in B^{\mathbf{N}}$ *and any* $x \in \mathbf{R}^N$ *consider the sequence* $\{z_n[b_.]\}$ *in* $\mathbf{R}^N$ *defined by*

$$\begin{cases} z_{n+1} = z_n + hf(z_n, F(z_n, b_n), b_n), \\ z_0 = x, \end{cases}$$

and define $\alpha^*[b_.]_n := F(z_n, b_n)$. *Then* $\alpha^* \in \Lambda$ *is optimal for the initial point* x, *i.e.*

$$v_h(x) = \sup_{b_.} J_h(x, \alpha^*[b_.], b_.).$$

Moreover

$$v_h(x) = J_h(x, \alpha^*[b_.^*], b_.^*)$$

for any $b_.^* \in B^{\mathbf{N}}$ *such that* $b_n^* \in G(z_n[b_.^*])$ *where*

$$G(y) := \operatorname*{argmax}_{b \in B} \ \min_{a \in A} v_h (y + hf(y, a, b)). \qquad \square$$

The proofs of these three results are not hard, and we refer the reader to [BF1], [BS3], [BCD].

In view of the rather simple theory of discrete time games, we use them to approximate the differential game presented in Section 3. Next theorem is taken from [BS3] and extends previous work of [BF1] for the case of a single player. The pioneering result of this kind is due to Capuzzo Dolcetta [CD] for the infinite horizon problem where, however, the approximating value functions v_h are equicontinuous, so a simpler proof can be given, while here each v_h is discontinuous.

Theorem 4.11. *Assume the basic hypotheses of Section 3 and* $\Omega := \mathbf{R}^N \setminus \mathcal{T}$ *bounded with Lipschitz boundary. If the lower (resp. upper) value function* V *(resp. U) of the differential game is continuous, then* v_h *converge to* V *(resp. u_h converge to U) uniformly as* $h \to 0$. $\qquad \square$

The reader is referred to [BS3] for a similar statement in the case of unbounded Ω, and to Section 5 below for some extensions of this result to the case of discontinuous value function. An immediate consequence of Theorem 4.11 and Corollary 3.5 is the equivalence between the Definition 3.1 of the value of the game and the following definition due to Fleming [F]

$$(4.8) \qquad w(x) := \lim_{h \to 0} v_h(x) = \lim_{h \to 0} u_h(x).$$

Corollary 4.12. *Assume all the hypotheses of Theorem 4.11 and the Isaacs' condition (3.5). Then the two limits in (4.8) coincide and* $w = U = V$.

Sketch of proof of Theorem 4.11. This proof is based on the ideas described in Part A of this section. We consider the upper and the lower weak limits of v_h

$$\bar{v}(x) := \limsup_{h \to 0}{}^* v_h(x)$$

$$:= \lim_{\delta \to 0} \sup\{v_h(y) \ : \ 0 < h < \delta, \ y \in \overline{\Omega}, \ |y - x| < \delta\}.$$

and

$$\underline{v}(x) := \liminf_{h \to 0}{}_* v_h(x)$$

$$:= \lim_{\delta \to 0} \inf\{v_h(y) \ : \ 0 < h < \delta, \ y \in \overline{\Omega}, \ |y - x| < \delta\}.$$

Note that $0 \le v_h \le 1$, so $0 \le \underline{v} \le \bar{v} \le 1$. We claim that $\bar{v}$ and $\underline{v}$ are, respectively, a sub and a supersolution of

(4.9)
$$\begin{cases} u + H(x, Du) = 1 & \text{in } \Omega, \\ u = 0 \quad \text{or} \quad u + H(x, Du) = 1 & \text{on } \partial\Omega, \end{cases}$$

in the sense of Definition 4.1, where

$$H(x, p) = \min_{b \in B} \max_{a \in A}\{-f(x, a, b) \cdot p\}.$$

Then we can follow the argument of Corollary 4.7: since V solves 4.9 (Theorem 3.3) and $V = 0$ on $\partial\Omega$, the comparison theorem 4.6 implies

$$\bar{v} \le V \le \underline{v},$$

so $\bar{v} = V = \underline{v}$ which gives the conclusion.

Next we prove the claim in a simpler case, namely the undiscounted minimum time problem. The proof for the discounted problem is essentially the same, but with longer calculations.

The value function w_h of the undiscounted discrete time problem satisfies the following Dynamic Programming Principle

(4.10)
$$w_h(x) = \sup_{b \in B} \inf_{a \in A} w_h(x + hf(x, a, b)) + h,$$

and we will prove that the upper weak limit $\bar{w} = \limsup_{h \to 0}{}^* w_h$ is a subsolution of

(4.11)
$$H(x, Du) = 1 \quad \text{in } \Omega.$$

We take $\varphi \in C^1(\overline{\Omega})$ such that $x \in \Omega$ is a strict maximum point of $\bar{w} - \varphi$ (the case that $x \in \partial\Omega$ can be treated similarly, as in the proof of Proposition 4.4). Let us assume for simplicity that w_h is upper semicontinuous, so that we can apply Lemma 4.5 (in the general case we should replace w_h with its upper semicontinuous envelope). Then there exist $h_n \to 0$ and $x_n \to x$ such that $w_{h_n} - \varphi$ has a maximum at x_n. We denote

$$y_n := x_n + h_n f(x_n, a, b), \qquad w_n := w_{h_n}.$$

From (4.10) and the property of x_n we get

$$\inf_{b \in B} \sup_{a \in A} \frac{\varphi(x_n) - \varphi(y_n)}{h_n} \leq \inf_{b \in B} \sup_{a \in A} \frac{w_n(x_n) - w_n(y_n)}{h_n} \leq 1.$$

By the differentiability of φ we obtain

$$\inf_{b \in B} \sup_{a \in A} \{-D\varphi(x_n) \cdot f(x_n, a, b) + o(h_n)\} \leq 1.$$

Finally we let $n \to \infty$ and get

$$H(x, D\varphi(x)) \leq 1,$$

which proves that $\bar{w}$ is a subsolution of (4.11). ∎

There are several related results obtained by methods of the theory of viscosity solutions. Explicit estimates of the rate of convergence of the form

$$\sup |v_h - v| \leq Ch^{\frac{\gamma}{2}}$$

where obtained in [Sou] with $\gamma = 1$ for the finite horizon problem with Lipschitz continuous value v, and in [CDI] for the infinite horizon problem with one player if v is Hölder continuous with exponent γ (here and in the following we are indicating by v the value function and v_h its semidiscrete approximation, no matter what control problem we are considering). The better estimate

$$\sup |v_h - v| \leq Ch^\gamma$$

was proved in [CDI] with $\gamma = 1$ for v semiconcave, and in [BF2] for the minimal time function (one player) if it is γ-Hölder continuous and there is a bound on the total variation of the optimal controls. The Euler approximation in (DS) can be replaced by higher order schemes to obtain better estimates, see [FaF]. Results on the rate of convergence for other approximation schemes are given in [BSa1], [BSa2], [LT].

From the point of view of applications the next natural questions to ask are whether the optimal feedbacks for the discrete time problem constructed in Proposition 4.10 are close to be optimal for the original differential game, and converge to optimal controls for this game, as $h \to 0$. Some answers for the infinite horizon problem with a single player can be found in [CD], [CDI], [BCD] under suitable assumptions and they are of the following kind. Let $a_\cdot^h \in A^{\mathbf{N}}$ be an optimal control for the discrete time minimization problem with step $h > 0$ and initial point x, i.e.

$$v_h(x) = J_h(x, a_\cdot^h),$$

and extend it to a piecewise constant control $a^h(\cdot) \in \mathcal{A}$ in the obvious way

$$a^h(t) := a^h_{[t/h]}.$$

Then $a^h(\cdot)$ is a minimizing sequence for the continuous time problem, i.e.

$$\lim_{h \to 0} J(x, a^h(\cdot)) = v(x);$$

moreover there is a subsequence $h_n \to 0$ such that, on $[0, T]$, $a^{h_n}(\cdot)$ converge weakly star in L^∞ to an optimal relaxed control $a^*(\cdot)$, and the corresponding trajectories $y_x(\cdot, a^{h_n})$ converge uniformly to $y_x(\cdot, a^*)$, for all $T > 0$.

It is important to mention that the semidiscrete approximation scheme presented here can be further discretized with respect to space in order to get an algorithm which is fit for numerical computations. This was first done by Falcone [Fa] for the infinite horizon problem with one player; he also proved estimates of the rate of convergence and found an acceleration method which speeds up considerably the computations. We refer the reader to the survey paper [CDF] and the books [FS], [BCD], [BPR] for general presentations of numerical methods for viscosity solutions of first and second order PDEs, to [CFa2] for problems with state constraints, to [CFLS] for an improvement of the algorithm using domain decomposition methods, to [CFa1] for estimates in the infinite horizon stochastic control problem, and to [BSou] for the convergence of an approximation scheme for stochastic games.

A fully discrete scheme for the discounted minimum time differential game was studied in [A] for the special case of a lion pursuing a slower man in a bounded set, and in [BFS] for the general problem discussed here. In this paper the proof of convergence is based on the same ideas as the proof of Theorem 4.11 if the value function is continuous. The case of discontinuous value function is studied in [BFS] and [Sor2] for problems with one player, and in [BBF] for games, by using some results of the theory of discontinuous viscosity solutions, see next Section 5.

Other recent numerical methods for optimal control problems can be found in [KD], [P1] [GT], [CQS1], and for differential games in [Su3], [TG], [CQS2], [P2].

5. Well-posedness of the Dirichlet problem among discontinuous functions.

As we underlined at the end of Sections 1 and 3, the value functions of control problems and games involving the exit time of the trajectories from a given domain are not continuous in general, no matter how smooth the data are. Since viscosity solutions are continuous by definition, we have to give a weaker notion of solution if we want to characterize the value function as the unique solution of the natural boundary value problem for the HJB or HJI equation, e.g. (DP) and (3.3) for the discounted minimum time.

Also the Dirichlet problem for second order degenerate elliptic equations has sometimes a natural candidate solution which is discontinuous. Consider for instance the linear equation

(5.1)
$$\begin{cases} -\frac{1}{2}a_{ij}(x)u_{x_ix_j} - f(x)\cdot Du + u = 1 & \text{in } \Omega, \\ u = 0 & \text{on } \partial\Omega. \end{cases}$$

If this problem has a classical solution u, then by stochastic calculus one finds the representation formula

(5.2)
$$u(x) = E(1 - e^{-t_x})$$

where E denotes the expectation and t_x is the first exit time from Ω of the solution of the stochastic differential equation

(5.3)
$$\begin{cases} dy_t = f(y_t)dt + a^{\frac{1}{2}}(y_t)dW_t, \\ y_0 = x, \end{cases}$$

where $a^{\frac{1}{2}}$ is the square root or the matrix (a_{ij}) and W_t is a Brownian motion. If in a neighbourhood of some point $x \in \partial\Omega$ the diffusion term vanishes, i.e. $a_{ij} \equiv 0$, and the solution of (5.3) enters Ω for small $t > 0$, then it is easy to see from (5.2) that u is discontinuous at x, and it is not hard to give examples where the discontinuity also propagates in Ω.

In this section we discuss three different approaches to the well-posedness of the Dirichlet problem among discontinuous functions, following the chronological order. The first two are restricted at the moment to first order equations with convex Hamiltonian with respect to the p variables, so they apply only to deterministic optimal control problems, the third works for fully nonlinear nonconvex degenerate elliptic equations in the same generality as in [CIL] and [C].

Approach 1: Ishii's definition and the "complete solution".

In [I2] Ishii proposed a notion of viscosity solution which does not imply continuity automatically, by taking the upper and lower semicontinuous envelopes of a locally bounded function u

$$u^*(x) := \lim_{r \to 0} \sup\{u(y) \; : \; |x - y| \leq r\},$$

$$u_*(x) := \lim_{r \to 0} \inf\{u(y) \; : \; |x - y| \leq r\},$$

and saying that u solves a given PDE if u^* and u_* are, respectively, a viscosity sub and supersolution of the PDE. This is a very weak notion of solution: note that the characteristic function of the rationals solves $u' = 0$ in this sense.

This definition is used in [I3] to study deterministic control problems involving the exit times from a domain, and by Barles and Perthame [BP1] to prove the first uniqueness theorem among discontinuous solutions for the HJB equations of such problems. Here we limit ourselves to the discounted minimum time problem, and give a result by the author and Staicu [BSt] which is also fit for proving the convergence of approximation schemes.

We consider the boundary value problem

(5.5a)
$$u + H(x, Du) = 1 \quad \text{in} \quad \mathbb{R}^N \setminus \mathcal{T} =: \Omega,$$

(5.5b)
$$u \geq 0 \quad \text{on} \quad \partial \mathcal{T},$$

(5.5c)
$$u \leq 0 \quad \text{or} \quad u + H(x, Du) \leq 1 \quad \text{on} \quad \partial \mathcal{T},$$

where $\mathcal{T}$ is a given closed set and

(5.6)
$$H(x, p) := \max_{a \in A}\{-f(x, a) \cdot p\}$$

with f and A satisfying the basic assumptions of Section 1.

Definition 5.1. *A locally bounded function* $u : \overline{\Omega} \to \mathbb{R}$ *is a* non-continuous *viscosity solution of (5.5), briefly* nc-solution, *if* u^* *is a subsolution of (5.5a) and (5.5c) (see definition (4.1)), and* u_* *is a supersolution of (5.5a) and (5.5b). If, in addition,* u^* *is the maximal subsolution and* u_* *is the minimal supersolution, we say that* u *is a* complete solution *of (5.5).*

To motivate the name, let us note that, if a complete solution w exists, then its completed graph contains the graphs of all nc-solutions u, i.e.

$$w_* \leq u_* \leq u \leq u^* \leq w^*.$$

It is not hard to deduce from the Dynamic Programming Principle that the discounted minimum time function V is always a nc-solution of (5.5), and that also

$$\hat{v}(x) := \inf_{a(\cdot)} \left\{ \int_0^t e^{-s}ds \; : \; y_x(t,a) \in int\mathcal{T} \right\},$$

is such, see [I3]. Note that $\hat{v}$ is the discounted minimum time to reach the open target $int\mathcal{T}$.

Theorem 5.2. *Assume the basic hypotheses of Section 1, (5.6) and $\mathcal{T} = \overline{int\mathcal{T}}$. Then $\hat{v} = \hat{v}^*$ is the maximal bounded subsolution of (5.5a) (5.5c) and $\hat{v}_* = V_*$ is the minimal bounded supersolution of (5.5a) (5.5b). In particular $\hat{v}$ is the unique bounded upper semicontinuous complete solution of (5.5).* □

The proof of this result is a combination of control theoretic and PDE arguments, see [BSt], [BCD]. The following easy consequence is essentially the uniqueness theorem in [BP1].

Corollary 5.3. *Under the assumptions of Theorem 5.2, V_* is the unique bounded lower semicontinuous nc-solution of (5.5).* □

The existence of a complete solution allows to extend the method of weak limits described in Section 4A to the case where a continuous solution does not exist. Next result is an example of application of Theorem 5.2 to the semidiscrete approximation of Section 4B, in the special case of a single player, i.e. B is a singleton. In this case the approximate value function is

(5.7) $$v_h(x) = \inf_{a. \in A^N} \left(1 - e^{-hn_h(x,a.)} \right),$$

where n_h is the first step when the solution of (DS) reaches the target $\mathcal{T}$.

Corollary 5.4. *Under the assumptions of Theorem 5.2, v_h converge to $\hat{v}$, as $h \to 0$, uniformly on every compact set where $\hat{v}$ is continuous.*

Proof. As in the proof of Theorem 4.11 we consider the weak limits

$$\bar{v}(x) = \limsup_{(h,y)\to(0,x)} v_h(y),$$

$$\underline{v}(x) = \liminf_{(h,y)\to(0,x)} v_h(y),$$

and recall that they are, respectively, a subsolution of (5.5a) and (5.5c) and a supersolution of (5.5a) and (5.5b). Then Theorem 5.2 gives

$$\hat{v}_* \le \underline{v} \le \bar{v} \le \hat{v}^*,$$

so $\hat{v} = \underline{v} = \bar{v}$ at every point where $\hat{v}$ is continuous, which gives the conclusion. ∎

Approach 2: Bilateral supersolutions.

Here we consider the Dirichlet boundary value problem

(5.8a)
$$u + H(x, Du) = 1 \quad \text{in} \quad \Omega,$$

(5.8b)
$$u = 0 \quad \text{on} \quad \partial\Omega,$$

with H given by (5.6) and $\Omega = \mathbb{R}^N \setminus \mathcal{T}$, $\mathcal{T}$ closed. We saw in Theorem 2.9 that for continuous functions and convex Hamiltonians the notion of viscosity solution is equivalent to that of bilateral supersolution. Since this definition uses only the subdifferential, and not the superdifferential, it extends naturally to lower semicontinuous functions, as it was first observed by Barron and Jensen [BJ2].

Definition 5.5. *A function $u \in LSC(\Omega)$ is a bilateral supersolution of (5.8a) if*
$$u(x) + H(x, p) = 1 \quad \forall x \in \Omega, \ p \in D^- u(x). \qquad \square$$

The correct weak way to interpret the Dirichlet boundary condition (5.8b) in this context was proposed by Soravia [Sor2].

Definition 5.6. *A bilateral supersolution $u \in LSC(\mathbb{R}^N)$ of (5.8a) is a bilateral supersolution of the BVP (5.8) if $u \equiv 0$ in $\mathbb{R}^N \setminus \Omega$ and*
$$u(x) + H(x, p) \leq 1 \quad \forall x \in \partial\Omega, \ p \in D^- u(x). \qquad \square$$

Next result is a characterization of the discounted minimum time function V as the unique bilateral supersolution of (5.8). Its proof can be found in [Sor2] and [BCD], where more general problems involving nonconstant running and terminal costs are also considered.

Theorem 5.7. *Assume (5.6) and the basic hypotheses on f of Section 1. Then V_* is the unique bounded bilateral supersolution of (5.8).* $\qquad \square$

Note that Theorem 5.7 works with no assumptions on $\Omega = \mathbb{R}^N \setminus \mathcal{T}$; in particular it applies to the case $int\mathcal{T} = \emptyset$, which is interesting for the applications, where Theorem 5.2 does not work. On the other hand the stability result one gets in this context is weaker than the one obtained with the first approach, Corollary 5.4. We give a convergence theorem for the semidiscrete scheme of Section 4B, where the approximating v_h are given by (5.7). The corresponding result for a fully discrete scheme can be found in [Sor2].

Corollary 5.8. *Under the assumptions of Theorem 5.7*

$$V_*(x) = \underline{v}(x) := \liminf_{(h,y)\to(0,x)} v_h(y).$$

Sketch of proof. We saw in the proof of Theorem 4.11 that the lower weak limit $\underline{v}$ is a supersolution of (5.8a). Similarly one shows that $\underline{v}$ is a supersolution of

$$-u - H(x, Du) = -1 \quad \text{in} \quad \mathbb{R}^N,$$

and it is easy to see that $\underline{v} \equiv 0$ in $\mathcal{T} = \mathbb{R}^N \setminus \Omega$. Therefore $\underline{v}$ is a bilateral supersolution of (5.8) and we conclude by Theorem 5.7. ∎

It is worth mentioning that Barles [Ba1] simplified the methods of [BJ2] for studying bilateral supersolutions, and that Frankowska [Fr2] found equivalent definitions and a new proof of uniqueness for the finite horizon problem by using methods of nonsmooth analysis and viability theory.

Approach 3: Envelope solutions.

Here we consider the Dirichlet problem

$$(DP) \qquad \begin{cases} F(x, u, Du, D^2u) = 0 & \text{in } \Omega, \\ u = g & \text{on } \partial\Omega, \end{cases}$$

where Ω is an arbitrary open set, $F : \Omega \times \mathbb{R} \times \mathbb{R}^N \times S(N) \to \mathbb{R}$ is continuous, $S(N)$ is the set of real symmetric $N \times N$ matrices, and F satisfies

$$F(x, r, p, X) \geq F(x, s, p, Y) \quad \text{if} \quad r \geq s \quad \text{and} \quad Y - X \geq 0.$$

The definitions of sub and supersolution of (DP) are obvious extensions of Definition 1.7, see e.g. [C], [CIL]. We will assume that a Comparison Principle holds for (DP), namely

$$(ComP) \quad \begin{cases} \text{if } w \in BUSC(\overline{\Omega}) \text{ and } W \in BLSC(\overline{\Omega}) \text{ are, respectively, a sub and} \\ \text{a supersolutions of the PDE in (DP), such that } w \leq W \text{ on } \partial\Omega, \\ \text{then } w \leq W \text{ in } \Omega, \end{cases}$$

where $BUSC(\overline{\Omega})$ (resp. $BLSC(\overline{\Omega})$) denotes the set of bounded lower (resp. upper) semicontinuous functions $\overline{\Omega} \to \mathbb{R}$. We denote with $\mathcal{S}$ (resp. $\mathcal{Z}$) the set of bounded subsolutions (resp. supersolutions) of (DP), namely, viscosity subsolutions $w \in BUSC(\overline{\Omega})$ (resp. supersolutions $W \in BLSC(\overline{\Omega})$) of the PDE in (DP) such that

$w \leq g$ (resp. $W \geq g$) on $\partial\Omega$. Note that, when a viscosity solution $u \in C(\overline{\Omega})$ exists, (ComP) implies the characterization

$$u(x) = \min_{W \in \mathcal{Z}} W(x) = \max_{w \in \mathcal{S}} w(x).$$

This motivates the following definition.

Definition 5.9. *An* envelope solution, *briefly* e-solution, *of* (DP) *is a function* $u : \overline{\Omega} \to \mathbb{R}$ *such that*

$$u(x) = \inf_{W \in \mathcal{Z}} W(x) = \sup_{w \in \mathcal{S}} w(x).$$

Note that if an e-solution exists it is obviously unique. Here is the existence theorem.

Theorem 5.10. *Assume* $\mathcal{S} \neq \emptyset$, $\mathcal{Z} \neq \emptyset$ *and* (ComP).

(i) *If there exists a bounded continuous subsolution* $\underline{u}$ *of* (DP) *such that* $\underline{u} = g$ *on* $\partial\Omega$, *then*

$$u(x) = \min_{W \in \mathcal{Z}} W(x)$$

is the e-solution.

(ii) *If there is a bounded continuous supersolution* $\tilde{u}$ *of* (DP) *such that* $\tilde{u} = g$ *on* $\partial\Omega$, *then*

$$u(x) = \max_{w \in \mathcal{S}} w(x)$$

is the e-solution.

Sketch of proof of (i). In view of the Comparison Principle (ComP) it is enough to prove the existence of $U \in \mathcal{Z}$ such that

$$U(x) \leq u(x) := \sup_{w \in \mathcal{S}} w(x).$$

The construction of such U is the following. We denote

$$\Omega_n := \left\{ x \in \Omega \; : \; dist(x, \partial\Omega) > \frac{1}{n} \right\},$$

define

$$u_n(x) := \sup\{w(x) \; : \; w \in \mathcal{S}, \; w = \underline{u} \text{ in } \overline{\Omega} \setminus \Omega_n\},$$

observe that the sequence u_n is nondecreasing and then define

$$U(x) := \lim_n (u_n)_*(x).$$

The proof that $U \geq g$ on $\partial\Omega$ is not hard by using the construction of U and the continuity of $\underset{\sim}{u}$. To obtain that $U \in \mathcal{Z}$ we need two more steps.

1. We show that $(u_n)_*$ is a supersolution of

$$(5.9) \qquad\qquad F(x, u, Du, D^2u) = 0$$

in Ω_n : this is a variant of Perron's method as adapted to viscosity solutions by Ishii [I2], see e.g. [CIL], [C], [BCD], [BB].

2. We observe that a nondecreasing pointwise limit of supersolutions of (5.9) is a supersolution of (5.9), so U is such in all Ω, and this completes the proof. ∎

The definition of e-solution and Theorem 5.10 are taken from [BCD], which deals with first order equations, and [BB]. For first order equations, e-solutions are equivalent to Subbotin's (generalized) minimax solutions [Su1] [Su2], and indeed the existence theorem for such equations was first proved by him using different methods.

Theorem 5.10 applies to a class of Hamilton-Jacobi-Bellman-Isaacs equations arising in stochastic optimal control and differential games. They are equations of the form

$$\sup_{a} \inf_{b} \mathcal{L}^{a,b} u(x) = 0,$$

or

$$\inf_{b} \sup_{a} \mathcal{L}^{a,b} u(x) = 0,$$

where

$$\mathcal{L}^{a,b} u(x) := -\alpha_{ij}(x, a, b)u_{x_i x_j} - f(x, a, b) \cdot Du + c(x, a, b)u - l(x, a, b)$$

and the matrix $\alpha = (\alpha_{ij})$ can be written as $\alpha = \sigma\sigma^t$ for some $N \times N$ matrix $\sigma(x, a, b)$ (so α is nonnegative definite). These equations satisfy (ComP) if, for instance, Ω is bounded, all the coefficients of the linear operators $\mathcal{L}^{a,b}$ are bounded and continuous with respect to x, uniformly in a and b, α_{ij} and f are also lipschitzean in x uniformly in a and b, and

$$c(x, a, b) \geq K > 0, \quad \text{for all } x, a, b,$$

see e.g. [CIL], [C]. If we consider the Dirichlet Problem (DP) for these equations with $g \equiv 0$, then the assumptions of part (i) of Theorem 5.10 are satisfied by taking $\underset{\sim}{u} \equiv 0$, provided that

$$l(x, a, b) \geq 0 \quad \text{for all } x, a, b.$$

In particular Theorem 5.10 applies to the Dirichlet problem (5.8) associated to the discounted minimum time differential game with target $\mathcal{T} = \mathbb{R}^N \setminus \Omega$, where the Hamiltonian is

$$(5.10) \qquad H(x,p) := \min_{b \in B} \max_{a \in A}\{-f(x,a,b) \cdot p\}.$$

A natural question is whether the e-solution of (5.8) is the lower value of the game in some sense. The answer to this question is rather technical and we refer the interested reader to [BBF] and [Su2]. However the answer is simple in the special case of a single player, i.e. H given by (5.6), and it gives a connection with the other two approaches described in this section.

Theorem 5.11. *Under the basic assumptions of Section 1, the e-solution of (5.8), with H given by (5.6), is V_*, where V is the discounted minimum time function.* □

The proof of this result can be found in [BSt] or [BCD]. Note that no assumptions on $\Omega = \mathbb{R}^N \setminus \mathcal{T}$ are needed in Theorem 5.11, as in Theorem 5.7.

The last result of this section describes a stability property of e-solutions studied in [BBF]. As for the first two approaches, we restrict ourselves to the semi-discrete scheme of Section 4B. Here, however, the Hamiltonian can be the non-convex one given by (5.10), and the open set Ω is approximated from the interior. We define for $\varepsilon > 0$

$$\mathcal{T}_\varepsilon := \{x \in \mathbb{R}^N : dist(x, \mathcal{T}) \le \varepsilon\},$$

and consider the value function v_h^ε of the discrete-time minorant game with target $\mathcal{T}_\varepsilon$.

Theorem 5.12. *Under the basic assumptions of Section 3, v_h^ε converge to the e-solution u of (5.8), with H given by (5.10), as $(\varepsilon, h) \to (0,0)$ with h linked to ε, in the following sense: for all $x \in \overline{\Omega}$ and $\gamma > 0$ there exists a function $\tilde{h} :]0, +\infty[\to]0, +\infty[$ and $\bar{\varepsilon} > 0$ such that*

$$|v_h^\varepsilon(y) - u(x)| \le \gamma \quad \forall y : |x - y| \le \tilde{h}(\varepsilon)$$

and for all $\varepsilon \le \bar{\varepsilon}$, $h \le \tilde{h}(\varepsilon)$. □

The proof can be found in [BBF], as well as a similar convergence result for a fully discrete scheme.

We end this section by referring the reader interested in discontinuous viscosity solutions to [Ba2] for the first two approaches, to [Su2], [BB] and [G] for the third approach, and to [BCD] for a detailed comparison among the three approaches.

References

[A] B. Alziary de Roquefort, *Jeux différentiels et approximation numérique de fonctions valeur*, RAIRO Modél. Math. Anal. Numér. 25, pp. 517-560, 1991.

[BBCD] F. Bagagiolo, M. Bardi, I. Capuzzo Dolcetta *A viscosity solutions approach to some asymptotic problems in optimal control*, in Proceedings of the Conference "PDE's methods in control, shape optimization and stochastic modelling", Pisa 1994, J.P. Zolesio ed., Marcel Dekker, to appear.

[B] M. Bardi, *A boundary value problem for the minimum time function*, SIAM J. Control Optim. 26, pp. 776-785, 1989.

[BB] M. Bardi, S. Bottacin, *Discontinuous solutions of degenerate elliptic boundary value problems*, preprint 22, Dipartimento di Matematica, Università di Padova, 1995.

[BBF] M. Bardi, S. Bottacin, M. Falcone, *Convergence of discrete schemes for discontinuous value functions of pursuit-evasion game*, in "New Trends in Dynamic Games and Application", G.J. Olsder ed., pp. 273-304, Birkhäuser, Boston, 1995.

[BCD] M. Bardi, I. Capuzzo Dolcetta, *Optimal control and viscosity solutions of Hamilton-Jacobi-Bellman equations*, Birkhäuser, Boston, to appear.

[BF1] M. Bardi, M. Falcone, *An approximation scheme for the minimum time function*, SIAM J. Control Optim. 28, pp. 950-965, 1990.

[BF2] M. Bardi, M. Falcone, *Discrete approximation of the minimal time function for systems with regular optimal trajectories*, in "Analysis and Optimization of Systems", A. Bensoussan and J.L. Lions eds., pp. 103-112, Lect. Notes Control Info. Sci. 144, Springer-Verlag, 1990.

[BFS] M. Bardi, M. Falcone, P. Soravia, *Fully discrete schemes for the value function of pursuit-evasion games*, in "Advances in dynamic games and applications", T. Basar and A. Haurie eds., pp. 89-105, Birkhäuser, 1993.

[BPR] M. Bardi, T. Parthasarathy, T.E.S. Raghavan eds. *Stochastic and differential games: theory and numerical methods*, Birkhäuser, Boston, to appear.

[BSa1] M. Bardi, C. Sartori, *Approximations and regular perturbations of optimal control problems via Hamilton-Jacobi theory*, Appl. Math. Optim. 24, pp. 113-128, 1991.

[BSa2] M. Bardi, C. Sartori, *Convergence results for Hamilton-Jacobi-Bellman equations in variable domains*, Differential Integral Equations 5, pp. 805-816, 1992.

90

[BS0] M. Bardi, P. Soravia, *A PDE framework for differential games of pursuit-evasion type*, in "Differential games and applications", T. Basar and P. Bernhard eds., pp. 62-71, Lect. Notes Control Info. Sci. 119, Springer-Verlag, 1989.

[BS1] M. Bardi, P. Soravia, *Hamilton-Jacobi equations with singular boundary conditions on a free boundary and applications to differential games*, Trans. Amer. Math. Soc. 325, pp. 205-229, 1991.

[BS2] M. Bardi, P. Soravia, *A comparison result for Hamilton-Jacobi equations and applications to some differential games lacking controllability*, Funkcial. Ekvac. 37, pp. 19-43, 1994.

[BS3] M. Bardi, P. Soravia, *Approximation of differential games of pursuit-evasion by discrete-time games*, in "Differential Games - Developments in modelling and computation", R.P. Hamalainen and H.K. Ethamo eds., pp. 131-143, Lect. Notes Control Info. Sci. 156, Springer-Verlag, 1991.

[BSt] M. Bardi, V. Staicu, *The Bellman equation for time-optimal control of non-controllable nonlinear systems*, Acta Applic. Math. 31, pp. 201-223, 1993.

[Ba0] G. Barles, *An approach of deterministic control problems with unbounded data*, Ann. Inst. H. Poincaré Anal. Nonlin. 7, pp.235-258, 1990.

[Ba1] G. Barles, *Discontinuous viscosity solutions of first order Hamilton-Jacobi equations: A guided visit*, Nonlinear Anal. T.M.A. 20, pp. 1123-1134, 1993.

[Ba2] G. Barles, *Solutions de viscosité des équations de Hamilton-Jacobi*, Springer-Verlag, Paris, 1994.

[BP1] G. Barles, B. Perthame, *Discontinuous solutions of deterministic optimal stopping time problems*, RAIRO Modél. Math. Anal. Numér. 21, pp. 557-579, 1987.

[BP2] G. Barles, B. Perthame, *Exit time problems in optimal control and vanishing viscosity method*, SIAM J. Control Optim. 26, pp. 1133-1148, 1988.

[BSou] G. Barles, P.E. Souganidis, *Convergence of approximation schemes for fully nonlinear systems*, Asymptotic Anal. 4, pp. 271-283, 1991.

[Br] E. N. Barron, *Differential games with maximum cost*, Nonlinear Anal. T. M. A. 14, pp. 971-989, 1990.

[BI] E.N. Barron, H. Ishii, *The Bellman equation for minimizing the maximum cost*, Nonlinear Anal. T.M.A. 13, pp. 1067-1090, 1989.

[BJ1] E.N. Barron, R. Jensen, *Total risk aversion, stochastic optimal control, and differential games*, Appl. Math. Optim. 19, pp. 313-327, 1989.

[BJ2] E.N. Barron, R. Jensen, *Semicontinuous viscosity solutions of Hamilton-Jacobi equations with convex Hamiltonians*, Comm. Partial Differential Equations 15, pp. 1713-1742, 1990.

[BJM] E.N. Barron, R. Jensen, J.L. Menaldi, *Optimal control and differential games with measures*, Nonlinear Anal. T.M.A. 21, pp. 241-268, 1993.

[BaBe] T. Basar, P. Bernhard, H^∞-*optimal control and related minimax design problems*, Birkhäuser, Boston, 1991.

[Be1] L.D. Berkovitz, *Optimal feedback controls*, SIAM J. Control Optim. 27, pp. 991-1006, 1989.

[Be2] L.D. Berkovitz, *A theory of differential games*, in "Advances in dynamic games and applications", T. Basar and A. Haurie eds., pp. 3-22, Birkhäuser, 1993.

[BiS] R.M. Bianchini, G. Stefani, *Time-optimal problem and time-optimal map*, Rend. Sem. Mat. Univ. Pol. Torino 48, pp. 401-429, 1990.

[Bo] S. Bortoletto, *The Bellman equation for constrained deterministic optimal control problems*, Differential Integral Equations 6, pp. 905-924, 1993.

[CFa1] F. Camilli, M. Falcone, *An approximation scheme for the optimal control of diffusion processes*, RAIRO Modél. Math. Anal. Numér. 29, pp. 97-122, 1995.

[CFa2] F. Camilli, M. Falcone, *Approximation of optimal control problems with state constraints: estimates and applications*, in "Nonsmooth analysis and geometric methods in deterministic optimal control", V. Jurdjevic, B.S. Mordukhovic, R.T. Rockafellar and H.J.Sussman eds., I.M.A. Volumes in Applied Mathematics, Springer-Verlag, to appear.

[CFLS] F. Camilli, M. Falcone, P. Lanucara, A. Seghini, *A domain decomposition method for Bellman equations*, in "Domain Decomposition methods in Scientific and Engineering Computing", D.E. Keyes and J. Xu eds., pp. 477-483, Contemporary Mathematics 180, A.M.S., 1994.

[CDP] P. Cannarsa, G. Da Prato, *Nonlinear optimal control with infinite horizon for distributed parameter systems and stationary Hamilton-Jacobi equations*, SIAM J. Control Optim. 27, pp. 861-875, 1989.

[CF] P. Cannarsa, H. Frankowska, *Some characterizations of optimal trajectories in control theory*, SIAM J. Control Optim. 29, pp. 1322-1347, 1991.

[CGS] P. Cannarsa, F. Gozzi, H. M. Soner, *A dynamic programming approach to nonlinear boundary control problems of parabolic type*, J. Funct. Anal. 117, pp. 25-61, 1993.

[CS] P. Cannarsa, C. Sinestrari, *Convexity properties of the minimum time function*, Calc. Var. 3, pp. 273-298, 1995.

[CD] I. Capuzzo Dolcetta, *On a discrete approximation of the Hamilton-Jacobi equation of dynamic programming*, Appl. Math. Optim. 10, pp. 367-377, 1983.

[CDE] I. Capuzzo Dolcetta, L.C. Evans, *Optimal switching for ordinary differential equations*, SIAM J. Control Optim. 22, pp.143-161, 1984.

[CDF] I. Capuzzo Dolcetta, M. Falcone, *Viscosity solutions and discrete dynamic programming*, Ann. Inst. H. Poincaré Anal. Non Lin. 6 (Supplement), pp. 161-183, 1989.

[CDI] I. Capuzzo Dolcetta, H. Ishii, *Approximate solutions of the Bellman equation of deterministic control theory*, Appl. Math. Optim. 11, pp. 161-181, 1984.

[CDL] I. Capuzzo Dolcetta, P.L. Lions, *Hamilton-Jacobi equations with state constraints*, Trans. Amer. Math. Soc. 318, pp. 643-683, 1990.

[Ca] C. Caratheodory, *Calculus of variations and partial differential equations of the first order*, 2nd English edn., Chelsea, New York, 1982.

[CQS1] P. Cardaliaguet, M. Quincampoix, P. Saint-Pierre, *Temps optimaux pour des problèmes de controle avec contraintes et sans controlabilité locale*, C. R. Acad. Sci. Paris 318, Ser. I, pp. 607-612, 1994.

[CQS2] P. Cardaliaguet, M. Quincampoix, P. Saint-Pierre, *Set-valued numerical analysis for optimal control and differential games*, in [BPR].

[Cl1] F.H. Clarke, *Optimization and nonsmooth analysis*, Wiley, New York, 1983.

[Cl2] F.H. Clarke, *Methods of Dynamic and Nonsmooth Optimization*, CBMS-NSF Reg. Conf. Ser. Appl. Math. 57, S.I.A.M., Philadelphia, 1989.

[C] M.G. Crandall, *Viscosity Solutions: a Primer*, in "Viscosity solutions and applications", I. Capuzzo Dolcetta and P.L. Lions eds., Lecture Notes in Mathematics, Springer-Verlag, 1997.

[CEL] M.G. Crandall, L.C. Evans, P.L. Lions, *Some properties of viscosity solutions of Hamilton-Jacobi equations*, Trans. Amer. Math. Soc. 282, pp. 487-502, 1984.

[CIL] M.G. Crandall, H. Ishii, P.L. Lions, *User's guide to viscosity solutions of second order partial differential equations*, Bull. Amer. Math. Soc. 27, pp. 1-67, 1992.

[CL1] M.G. Crandall, P.L. Lions, *Viscosity solutions of Hamilton-Jacobi equations*, Trans. Amer. Math. Soc. 277, pp. 1-42, 1983.

[CL2] M.G. Crandall, P.L. Lions, *Viscosity solutions of Hamilton-Jacobi equations in infinite dimensions, Part VI*, in "Evolution equations, control theory, and biomathematics", P. Clément and G. Lumer eds., Lect. Notes Pure Appl. Math. 155, Marcel Dekker, New York, 1994.

[CL3] M.G. Crandall, P.L. Lions, *Viscosity solutions of Hamilton-Jacobi equations in infinite dimensions, Part VII*, J. Func. Anal. 125, pp. 111-148, 1994.

[D] F. Da Lio, *Equazioni di Bellman per problemi di controllo ottimo illimitato*, Thesis, Università di Padova, July 1994.

[E] L. C. Evans, *Nonlinear systems in optimal control theory and related topics*, in "Systems of nonlinear PDEs", J.M. Ball ed., pp. 95-113, D. Reidel, Dordrecht, 1983.

[EI] L.C. Evans, H. Ishii, *Differential games and nonlinear first order PDE in bounded domains*, Manuscripta Math. 49, pp. 109-139, 1984.

[EJ] L.C. Evans, M.R. James, *The Hamilton-Jacobi-Bellman equation for time optimal control*, SIAM J. Control Optim. 27, pp. 1477-1489, 1989.

[ES] L.C. Evans, P.E. Souganidis, *Differential games and representation formulas for solutions of Hamilton-Jacobi equations*, Indiana Univ. Math. J. 33, pp. 773-797, 1984.

[Fa] M. Falcone, *A numerical approach to the infinite horizon problem of deterministic control theory*, Appl. Math. Optim. 15, pp.1-13, 1987 (Corrigenda in vol. 23, pp. 213-214, 1991).

[FaF] M. Falcone, R. Ferretti, *Discrete time high-order schemes for viscosity solutions of Hamilton-Jacobi-Bellman equations*, Numerische Mathematik 67, pp. 315-344, 1994.

[F] W.H. Fleming, *The convergence problem for differential games*, J. Math. Anal. Appl. 3, pp. 102-116, 1961.

[FME] W.H. Fleming, W.M. McEneaney, *Risk sensitive optimal control and differential games*, in Lecture Notes on Control and Information Sciences 184, pp. 185-197, Springer, 1992.

[FS] W.H. Fleming, H.M. Soner, *Controlled Markov processes and viscosity solutions*, Springer, New York, 1993.

[FSou] W.H. Fleming, P.E. Souganidis, *On the existence of value function of two-players, zero-sum stochastic differential games*, Indiana Univ. Math. J. 38, pp. 293-314, 1989.

[Fr1] H. Frankowska, *Optimal trajectories associated with a solution of the contingent Hamilton-Jacobi equation*, Appl. Math. Optim. 19, pp.291-311, 1989.

[Fr2] H. Frankowska, *Lower semicontinuous solutions of Hamilton-Jacobi-Bellman equations*, SIAM J. Control Optim. 31, pp. 257-272, 1993.

[G] P. Goatin, *Sul problema di Dirichlet con condizioni al bordo generalizzate per equazioni ellittiche degeneri nonlineari*, Thesis, Università di Padova, November 1995.

[GT] R.L.V. Gonzalez, M.M. Tidball, *Sur l'ordre de convergence des solutions discrétisées en temps et en espace de l'équation de Hamilton-Jacobi*, C. R. Acad. Sci. Paris 314, Sér. I, pp. 479-482, 1992.

[Is] R. Isaacs, *Differential games*, Differential games, Wiley, New York, 1965.

[I1] H. Ishii, *Uniqueness of unbounded viscosity solutions of Hamilton-Jacobi equations*, Indiana Univ. Math. J. 33, pp. 721-748, 1984.

[I2] H. Ishii, *Perron's method for Hamilton-Jacobi equations*, Duke Math. J. 55, pp. 369-384, 1987.

[I3] H. Ishii, *A boundary value problem of the Dirichlet type for Hamilton-Jacobi equations*, Ann. Sc. Norm. Sup. Pisa (IV) 16, pp. 105-135, 1989.

[I4] H. Ishii, *Viscosity solutions for a class of Hamilton-Jacobi equations in Hilbert spaces*, J. Func. Anal. 105, pp. 301-341, 1992.

[J] M.R. James, *Asymptotic analysis of nonlinear stochastic risk-sensitive control and differential games*, Math. Control Signals Systems 5, pp. 401-417, 1992.

[Ko] S. Koike, *On the state constraint problem for differential games*, Indiana Univ. Math. J. 44, pp. 467-487, 1995.

[KK] A.N. Krasovskii, N.N. Krasovskii, *Control under lack of information*, Birkhäuser, Boston, 1995.

[KS] N.N. Krasovskii, A.I. Subbotin, *Game theoretical control problems*, Springer, New York, 1988.

[K] S.N. Kruzkov, *Generalized solutions of the Hamilton-Jacobi equations of eikonal type I*, Math. USSR Sbornik 27, pp. 406-445, 1975.

[KD] H.J. Kushner, P.G. Dupuis, *Numerical methods for stochastic control problems in continuous time*, Springer-Verlag, New York, 1992.

[Le] J. Lewin, *Differential games*, Springer, London, 1994.

[L1] P.L. Lions, *Generalized solutions of Hamilton-Jacobi equations*, Pitman, Boston, 1982.

[L2] P.L. Lions, *Optimal control of diffusion processes and Hamilton-Jacobi-Bellman equations. Part 1: The dynamic programming principle and applications, Part 2: Viscosity solutions and uniqueness*, Comm. Partial. Differential Equations 8, (1983), 1101-1174 and 1229-1276.

[L3] P.L. Lions, *Neumann type boundary conditions for Hamilton-Jacobi equations*, Duke J. Math. 52, pp. 793-820, 1985.

[L4] P.L. Lions, *Viscosity solutions of fully nonlinear second-order equations and optimal stochastic control in infinite dimensions. Part III*, J. Func. Anal. 86, pp. 1-18, 1989.

[LS] P.L. Lions, P.E. Souganidis, *Differential games and directional derivatives of viscosity solutions of Bellman's and Isaacs' equations I and II*, SIAM J. Control Optim. 23 and 24, pp. 566-583 and 1086-1089, 1985 and 1986.

[LT] P. Loreti, E. Tessitore, *Approximation and regularity results on constrained viscosity solutions of Hamilton-Jacobi-Bellman equations*, J. Math. Systems Estimation Control 4, pp. 467-483, 1994.

[ME] W.M. McEneaney, *Uniqueness for viscosity solutions of nonstationary HJB equations under some a priori conditions (with applications)*, SIAM J. Control Optim. 33, pp. 1560-1576, 1995.

[Mi1] S. Mirica, *Inégalité différentielles impliquant l'équation de Bellman-Isaacs et ses généralisations dans la théorie du controle optimal*, Anal. Univ. Bucuresti Mat. Anul 37, pp.25-35, 1988.

[Mi2] S. Mirica, *Some generalizations of the Bellman-Isaacs equation in deterministic optimal control*, Studii Cercetari Mat. 42, pp.437-447, 1990.

[Mi3] S. Mirica, *Nonsmooth fields of extremals and constructive Dynamic Programming in optimal control*, preprint 18, Dipartimento di Matematica, Università di Padova, 1993.

[Mi4] S. Mirica, *Optimal feedback control in closed form via Dynamic Programming*, preprint 1993.

[M] M. Motta, *On nonlinear optimal control problems with state constraints*, SIAM J. Control Optim. 33, pp. 1411-1424, 1995.

[MR1] M. Motta, F. Rampazzo, *Dynamic Programming for nonlinear systems driven by ordinary and impulsive controls*, SIAM J. Control Optim. to appear, 1996.

[MR2] M. Motta, F. Rampazzo, *The value function of a slow growth control problem with state constraints*, J. Math. Systems Estimation Control to appear.

[P1] H. J. Pesch, *Offline and online computation of optimal trajectories in the aerospace field*, in "Applied Mathematics in Aerospace Science and Engeneering", pp. 165-219, A. Miele and A. Salvetti eds., Plenum, New York, 1994.

[P2] H. J. Pesch, *Solving optimal control and pursuit-evasion game problems of high complexity*, in "Computational Optimal Control", pp. 43-64, R. Bulirsch and D. Kraft eds., Birkhäuser, Basel, 1994.

[R] F. Rampazzo, *Differential games with unbounded versus bounded controls*, preprint 30, Dipartimento di Matematica , Università di Padova, 1995.

[RV] J. D. L. Rowland, R. B. Vinter, *Construction of optimal feedback controls*, Systems Control Lett. 16, pp. 357-367, 1991.

[S1] H.M. Soner, *Optimal control problems with state-space constraints I and II*, SIAM J. Control Optim. 24, pp. 551-561 and pp. 1110-1122, 1986.

[S2] H.M. Soner, *Controlled Markov Processes, Viscosity Solutions and Applications to Mathematical Finance*, in "Viscosity solutions and applications", I. Capuzzo Dolcetta and P.L. Lions eds., Lecture Notes in Mathematics, Springer-Verlag, 1997.

[Sor0] P. Soravia, *The concept of value in differential games of survival and viscosity solutions of Hamilton-Jacobi equations*, Differential Integral Equations 5, pp. 1049-1068, 1992.

[Sor1] P. Soravia, *Pursuit-evasion problems and viscosity solutions of Isaacs equations*, SIAM J. Control Optim. 31, pp. 604-623, 1993.

[Sor2] P. Soravia, *Discontinuous viscosity solutions to Dirichlet problems for Hamilton-Jacobi equations with convex Hamiltonians*, Comm. Partial Differential Equations 18, pp. 1493-1514, 1993.

[Sor3] P. Soravia, H_∞ *control of nonlinear systems: differential games and viscosity solutions*, SIAM J. Control Optim., to appear.

[Sor4] P. Soravia, *Differential games and viscosity solutions to study the H_∞ control of nonlinear, partially observed systems*, Preprint volume of the 6th International Symposium on Dynamic Games and Applications, M. Breton and G. Zaccour eds., Montreal, 1994.

[Sor5] P. Soravia, *Estimates of convergence of fully discrete schemes for the Isaacs equation of pursuit-evasion differential games via maximum principle*, preprint, Dipartimento di Matematica, Università di Padova, 1995.

[Sor6] P. Soravia, *Stability of dynamical systems with competitive controls: the degenerate case*, J. Math. Anal. Appl. 191, pp. 428-449, 1995.

[Sor7] P. Soravia, *Optimality principles and representation formulas for viscosity solutions of Hamilton-Jacobi equations, I and II*, preprints, Dipartimento di Matematica, Università di Padova, 1995.

[Sor8] P. Soravia, *Equivalence between nonlinear H_∞ control problems and existence of viscosity solutions of Hamilton-Jacobi-Isaacs equations*, preprint, Dipartimento di Matematica, Università di Padova, 1995.

[Sou] P.E. Souganidis, *Max-min representations and product formulas for the viscosity solutions of Hamilton-Jacobi equations with applications to differential games*, Nonlinear Anal. T.M.A. 9, pp. 217-257, 1985.

[Su1] A.I. Subbotin, *Discontinuous solutions of a Dirichlet type boundary value problem for first order partial differential equations*, Russian J. Numer. Anal. Math. Modelling 8, pp. 145-164, 1993.

[Su2] A.I. Subbotin, *Generalized solutions of first order PDEs: The Dynamic Optimization Perspective*, Birkhäuser, Boston, 1995.

[Su3] A.I. Subbotin, *Constructive theory of positional differential games and generalized solutions to Hamilton-Jacobi equations*, in [BPR].

[Sua] N.N. Subbotina, *The maximum principle and the superdifferential of the value function*, Prob. Control Inform. Th. 18, pp. 151-160, 1989.

[T] D. Tataru, *Viscosity solutions for Hamilton-Jacobi equations with unbounded nonlinear term: a simplified approach*, J. Differential Equations 111, pp.123-146, 1994.

[TG] M.M. Tidball and R.L.V. Gonzalez, *Zero sum differential games with stopping times: some results about their numerical resolution*, in "Advances in dynamic games and applications", T. Basar and A. Haurie eds., pp. 106-124, Birkhäuser, 1993.

[Y1] J. Yong, *A zero-sum differential game in a finite duration with switching strategies*, SIAM J. Control Optim. 28, pp. 1234-1250, 1990.

[Y2] J. Yong, *Zero-sum differential games involving impulse controls*, Appl. Math. Optim. 29, pp.243-261, 1994.

[Z1] X.Y. Zhou, *Maximum principle, dynamic programming, and their connection in deterministic control*, J. Optim. Th. Appl. 65, pp. 363-373, 1990.

[Z2] X.Y. Zhou, *Verification theorems within the framework of viscosity solutions*, J. Math. Anal. Appl. 177, pp. 208-225, 1993.

Regularity for Fully Nonlinear Elliptic Equations and Motion by Mean Curvature

LAWRENCE C. EVANS

Department of Mathematics
University of California
Berkeley, California 94720-3840 USA

Introduction

These are the notes for a series of six lectures I gave at Montecatini Terme, Italy, during June 12–17, 1995, as part of a CIME program on "Viscosity Solutions and Applications".

Each speaker presented a minicourse of six lectures on a variety of topics. As I was asked to speak on two essentially distinct subjects (fully nonlinear elliptic PDE and mean curvature motion), I in fact presented two microcourses, or more precisely nanocourses, of three lectures each.

Part I. Regularity for fully nonlinear elliptic PDE of second order

    A. Introduction, viscosity solutions, linear estimates

    B. $C^{1,\alpha}$ estimates

    C. $C^{2,\alpha}$ estimates

Part II. Motion by mean curvature

    D. Introduction, distance function

    E. Level set method

    F. Geometric properties

For these notes I have changed the oral presentation somewhat, mostly to be consistent with the other speakers' lectures.

I am pleased to thank I. Capuzzo-Dolcetta and P.-L. Lions for inviting me to speak.

Part I
Regularity Theory for Fully Nonlinear Elliptic PDE of Second Order

A. Introduction, viscosity solutions, linear estimates

1. Introduction. This and the next two lectures concern regularity for *fully nonlinear second-order elliptic PDE* and in particular for the model problem:

$$(1) \qquad F(D^2 u) = f \qquad \text{in } U .$$

Here U denotes an open subset of $\mathbb{R}^n$, $f : U \to \mathbb{R}$ is given, $u : \overline{U} \to \mathbb{R}$ is the unknown, and

$$D^2 u = \begin{pmatrix} u_{x_1 x_1} & \cdots & u_{x_1 x_n} \\ & \ddots & \\ u_{x_n x_1} & & u_{x_n x_n} \end{pmatrix}$$

is the Hessian matrix of u. We are given the nonlinearity

$$F : \mathbb{S}^n \to \mathbb{R} ,$$

$\mathbb{S}^n$ denoting the space of real, $n \times n$ symmetric matrices.

The overall problem is to study solutions u of the PDE (1), subject say to the boundary condition

$$(2) \qquad u = 0 \quad \text{on} \quad \partial U .$$

As the left-hand side of (1) is in general a nonlinear combination of the various second derivatives of u, this PDE is called *fully nonlinear*.

We will require of course various assumptions concerning F, the most important of which is a kind of (reverse) monotonicity condition with respect to matrix ordering.

Definition. (i) We say that F is *elliptic* provided

$$(3) \qquad A \geq B \quad \text{in} \quad \mathbb{S}^n$$

implies

$$(4) \qquad F(B) \geq F(A) \qquad (A, B \in \mathbb{S}^n) .$$

(ii) We say F is **uniformly elliptic** if there exist constants $0 < \theta \le \Theta < \infty$ such that (3) implies

(5) $$\theta\|A - B\| \le F(B) - F(A) \le \Theta\|A - B\| .$$

Notation. (a) Condition (3) means $C = A - B$ is a nonnegative definite symmetric $n \times n$ matrix.

(b) In (5), we take the norm

$$\|A - B\| = \|C\| = \sup_{|\xi|=1} |\xi \cdot C\xi|$$

(c) If $\xi \in \mathbb{R}^n$, $\xi = (\xi_1, \ldots, \xi_n)$, then

$$\xi \otimes \xi = \begin{pmatrix} \xi_1\xi_1 & \cdots & \xi_1\xi_n \\ & \ddots & \\ \xi_n\xi_1 & \cdots & \xi_n\xi_n \end{pmatrix} \in \mathbb{S}^n$$

$\square$

Examples of nonlinear elliptic PDE arise in stochastic control theory (cf. Soner's lectures) and in geometry (cf. Lecture D below).

We will assume hereafter that F, f are smooth. We can consequently reinterpret (4),(5) in differential form as follows. Fix a matrix $R \in \mathbb{S}^n$ and a vector $\xi \in \mathbb{R}^n$. Taking

$$B = R , \qquad A = R + \lambda(\xi \otimes \xi) \qquad (\lambda > 0) ,$$

we deduce from (4) that

$$\frac{1}{\lambda} \left[F(R + \lambda(\xi \otimes \xi)) - F(R) \right] \le 0$$

Upon letting $\lambda \to 0^+$ we conclude:

(6) $$-\frac{\partial F}{\partial r_{ij}}(R)\xi_i\xi_j \ge 0 \qquad (R \in \mathbb{S}^n, \ \xi \in \mathbb{R}^n) .$$

Similarly (5) implies

(7) $$\theta|\xi|^2 \le -\frac{\partial F}{\partial r_{ij}}(R)\xi_i\xi_j \le \Theta|\xi|^2 \qquad (R \in \mathbb{S}^n, \ \xi \in \mathbb{R}^n) .$$

Hence the strict monotonicity condition of uniform ellipticity (5) means that the "linearization" of F about any fixed matrix $R \in \mathbb{S}^n$ is a uniformly elliptic operator. Our study of regularity for solutions of the fully nonlinear PDE (1) will therefore profit from good estimates for linear elliptic PDE.

2. Viscosity solutions. A main concern of Lectures A–C will be the possibility of estimating a smooth solution of (1),(2). It will be helpful however to first introduce an appropriate notion of a weak solution.

Motivation. We intend to find a way to interpret a merely continuous function $u : \overline{U} \to \mathbb{R}$ as solving (1). Since u will not necessarily have second derivatives, we must somehow "move" the second derivatives onto a smooth test function ϕ. The idea is to assume temporarily that u is a smooth solution of (1) and then to use the maximum principle.

So let ϕ be a smooth function and suppose that at some point $x_0 \in U$ we have

$$u(x_0) = \phi(x_0)$$

with

$$u(x) \leq \phi(x) \qquad \text{for all } x \text{ near } x_0 .$$

Thus $u - \phi$ has a local maximum at x_0. Geometrically the graph of u is touched from above at x_0 by the graph of ϕ. As we are temporarily supposing u to be smooth, we deduce

$$D^2 u(x_0) \leq D^2 \phi(x_0) .$$

Hence the ellipticity condition (4) implies

$$F(D^2 u(x_0)) \geq F(D^2 \phi(x_0)) .$$

But then the PDE (1) gives

$$F(D^2 \phi(x_0)) \leq f(x_0) .$$

The opposite inequality would occur if the graph of ϕ were to touch the graph of u from below at x_0.

These considerations motivate the following:

Definitions. (i) We say $u \in C(\overline{U})$ is a *subsolution of (1) in the viscosity sense* if

$$F(D^2 \phi(x_0)) \leq f(x_0)$$

whenever ϕ is smooth and $u - \phi$ has a local maximum at a point $x_0 \in U$.

(ii) We say $u \in C(\overline{U})$ is a *supersolution of (1) in the viscosity sense* if

$$F(D^2 \phi(x_0)) \geq f(x_0)$$

whenever ϕ is smooth and $u - \phi$ has a local minimum at a point $x_0 \in U$.

(iii) We say u is a *viscosity solution* of (1) if u is both a viscosity subsolution and supersolution.

The great advantage of this definition is that it allows us to interpret a nonsmooth function u as solving (1). The point is that by weakening what we mean by a solution of (1),(2) we are more likely to be able to construct a solution.

The accompanying lectures of M. G. Crandall explain in more detail the stability, uniqueness and existence properties of viscosity solutions.

I will in these notes mostly provide various formal calculations leading to estimates of $|u|$, $|Du|$, $|D^2u|$, etc. under the assumption that u is a smooth solution of (1),(2). Work, sometimes lots of work, is needed to obtain the same estimate for viscosity solutions, and for this I refer readers to the very clear book [C-C] of Cabre and Caffarelli. I think however it is extremely important for students to understand also that unjustified, formal calculations can serve as guides for rigorous proofs.

Example: Jensen's regularizations. To illustrate this point, let me show you the following heuristic calculation. Suppose u is a smooth solution of

$$(8) \qquad\qquad F(D^2u) = 0 \qquad \text{in } \mathbb{R}^n \,,$$

F being smooth and elliptic. Consider the Hamilton-Jacobi equation

$$(9) \qquad \begin{cases} v_t + H(Dv) = 0 & \text{in } \mathbb{R}^n \times (0, \infty) \\ v = u & \text{on } \mathbb{R}^n \times \{t = 0\} \,, \end{cases}$$

where $H : \mathbb{R}^n \to \mathbb{R}$ is smooth, convex. I will show that the effect of the flow (9) is to convert u into a supersolution. Indeed, set

$$w = F(D^2v) \,.$$

Then

$$w_t = F_{r_{ij}}(D^2v)v_{x_ix_jt}$$
$$w_{x_k} = F_{r_{ij}}(D^2v)v_{x_ix_jx_k} \,, \qquad (1 \le k \le n)$$

Now (9) implies

$$v_{x_it} + H_{p_k}(Dv)v_{x_kx_i} = 0 \qquad (1 \le i \le n)$$
$$v_{x_ix_jt} + H_{p_k}(Dv)v_{x_kx_ix_j} + H_{p_kp_l}(Dv)v_{x_kx_i}v_{x_lx_j} = 0 \qquad (1 \le i, j \le n)$$

Thus

$$(10) \quad \begin{aligned} w_t + H_{p_k} w_{x_k} &= F_{r_{ij}}(-H_{p_k} v_{x_k x_i x_j} - H_{p_k p_l} v_{x_k x_i} v_{x_l x_j}) \\ &\quad + H_{p_k}(F_{r_{ij}} v_{x_i x_j x_k}) \\ &= -H_{p_k p_l} F_{r_{ij}} v_{x_k x_i} v_{x_l x_j} \geq 0 , \end{aligned}$$

as F is elliptic and H is convex. Since $w = F(D^2 u) = 0$ at $t = 0$, we deduce from (10) that

$$w = F(D^2 v) \geq 0 \quad \text{in } \mathbb{R}^n \times (0, \infty) .$$

In particular, for each $t > 0$, $v(\cdot, t)$ is a supersolution.

As a special case, let $H(p) = \frac{|p|^2}{2}$. Then the (unique viscosity) solution of (9) turns out to be

$$v(x, t) = \min_{y \in \mathbb{R}^n} \left(u(y) + \frac{|x - y|^2}{2t} \right)$$

Consequently we can expect upon letting $t = \varepsilon$ that

$$u_\varepsilon(x) = \min_{y \in \mathbb{R}^n} \left(u(y) + \frac{|x - y|^2}{2\varepsilon} \right)$$

will satisfy

$$F(D^2 u_\varepsilon) \geq 0 \quad \text{in } \mathbb{R}^n .$$

This is indeed true, provided we interpret everything in the viscosity sense — see Jensen-Lions-Souganidis [J-L-S]. My point here is that the formal calculations lead us to guess the correct result.

Taking $H(p) = (1 + |p|^2)^{\frac{1}{2}}$, we deduce

$$\tilde{u}_\varepsilon = \text{function whose graph in } \mathbb{R}^{n+1} \text{ is at a}$$
$$\text{distance } \varepsilon \text{ below the graph of } u$$

is also a supersolution. This was Jensen's original observation [J]. □

3. **Review of linear estimates.** As noted above, our estimates for the fully nonlinear PDE will depend upon deep estimates for linear elliptic PDE. We therefore review here the basic estimates for *nondivergence structure* equations of the form

$$Lu = -a_{ij}(x) u_{x_i x_j} = f \quad \text{in } U ,$$

where the coefficients $a_{ij} \in L^\infty(U)$ $(1 \leq i, j \leq n)$ satisfy the uniform ellipticity condition

(12) $$\theta|\xi|^2 \leq a_{ij}(x)\xi_i\xi_j \leq \Theta|\xi|^2 \qquad (x \in U, \ \xi \in \mathbb{R}^n)$$

for constants $0 < \theta \leq \Theta < \infty$. We assume the u, f, ∂U, etc., are smooth. It will be extremely important for our applications that the following estimates do not depend in any way on the smoothness or even continuity of the a_{ij}.

Estimate 1: (Sup-norm bound)

(13) $$\|u\|_{L^\infty(U)} \leq \|u\|_{L^\infty(\partial U)} + C\|f\|_{L^\infty(U)} .$$

Estimate 2: (Estimate of gradient on ∂U). If $u = 0$ on ∂U, then

(14) $$\|Du\|_{L^\infty(\partial U)} \leq C(\|u\|_{L^\infty(U)} + \|f\|_{L^\infty(U)})$$

Estimate 3: (Alexandroff-Bakelman-Pucci). We have

(15) $$\sup_U u \leq \sup_{\partial U} u + C\|f\|_{L^n(U)}$$

Estimate 4: (Krylov-Safonov Hölder estimates). (i) For each $V \subset\subset U$ there exists $\alpha > 0$ such that

(16) $$\|u\|_{C^{0,\alpha}(V)} \leq C(\|u\|_{L^\infty(U)} + \|f\|_{L^n(U)})$$

(ii) If $u|_{\partial U}$ is Hölder continuous with exponent β, there exists $0 < \alpha \leq \beta$ such that

(17) $$\|u\|_{C^{0,\alpha}(U)} \leq C(\|u\|_{C^{0,\beta}(\partial U)} + \|f\|_{L^n(U)})$$

Estimate 5: (Krylov-Safonov Harnack inequality). If u solves (1) and $u \geq 0$, then for each $V \subset\subset U$ there exists a constant C such that

(18) $$\sup_V u \leq C(\inf_V u + \|f\|_{L^n(U)}) .$$

These inequalities are proved in [G-T] for sufficiently smooth solutions u of $Lu = f$. The various constants C depend on n, θ, Θ, the smoothness of ∂U, etc., but not on u and f.

See [C-C] for viscosity solution interpretations and proofs of the foregoing.

B. $C^{1,\alpha}$ estimates

In this lecture we obtain *a priori* $C^{1,\alpha}$ estimates for a smooth solution of our fully nonlinear PDE

1. A boundary Hölder estimate. We will require first a rather delicate linear estimate at the boundary. We reproduce the following recent proof of Caffarelli, taken from [G-T 2].

Notation. $B^+(0,r) = \{x \in \mathbb{R}^n \mid |x| \le r, x_n \ge 0\}.$

Theorem (Krylov). *Let u solve*

$$(1) \qquad \begin{cases} -a_{ij}u_{x_ix_j} = f & \text{in } B^+(0,1) \\ u = 0 & \text{on } \partial B^+(0,1) \cap \{x_n = 0\} \ . \end{cases}$$

Then for $0 < r \le 1$,

$$(2) \qquad \underset{B^+(0,r)}{\text{osc}} \left(\frac{u}{x_n}\right) \le C\, r^\alpha \left(\underset{B^+(0,1)}{\text{osc}} \left(\frac{u}{x_n}\right) + \|f\|_{L^\infty}\right) ,$$

the constants $0 < \alpha < 1$, $C > 0$ depending only on n, θ, Θ, etc.

OUTLINE OF PROOF. **1.** An argument using barriers (cf. Estimate 1 above) shows

$$v = \frac{u}{x_n}$$

is locally bounded in $B^+(0,1)$. Assume for the moment $u \ge 0$.

2. Write $x = (x', x_n)$, $(x' = (x_1,\dots,x_{n-1}) \in \mathbb{R}^{n-1})$ and define

$$C_{r,\delta} = \{x \mid |x'| \le r, \ 0 \le x_n \le \delta r\} \ .$$

We *claim* that there exists $\delta > 0$ such that

$$(3) \qquad \inf_{\substack{|x'|\le r \\ x_n = \delta r}} v \le C \left(\inf_{C_{r/2,\delta}} v + r\|f\|_{L^\infty}\right)$$

To see this assume without loss $r = 1$,

$$\inf_{\substack{|x'|\le 1 \\ x_n = \delta}} v = 1 \ .$$

Define

$$w(x) = \left[(1 - |x'|^2) + (1 + \|f\|_{L^\infty})\frac{(x_n - \delta)}{\delta^{\frac{1}{2}}}\right] x_n$$

and check $Lw \leq f$ if $\delta > 0$ is small enough and $0 \leq x_n \leq \delta$. Also $w \leq u$ on $\partial C_{1,\delta}$. Thus the maximum principle implies

$$w \leq u \quad \text{on} \quad C_{1,\delta} \ ;$$

and so on $C_{\frac{1}{2},\delta}$,

$$v = \frac{u}{x_n} \geq (1 - |x'|^2) + (1 + \|f\|_{L^\infty})\frac{x_n - \delta}{\delta^{\frac{1}{2}}}$$
$$\geq \frac{1}{2} - C\|f\|_{L^\infty} \quad \text{for small} \ \delta > 0 \ .$$

This gives (3) (for $r = 1$, $\inf_{\substack{|x'| \leq 1 \\ x_n = \delta}} v = 1$).

3. Now set

$$D_{r,\delta} = \left\{ x \mid |x'| \leq r, \ \frac{\delta r}{2} \leq x_n \leq \frac{3\delta r}{2} \right\} \ .$$

Note $D_{r,\delta} \subset C_{2r,\delta}$. Observe also

$$\frac{2u}{3\delta r} \leq v = \frac{u}{x_n} \leq \frac{2u}{\delta r}$$

on $D_{r,\delta}$. Thus the Harnack inequality (Estimate 5) gives

$$\sup_{D_{r,\delta}} u \leq C(\inf_{D_{r,\delta}} u + r^2\|f\|_{L^\infty}) \ .$$

Hence

$$\begin{aligned}
\sup_{D_{r,\delta}} v &\leq C(\inf_{D_{r,\delta}} v + r\|f\|_{L^\infty}) \\
&\leq C(\inf_{\substack{|x'| \leq r \\ x_n = \delta r}} v + r\|f\|_{L^\infty}) \\
&\leq C(\inf_{C_{r/2,\delta}} v + r\|f\|_{L^\infty}) \ , \quad \text{by (3)}
\end{aligned}$$

(4)

4. Now drop the assumption $u \geq 0$. Set

$$M_{2r} = \sup_{C_{2r,\delta}} v \ , \qquad m_{2r} = \inf_{C_{2r,\delta}} v \ , \qquad \underset{C_{2r,\delta}}{\text{osc}} \ v = M_{2r} - m_{2r} \ .$$

Apply (4) to $M_{2r} - v \geq 0$, $v - m_{2r} \geq 0$ in place of v:

$$\sup_{D_{r,\delta}}(M_{2r} - v) \leq C[\inf_{C_{r/2,\delta}} (M_{2r} - v) + r\|f\|_{L^\infty}] \ ,$$
$$\sup_{D_{r,\delta}}(v - m_{2r}) \leq C[\inf_{C_{r/2,\delta}} (v - m_{2r}) + r\|f\|_{L^\infty}] \ .$$

Adding these inequalities we deduce:

$$M_{2r} - m_{2r} \leq \sup_{D_{r,\delta}}(M_{2r} - v) + \sup_{D_{r,\delta}}(v - m_{2r})$$

$$\leq C[(M_{2r} - M_{r/2}) + (m_{r/2} - m_{2r}) + r\|f\|_{L^\infty}]$$

Rearrange this inequality to discover

$$\operatorname*{osc}_{C_{r/2,\delta}} v \leq \eta \operatorname*{osc}_{C_{2r,\delta}} v + Cr\|f\|_{L^\infty}$$

for $\eta = (1 - \frac{1}{C}) < 1$. This inequality and a standard iteration yield (2). □

As a corollary we have:

Estimate 6: Let u solve $Lu = f$ in U, $u = 0$ on ∂U. Then

(5)
$$\|Du\|_{C^{0,\alpha}(\partial U)} \leq C\|f\|_{L^\infty(U)} ,$$

for constants $C, \alpha > 0$ depending only on n, θ, Θ, ∂U, etc.

The proof of (5) follows by locally straightening out ∂U to convert to the geometric setting of Theorem 1 and letting $x_n \to 0$.

2. $C^{1,\alpha}$ estimates for fully nonlinear elliptic equations. Let us return to the fully nonlinear PDE

(7)
$$\begin{cases} F(D^2u) = f & \text{in } U \\ u = 0 & \text{on } \partial U . \end{cases}$$

We will next show how the linear theory discussed above gives us *a priori* estimates on the solution u. The idea is to "linearize" in various ways.

a. First linearization. We may as well suppose $F(0) = 0$. Then (7) implies

$$f = F(D^2u) - F(0) = \int_0^1 \frac{d}{dt} F(tD^2u)dt$$

$$= \int_0^1 F_{r_{ij}}(tD^2u)dt\ u_{x_ix_j}$$

Setting

$$a_{ij}(x) = -\int_0^1 F_{r_{ij}}(tD^2u(x))dt ,$$

we see

(8)
$$\begin{cases} -a_{ij}u_{x_ix_j} = f & \text{in } U \\ u = 0 & \text{on } \partial U . \end{cases}$$

Then according to Estimates 1, 2 and 6 we have:

$$\|u\|_{C^{0,\alpha}(U)} \leq C\|f\|_{L^{\infty}(U)} ,$$
(9)

$$\|Du\|_{C^{0,\alpha}(\partial U)} \leq C\|f\|_{L^{\infty}(U)} .$$
(10)

b. Second linearization. We use a different trick to invoke once more the estimates above for linear equations. Let us fix $k \in \{1,\ldots,n\}$ and differentiate the PDE (7) with respect to x_k. We find

$$F_{r_{ij}}(D^2u)u_{x_k x_i x_j} = f_{x_k} \quad \text{in } U \quad (1 \leq k \leq n) .$$

Thus

$$-\tilde{a}_{ij}\tilde{u}_{x_i x_j} = \tilde{f} \quad \text{in } U ,$$
(11)

for

$$\tilde{u} = u_{x_k} , \qquad \tilde{f} = f_{x_k}$$
$$\tilde{a}_{ij} = -F_{r_{ij}}(D^2u) .$$

In view of (10) and Estimates 1–4, we deduce

$$\|\tilde{u}\|_{C^{0,\alpha}(U)} \leq C\|\tilde{f}\|_{L^{\infty}(U)}$$

for some $\alpha > 0$. As $\tilde{u} = u_{x_k}$ $(k = 1,\ldots,n)$, we conclude that

$$\|Du\|_{C^{0,\alpha}(U)} \leq C\|f\|_{C^{0,1}(U)} .$$
(12)

Let us summarize:

Theorem. *If u is a smooth solution of the uniformly elliptic, fully nonlinear PDE (7), we have the estimate*

$$\|u\|_{C^{1,\alpha}(U)} \leq C\|f\|_{C^{0,1}(U)}$$
(13)

for constants $C > 0$, $0 < \alpha < 1$, depending only on n, θ, Θ, ∂U, etc.

Remark. See [C-C, Chapters 5,9] for a proof of estimate (13) (and improvements) if u is only a viscosity solution of (7). $\quad\square$

C. $C^{2,\alpha}$ estimates

It is unknown in dimensions $n \geq 3$ whether the (unique) viscosity solution of

(1)
$$\begin{cases} F(D^2 u) = f & \text{in } U \\ u = 0 & \text{on } \partial U \end{cases}$$

belongs to $C^{2,\alpha}_{\text{loc}}$ or $C^{1,1}_{\text{loc}}$. This is a major unsolved problem.

To continue further we will therefore need to assume, in addition to uniform ellipticity, that

(2)
$$F \quad \text{is convex.}$$

Many, if not most, of the important examples of fully nonlinear elliptic PDE satisfy this condition. We will see that the "one-sided control" on $D^2 F$ (due to convexity) will give us "two-sided control" on $D^2 u$.

1. Sup-norm bounds on $D^2 u$.

Theorem. *There exists a constant C such that*

(3)
$$\|D^2 u\|_{L^\infty(U)} \leq C\|f\|_{C^{0,1}(U)}$$

OUTLINE OF PROOF. 1. Assume for simplicity

$$\begin{cases} F(D^2 u) = f & \text{in } B^+(0,1) \\ u = 0 & \text{on } \partial B^+(0,1) \cap \{x_n = 0\} \end{cases}$$

Let $k \in \{1, \ldots, n-1\}$ and differentiate with respect to x_k:

$$F_{r_{ij}}(D^2 u) u_{x_k x_i x_j} = f_{x_k} \quad \text{in} \quad B^+(0,1)$$

Let

(4)
$$\begin{cases} \tilde{u} = u_{x_k}, & \tilde{f} = f_{x_k}, \\ \tilde{a}_{ij} = -F_{r_{ij}}(D^2 u). \end{cases}$$

Then

(5)
$$\begin{cases} -\tilde{a}_{ij} \tilde{u}_{x_i x_j} = \tilde{f} & \text{in } B^+(0,1) \\ \tilde{u} = 0 & \text{on } \partial B^+(0,1) \cap \{x_n = 0\} \end{cases}$$

Then a local version of Estimates 1,2 provide us with the bound

$$\|u_{x_k x_n}\|_{L^\infty(\partial B^+(0,\frac{1}{2}) \cap \{x_n = 0\})} \leq C(\|Du\|_{L^\infty} + \|f\|_{C^{0,1}})$$

for $k = 1, \ldots n - 1$. We next employ the PDE $F(D^2u) = f$ to obtain a bound on $u_{x_n x_n}|_{x_n = 0}$. Indeed the uniform ellipticity implies

$$\Theta \geq -\frac{\partial F}{\partial r_{nn}}(D^2 u) \geq \theta$$

and so we can solve for $u_{x_n x_n}$ in terms of f, $u_{x_k x_l}$ ($k = 1, \ldots, n-1$, $l = 1, \ldots, n$).

Locally straightening ∂U to reduce to the case above gives us

(6)
$$\|D^2 u\|_{L^\infty(\partial U)} \leq C(\|u\|_{C^{0,1}(U)} + \|f\|_{C^{0,1}})$$
$$\leq C\|f\|_{C^{0,1}}$$

So far we have not used the convexity of F.

2. We use the maximum principle trick to extend (6) to an estimate of $D^2 u$ inside U. Let ξ denote some unit vector in $\mathbb{R}^n$ and differentiate the PDE (1) twice in the direction ξ. We obtain

$$F_{r_{ij}}(D^2 u) u_{\xi\xi x_i x_j} + F_{ij,kl}(D^2 u) u_{x_i x_j \xi} u_{x_k x_l \xi} = f_{\xi\xi}$$

for $u_{\xi\xi} = u_{x_i x_j} \xi_i \xi_j$. Since F is convex, the matrix $D^2 F$ is nonnegative. Hence

(7)
$$-\hat{a}_{ij}\hat{u}_{x_i x_j} \leq \hat{f}$$

where

$$\begin{cases} \hat{u} = u_{\xi\xi}, & \hat{f} = f_{\xi\xi} \\ \hat{a}_{ij} = -F_{r_{ij}}(D^2 u). \end{cases}$$

Thus $\hat{u}$ is a *sub*solution of a linear elliptic PDE, and consequently (cf. Estimate 1):

$$\sup_U \hat{u} \leq \max_{\partial U} \hat{u} + C\|\hat{f}\|_{L^\infty}.$$

Consequently (6),(8) imply

$$\sup_U u_{\xi\xi} \leq C\|f\|_{C^{1,1}(U)}.$$

This is a one-sided upper estimate for $u_{\xi\xi}$ in any direction ξ.

We next employ as follows the PDE $F(D^2 u) = f$ to obtain a two-sided estimate. Since F is convex and $F(0) = 0$,

$$F(R) \geq DF(0) : R.$$

Thus

$$F_{r_{ij}}(0) u_{x_i x_j} \leq f \quad \text{in} \quad U.$$

Rotating and stretching coordinates we deduce

$$-\Delta u \leq C\|f\|_{L^\infty} ,$$

and so

$$u_{x_1 x_1} \geq -\sum_{i=2}^{n} u_{x_i x_i} - C\|f\|_{L^\infty}$$
$$\geq -C\|f\|_{C^{1,1}} .$$

This inequality is valid with any $u_{\xi\xi}$ replacing $u_{x_1 x_1}$:

$$\inf_U u_{\xi\xi} \geq -C\|f\|_{C^{1,1}}$$

□

2. $C^{2,\alpha}$ estimates. We next modify the foregoing calculations to obtain a far more subtle bound on the Hölder modulus of continuity of $D^2 u$.

Theorem. *There exist constants $C > 0$, $0 < \alpha < 1$ such that*

(9)
$$\|u\|_{C^{2,\alpha}(\bar{U})} \leq C\|f\|_{C^{1,1}(U)}$$

OUTLINE OF PROOF. 1. Again for simplicity assume

$$\begin{cases} F(D^2 u) = f & \text{in } B^+(0,1) \\ u = 0 & \text{on } \partial B^+(0,1) \cap \{x_n = 0\} \end{cases}$$

Take $k \in \{1, \ldots, n-1\}$ and differentiate with respect to x_k to obtain (4),(5).

By Estimate 6 (from Lecture 2) we deduce

$$\|u_{x_k x_n}\|_{C^{0,\alpha}(\partial B^+(0,\frac{1}{2}) \cap \{x_n=0\})} \leq C\|f\|_{C^{0,1}(U)} ,$$

for $k = 1, \ldots, n-1$. We use the PDE $F(D^2 u) = f$ to obtain control on $u_{x_n x_n}|_{x_n=0}$.

Locally straightening ∂U to reduce to the case above gives us

(8)
$$\|D^2 u\|_{C^{0,\alpha}(\partial U)} \leq C\|f\|_{C^{0,1}(U)} .$$

This bond does not require the convexity of F.

2. We must extend (8) to give a bound on the Hölder modulus of $D^2 u$ within U. I will sketch next a derivation of interior Hölder estimates on $D^2 u$. These can be combined with the boundary estimate (8) to give global bounds on $\|D^2 u\|_{C^{0,\alpha}}$.

The idea is this: Using (7) we deduce from variants of the Krylov-Safonov estimates "one-sided" control on the oscillation of $\hat{u} = u_{\xi\xi}$ on balls. Then the PDE $F(D^2u) = f$ itself, which gives a functional relation among the various second derivatives, allows us to obtain "two-sided" control on the oscillation.

I will present from [C-C] a recent proof due to Caffarelli. We first record a technical fact from the linear theory:

$$(9) \qquad \begin{cases} \text{If } -\hat{a}_{ij}v_{x_ix_j} \geq 0 \text{ in } B(0,1) \text{ and } v \geq 0, \text{ then} \\ \inf_{B(0,\frac{1}{2})} v \geq C|\{v \geq 1\} \cap B(0,\frac{1}{2})|^\delta \end{cases}$$

for constants C, δ.

3. From now on we assume $B(0,1) \subset U$ and for simplicity assume as well $f \equiv 0$ on $B(0,1)$. Let us suppose

$$1 \leq \operatorname{diam} D^2u(B(0,1)) \leq 2$$

and that $D^2u(B(0,1)) \subset \mathbb{S}^n$ is covered by N balls $\tilde{B}_1, \tilde{B}_2, \tilde{B}_N$ of radius $\varepsilon > 0$. (Notation: $\tilde{B} = $ ball in $\mathbb{S}^n$.)

We *claim* that if $\varepsilon > 0$ is small enough, then

$$(10) \qquad \begin{cases} D^2u(B(0,\frac{1}{2})) \quad \text{can be covered by} \\ N-1 \text{ of the balls } \tilde{B}_1, \ldots, \tilde{B}_N . \end{cases}$$

To substantiate this claim, for each $i \in \{1, \ldots, N\}$ we first select $x_i \in B(0,1)$ so that

$$\tilde{B}_i \subset \tilde{B}(R_i, 2\varepsilon)$$

where $R_i = D^2u(x_i)$. Now let $\lambda > 0$ be a constant (to be selected later). Take ε so small that $2\varepsilon \leq \frac{\lambda}{2}$. Then

$$(11) \qquad \text{the balls } \{\tilde{B}(R_i, \frac{\lambda}{2})\}_{i=1}^N \text{ cover } D^2u(B(0,1)) \subset \mathbb{S}^n ,$$

and so

$$(12) \qquad \text{the sets } \{(D^2u)^{-1}(\tilde{B}(R_i, \frac{\lambda}{2}))\}_{i=1}^N \text{ cover } B(0,1) \subset \mathbb{R}^n .$$

We may assume that N is bounded above by a fixed constant (depending only on λ). Thus at least one of the sets in (12) intersects $B(0,\frac{1}{2})$ on a set of measure bounded away from zero. So without loss we may assume

$$(13) \qquad |(D^2u)^{-1}(\tilde{B}(R_1, \frac{\lambda}{2}) \cap B(0,\frac{1}{2})| \geq \eta$$

for some universal constant $\eta > 0$.

4. Since diam $D^2u(B(0,1)) \geq 1$, we may assume that

$$\|R_1 - R_2\| \geq \tfrac{1}{2} .$$

Now $F(R_1) = F(D^2u(x_1)) = 0$, $F(R_2) = F(D^2u(x_2)) = 0$. This implies — owing to the uniform ellipticity of F — that there exists a unit vector ξ such that

$$u_{\xi\xi}(x_2) \geq u_{\xi\xi}(x_1) + \lambda$$

for some positive constant λ depending only on θ, Θ (see [C-C]). We take this to be the constant λ in step 3. Hence

(14)
$$M = \sup_{B(0,1)} u_{\xi\xi} \geq u_{\xi\xi}(x_1) + \lambda .$$

Now if $x \in (D^2u)^{-1}(\tilde{B}(R_1, \tfrac{\lambda}{2}))$, then (14) implies

(15)
$$u_{\xi\xi}(x) \leq u_{\xi\xi}(x_1) + \frac{\lambda}{2} \leq M - \lambda + \frac{\lambda}{2} = M - \frac{\lambda}{2} .$$

Set

(16)
$$v = M - u_{\xi\xi} .$$

Then (13),(15) imply

(17)
$$|\{v \geq \frac{\lambda}{2}\} \cap B(0, \tfrac{1}{2})| \geq \eta .$$

Next recall from (7) that $-\hat{a}_{ij}\hat{u}_{x_ix_j} \leq 0$ in $B(0,1)$ for $\hat{u} = u_{\xi\xi}$, $\hat{f} \equiv 0$. Hence $-\hat{a}_{ij}v_{x_ix_j} \geq 0$, $v \geq 0$ in $B(0,1)$. Therefore (9) and (17) imply

$$\inf_{B(0,\frac{1}{2})} v \geq \sigma ,$$

σ an absolute constant. Hence

(18)
$$\max_{B(0,\frac{1}{2})} u_{\xi\xi} \leq M - \sigma .$$

5. Since the balls $\{\tilde{B}(R_i, 2\varepsilon)\}$ cover $D^2u(B(0,1))$, there exists $j \in \{1, \dots, N\}$ such that

(19)
$$M - u_{\xi\xi}(x_j) \leq 3\varepsilon .$$

Take ε so small that $6\varepsilon < \sigma$. Then (18),(19) imply

$$\text{(20)} \qquad D^2u(B(0,\tfrac{1}{2})) \cap \bar{B}(R_j, 2\varepsilon) = \emptyset .$$

Indeed if $x \in B(0, \tfrac{1}{2})$ and $D^2u(x) \in \bar{B}(R_j, 2\varepsilon)$, then

$$|u_{\xi\xi}(x) - u_{\xi\xi}(x_j)| \le 2\varepsilon .$$

But (19) implies

$$u_{\xi\xi}(x) \ge M - 5\varepsilon > M - \sigma ,$$

in contradiction to (18).

This verifies (20) and thus proves claim (10).

6. Next we show that there exists $\delta > 0$ such that

$$\text{(20)} \qquad \begin{cases} \text{diam } D^2u(B(0,1)) = 2 \\ \text{implies} \\ \text{diam } D^2u(B(0,\delta)) \le 1 \end{cases}$$

To see this let us cover $D^2u(B(0,1))$ with N balls of radius ε, ε as above. Then

$$D^2u(B(0,\tfrac{1}{2}))$$

is covered by $N - 1$ balls. Thus

$$D^2u(B(0,\tfrac{1}{4}))$$

is covered with $N - 2$ balls and so

$$D^2u(B(0, \tfrac{1}{2^k}))$$

is covered with $N - k$ balls. Hence for some k

$$D^2u(B(0, \tfrac{1}{2^k})) \le 1 .$$

Interior Hölder continuity for D^2u follows upon our iterating estimate (20). $\qquad \square$

This proof is technically simpler than the original proofs of $C^{2,\alpha}$ estimates (done independently by Krylov and myself). I strongly suggest again that interested readers consult the book of Cabre-Caffarelli [C-C].

Part II
Mean Curvature Motion

Lecture D: Introduction, distance function

The next three lectures concern a completely different topic, the flow of hypersurfaces in $\mathbb{R}^n$ by mean curvature. We will first discuss the local existence of a classical solution of such an evolution, following the method of [E-S II]. This requires analysis of a fully nonlinear parabolic PDE (satisfied by the signed distance function) and so provides a kind of bridge with the mathematics discussed in Part I.

These are meant to be elementary, expository lectures, which serve as an introduction to Souganidis' more advanced talks.

1. Mean curvature motion

Let Γ_0 denote a smooth, $(n-1)$-dimensional hypersurface lying in $\mathbb{R}^n$. We propose to evolve Γ_0 in time, generating thereby a family

$$\{\Gamma_t\}_{t\geq 0}$$

of hypersurfaces, by moving the surface Γ_t at time t so that its normal velocity is the mean curvature vector. This is in fact the "geometric heat equation".

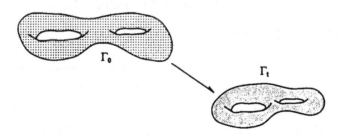

Geometric Notation.

$$v = \text{normal velocity}$$
$$\nu = \text{unit normal vector field to } \Gamma_t$$
$$\mathbf{H} = \text{mean curvature vector} = H\nu$$
$$H = \text{mean curvature, computed with respect to } \nu$$

From differential geometry we recall the formula

$$(1) \qquad\qquad H = -\mathrm{div}(\nu) \,,$$

where ν is extended arbitrarily off Γ_t and "div" is the usual divergence in $\mathbb{R}^n$.

Hence our geometric law of motion is

$$(2) \qquad\qquad \mathbf{v} = H\nu = -\mathrm{div}(\nu)\nu$$

2. The distance function

We now reinterpret (2) in terms of the signed distance function to Γ_t. Let us therefore for the moment assume that the geometric motion problem described in §1 has a smooth solution, existing for times $0 \leq t < t_\star$. We will assume that

$$\Gamma_0 = \partial U_0 \,,$$

U_0 denoting a bounded open set, and further that

$$\Gamma_t = \partial U_t \qquad 0 \leq t < t_\star$$

U_t denoting a bounded open set at time $t > 0$.

Definition. For each time $t \geq 0$, define

$$(3) \qquad\qquad d(x,t) = \begin{cases} \mathrm{dist}(x,\Gamma_t) & x \in \mathbb{R}^n - \bar{U}_t \\ -\mathrm{dist}(x,\Gamma_t) & x \in U_t \end{cases}$$

Then d is the **signed distance** function to $\{\Gamma_t\}_{0 \leq t < t_\star}$.

Note that even though we are assuming the surfaces $\{\Gamma_t\}_{0 \leq t < t_\star}$ are smooth, the mapping $x \mapsto d(x,t)$ need not be smooth. However, d is smooth in some sufficiently small region around the $\{\Gamma_t\}$, to which we hereafter restrict our attention.

Since d is determined by Γ_t, d "contains geometric information about Γ_t," which we now extract by differentiating d.

If x is sufficiently close to Γ_t and $d(x,t) > 0$, there exists a unique point $y \in \Gamma_t$ for which

$$d(x,t) = |x - y| \,.$$

Then

$$(4) \qquad\qquad \nu = Dd(x,t)$$

is the outer unit normal to Γ_t at y.

Information about the curvature of Γ_t is contained in $D^2 d$.

Notation. (a) The eigenvalues of $D^2 d$ will be written as $\lambda_1, \ldots, \lambda_n$.
(b) The *principal curvatures* of Γ_t at y will be denoted $\kappa_1, \ldots, \kappa_{n-1}$.

In particular, $H = \kappa_1 + \cdots + \kappa_{n-1}$. As in [G-T, §14.6], we have after a possible reordering

(5)
$$\lambda_i = \frac{-\kappa_i}{1 - \kappa_i d} \qquad (1 \leq i \leq n-1), \quad \lambda_n = 0 .$$

Inverting this relation we obtain

$$\kappa_i = \frac{\lambda_i}{\lambda_i d - 1} \qquad (1 \leq i \leq n-1) .$$

3. A PDE satisfied by the distance function

We propose next to convert the geometric law of motion (2) into a PDE satisfied by d. Now for a fixed point x such that $d(x,t) > 0$ we have

$$d_t = -\mathbf{v} \cdot \boldsymbol{\nu} = -H$$
$$= -\sum_{i=1}^{n-1} \kappa_i$$
$$= \sum_{i=1}^{n-1} \frac{\lambda_i}{1 - \lambda_i d}$$

The same formula is valid if $d(x,t) \leq 0$. Consequently,

(7)
$$d_t = F(D^2 d, d) \qquad \text{near } \Gamma ,$$

where

(8)
$$F(R, z) = f(\lambda, z)$$
$$= \sum_{i=1}^{n} \frac{\lambda_i}{1 - \lambda_i z}$$

and $\lambda = \lambda(R) = (\lambda_1, \ldots, \lambda_n)$ are the eigenvalues of R, ordered so that $\lambda_1 \leq \lambda_2 \leq \cdots \leq \lambda_n$. We will always stay so close to Γ_t that $\lambda_i d < 1$ $(i = 1, \ldots, n)$.

Remark. In particular, consider the region $\{d > 0\}$. Then if $\lambda_i \geq 0$, we have $\frac{\lambda_i}{1-\lambda_i d} \geq \lambda_i$. Furthermore if $\lambda_i \leq 0$, then $\frac{\lambda_i}{1-\lambda_i d} \geq \lambda_i$ as well. Thus

$$0 = d_t - \sum_{i=1}^n \frac{\lambda_i}{1 - \lambda_i d} \leq d_t - \sum_{i=1}^n \lambda_i = d_t - \Delta d .$$

Hence

(9) $$d_t - \Delta d \geq 0 \quad \text{in} \quad \{d \geq 0\} ;$$

and likewise,

(10) $$d_t - \Delta d \leq 0 \quad \text{in} \quad \{d \leq 0\} .$$

We deduce in particular

$$d_t = \Delta d \quad \text{on } \Gamma_t .$$

These observations, generalized to hold in the viscosity sense on all of $\mathbb{R}^n \times (0, \infty)$, are important in the study of moving fronts for reaction-diffusion equations. (See the lectures of Souganidis.) $\quad\square$

Let us next examine the nonlinearity (8). Note

$$\frac{\partial f}{\partial \lambda_i} (\lambda, z) = \frac{1}{(1 - \lambda_i z)^2} > 0 \quad (1 \leq i \leq n) .$$

Thus if $z \in \mathbb{R}$ is fixed and $A, B \in \mathbb{S}^n$ with $A \geq B$, we have

$$F(A, z) = f(\lambda(A), z) \geq f(\lambda(B), z) = F(B, z) .$$

Here we used the fact that $A \geq B$ implies

$$\lambda_i(A) \geq \lambda_i(B) \quad \text{for } i = 1, \ldots, n .$$

This follows from Courant's min-max representation formula for eigenvalues.

Consequently the fully nonlinear operator $-F(D^2 u, u)$ is elliptic. In particular then the *PDE* (7) *is a fully nonlinear parabolic PDE for d.*

4. Local existence of smooth solutions for mean-curvature flow

We next show that we can reverse the reasoning above and solve the PDE (7) in order to build a smooth solution $\{\Gamma_t\}_{0 \le t \le t_0}$ of the mean curvature flow, provided $t_0 > 0$ is sufficiently small.

Notation.

(a) $g(x) = \begin{cases} \text{dist}(x, \Gamma_0) & \text{if } x \in \mathbb{R}^n - \bar{U}_0 \\ -\text{dist}(x, \Gamma_0) & \text{if } x \in U_0 \end{cases}$

(b) $V = \{x \in \mathbb{R}^n \mid -\delta_0 < g(x) < \delta_0\}$

(c) $W = V \times (0, t_0)$, $\Sigma = \partial V \times [0, t_0]$.

We choose δ_0 so small that g is smooth within $\bar{V}$. We then consider the initial/boundary value problem

(12)
$$\begin{cases} v_t = F(D^2 v, v) & \text{in } W \\ |Dv|^2 = 1 & \text{on } \Sigma \\ v = g & \text{on } V \times \{t = 0\} \end{cases}$$

The point is that **if** $\{\Gamma_t\}_{0 < t \le t_0}$ is a classical mean curvature motion, then $v = d$ solves (12). On the other hand, [E-S II] shows that there exists a solution v of (12) if t_0 is sufficiently small, with

$$v \in C^{2+\alpha, \frac{2+\alpha}{2}}(\overline{W}) \cap C^\infty(W) .$$

The proof proceeds by looking for v in the form

$$v = g + th + w , \qquad \text{where } h = F(D^2 g, g) .$$

Then w satisfies a nonlinear PDE (with nonlinear boundary conditions), which is "close to linear" for small w. A contraction mapping argument yields the existence of w: see [E-S II].

It remains to show that $v(\cdot, t)$ is actually the signed distance from a surface Γ_t, which in turn is evolving by mean curvature flow. We require this

Lemma. *The function*

$$w = |Dv|^2 - 1$$

is identically equal to zero in $\overline{W}$.

PROOF. Clearly $w = 0$ on the parabolic boundary of W, owing to the boundary condition in (12). Let us also compute

$$v_{tx_k} = F_{r_{ij}}(D^2 v, v)v_{x_k x_i x_j} + F_z(D^2 v, v)v_{x_k} .$$

Hence

$$w_t = 2v_{x_k} v_{x_k t}$$

(13)
$$= 2F_{r_{ij}} v_{x_k} v_{x_k x_i x_j} + 2F_z |Dv|^2$$

$$= F_{r_{ij}} w_{x_i x_j} - 2F_{r_{ij}} v_{x_k x_i} v_{x_k x_j} + 2F_z |Dv|^2 .$$

Now $F(R, z) = f(\lambda, z)$, $\lambda = \lambda(R)$, and so

$$F_z(D^2 v, v) = f_z(\lambda(D^2 v), v) = \sum_{i=1}^n \frac{\lambda_i^2}{(1 - \lambda_i v)^2} .$$

In addition,

$$F_{r_{ij}}(D^2 v, v) v_{x_k x_i} v_{x_k x_j} = \sum_{i=1}^n \frac{\partial f}{\partial \lambda_i} (\lambda, v) \lambda_i^2$$

$$= \sum_{i=1}^n \frac{\lambda_i^2}{(1 - \lambda_i v)^2} = F_z(D^2 v, v) .$$

(We prove the second equality by rotating coordinates to diagonalizable $D^2 v$.) Then (13) implies

(14)
$$w_t = F_{r_{ij}} w_{x_i x_j} + 2F_z w \quad \text{in } W .$$

As $w = 0$ on the parabolic boundary of W, (14) and uniqueness imply $w \equiv 0$ in W. $\qquad\square$

Next **define**
$$\Gamma_t = \{x \in V \mid v(x, t) = 0\} \qquad (0 < t \le t_0)$$

Then the Lemma implies $v(\cdot, t)$ is the signed distance function to Γ_t, and the PDE (12) implies the surfaces $\{\Gamma_t\}_{0 < t \le t_0}$ move by curvature flow.

E. The level set method

The foregoing lecture provides us with a smooth solution of the mean curvature flow problem, at least for small times $0 \le t \le t_0$. Such a smooth flow definitely will cease to exist after a finite time however. If, for instance $\Gamma_0 = \partial B(0, R)$, a sphere of radius R, then
$$\Gamma_t = B(0, R(t)) \quad \text{if } 0 \le t < t^* ,$$
where $R(t) = (R^2 - 2(n-1)t)^{\frac{1}{2}}$, $t^* = R^2/2(n-1)$. The sphere consequently shrinks to the point 0 at time t^*, and so $\Gamma_{t^*} = \{0\}$ is not a smooth surface.

Even worse, a smooth flow $\{\Gamma_t\}_{0\leq t<t_*}$ existing on some interval $[0,t_*)$ can cease to be smooth in the limit $t \to t_*$, even if the Γ_t are not collapsing to a point. See for example the illustration

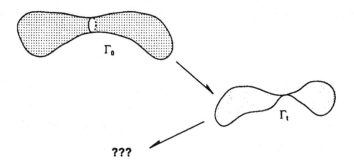

A fascinating question thus arises:

> How is it possible to define a "generalized" flow $\{\Gamma_t\}_{t>0}$ which extends the classical flow $\{\Gamma_t\}_{0<t<t_0}$ past the first time t_* of singularity?

The level set method, introduced by Osher-Sethian for numerical purposes, provides an answer.

1. Motivation for viscosity solution approach.

The theory of viscosity solutions is designed to resolve a similar problem for certain PDE (e.g. Hamilton-Jacobi equations), for which a classical solution exists only during a short initial time interval, after which singularities arise. We can expect that viscosity solution methods apply to those PDE for which there is — formally at least — a maximum principle or, more precisely, a comparison principle for solutions.

To see that there is some hope of applying such ideas in the geometric setting, consider the following heuristics. Let $\{\Gamma_t\}_{0\leq t\leq t_*}$, $\{\Delta_t\}_{0\leq t\leq t_*}$ be two smooth mean curvature flows, with

$$\begin{cases} \Delta_0 \text{ lying within the bounded region } U_0 \\ \Gamma_0 = \partial U_0 \end{cases}$$

Then Δ_t lies within U_t, where $\Gamma_t = \partial U_t$ for $0 \leq t \leq t_0$. In other words, *the flows* $\{\Gamma_t\}_{0\leq t\leq t_0}$, $\{\Delta_t\}_{0\leq t\leq t_0}$ *preserve the initial "ordering"*. This is a kind of analogue of the comparison principle for PDE. To see informally why Δ_t lies within U_t, suppose instead that there exists a first time $s \in [0,t]$ for which Δ_s intersects Γ_s at a point y. We then expect that the mean curvature of Δ_t at y is greater than the mean curvature of Γ_t at y. Consequently the flow will tend to push Δ_s back inside U_s. A careful proof uses the strong parabolic maximum principle.

2. Level set method. To gain further insight, recall from Lecture D that the (smooth) evolving surfaces $\{\Gamma_t\}_{0 \leq t \leq t_0}$ are the zero level sets of the signed distance function:

$$\Gamma_t = \{x \in \mathbb{R}^n \mid d(x,t) = 0\} \qquad (0 \leq t \leq t_0)$$

The key to the so-called *level set method* for studying generalized mean curvature flow is to consider instead of d a new function u with the property that — formally at least — *each level set of u evolves via mean curvature flow*. In other words, we seek a function u such that for each $\gamma \in \mathbb{R}$ the level sets

(1) $$\Gamma_t^\gamma = \{x \in \mathbb{R}^n \mid u(x,t) = \gamma\}$$

flow according to their mean curvature.

Following the general principle that we should always seek an "infinitesimal" version of any concept, let us now ask what PDE such a function u will satisfy. Let us assume for the moment u is smooth and $|Du| \neq 0$ in some region. Then for each $\gamma \in \mathbb{R}$, Γ_t^γ is a smooth hypersurface, with normal

(2) $$\nu = \frac{Du}{|Du|} .$$

The mean curvature of Γ_t^γ is then

(3) $$H = -\operatorname{div}(\nu) = -\operatorname{div}\left(\frac{Du}{|Du|}\right)$$

Now the normal velocity of the level surface Γ_t^γ in the direction ν is

$$\mathbf{v} = -\frac{u_t}{|Du|}\,\nu .$$

Thus our geometric law of motion

$$\mathbf{v} = H\nu$$

becomes

$$-\frac{u_t}{|Du|} = H = -\operatorname{div}\left(\frac{Du}{|Du|}\right) ,$$

and so

(4) $$u_t = |Du|\operatorname{div}\left(\frac{Du}{|Du|}\right) = \left(I - \frac{Du \otimes Du}{|Du|^2}\right) : D^2 u$$

along Γ_t^γ. Since the level surfaces fill up the region where $|Du| \neq 0$, we conclude the PDE (4) holds everywhere in this region.

In summary, if u is smooth, then

(5)
$$u_t = \left(\delta_{ij} - \frac{u_{x_i} u_{x_j}}{|Du|^2} \right) u_{x_i x_j}$$

provided $|Du| \neq 0$.

The idea is now to study the PDE (5) and to use it to **define** a generalized mean curvature motion existing for all time $t \geq 0$. Here is the procedure:

Construction of generalized mean curvature flow

Step 1: Given a compact set $\Gamma_0 \subset \mathbb{R}^n$, find a continuous bounded function $g : \mathbb{R}^n \to \mathbb{R}$ such that

$$\Gamma_0 = \{ x \mid g(x) = 0 \}$$

Step 2: Solve the initial value problem

(6)
$$\begin{cases} u_t = \left(\delta_{ij} - \frac{u_{x_i} u_{x_j}}{|Du|^2} \right) u_{x_i x_j} & \text{in } \mathbb{R}^n \times (0, \lambda) \\ u = g & \text{on } \mathbb{R}^n \times \{t = 0\} \end{cases}$$

Step 3: Then define

(7)
$$\Gamma_t = \{ x \mid u(x, t) = 0 \} \qquad (t \geq 0)$$

Assuming this all works we call

$$\{\Gamma_t\}_{t \geq 0}$$

the generalized mean curvature flow starting from Γ_0. This flow will exist for all times $t \geq 0$, although it may develop singularities (even if Γ_0 is a smooth hypersurface) and will certainly become the empty set after a finite time.

3. Justification of level set method

Considerable work is needed to justify the foregoing plan. The main issue is that we must show the PDE (6) has some kind of unique weak solution u, the uniqueness being necessary to justify the article "the" in "the generalized mean curvature flow". The overall problems are that the PDE (6) is degenerate parabolic and, worse, is undefined where $|Du| = 0$. And yet, since we definitely want to allow for the possibility that the level sets $\{\Gamma_t\}$ change topological type, the regions where $|Du| = 0$ are important.

The solution is to invoke viscosity solution methods to interpret the PDE (6), to solve this initial value problem, and to establish uniqueness ([E-S-I],[C-G-G]).

Following [E-S-I] we introduce the following

Definition. A bounded, continuous function u is a *viscosity solution* of (6) provided $u = g$ on $\mathbb{R}^n \times \{t = 0\}$, and for each $\phi \in C^\infty(\mathbb{R}^n \times (0,\infty))$ if $u - \phi$ has a local maximum (minimum) at a point $(x_0, t_0) \in \mathbb{R}^n \times (0,\infty)$, then

$$(8) \qquad \phi_t \underset{(\geq)}{\leq} \left(\delta_{ij} - \frac{\phi_{x_i}\phi_{x_j}}{|D\phi|^2} \right) \phi_{x_i,x_j} \quad \text{at } (x_0, t_0) \quad \text{if} \quad D\phi(x_0, t_0) \neq 0 \ ,$$

and

$$(9) \qquad \phi_t \underset{(\geq)}{\leq} (\delta_{ij} - \eta_i\eta_j)\phi_{x_i,x_j} \quad \text{at } (x_0, t_0) \quad \text{for some} \quad |\eta| \leq 1, \text{ if } D\phi(x_0, t_0) = 0 \ .$$

This definition is motivated by the proof of the following

Theorem 1. *There exists a viscosity solution of (6).*

PROOF. 1. We consider for $\varepsilon > 0$ the approximation

$$(10) \qquad \begin{cases} u_t^\varepsilon = \left(\delta_{ij} - \dfrac{u_{x_i}^\varepsilon u_{x_j}^\varepsilon}{|Du^\varepsilon|^2 + \varepsilon^2} \right) u_{x_i x_j}^\varepsilon & \text{in } \mathbb{R}^n \times (0,\infty) \\ u^\varepsilon = g & \text{on } \mathbb{R}^n \times \{t = 0\} \end{cases}$$

Problem (10) has a unique, smooth, bounded solution u^ε. We can differentiate the PDE (10) with respect to x_k ($k = 1, \ldots, n$) and t, and apply the maximum principle to derive the bounds

$$\sup_{0 < \varepsilon \leq 1} \|u^\varepsilon, Du^\varepsilon, u_t^\varepsilon\|_{L^\infty} \leq C \ ,$$

the constant C depending only on $\|g\|_{C^{1,1}}$. Consequently we have for some sequence $\varepsilon_j \to 0$:

$$u^{\varepsilon_j} \to u \quad \text{locally uniformly on } \mathbb{R}^n \times [0,\infty) \ ,$$

where u is bounded, Lipschitz.

2. We will now demonstrate that u is a viscosity solution of (6). So let $\phi \in C^\infty(\mathbb{R}^n \times (0,\infty))$ and suppose $u - \phi$ has a (strict) local maximum at (x_0, t_0). Then

$$(11) \qquad u^{\varepsilon_j} - \phi \quad \text{has a local maximum at} \quad (x_j, t_j)$$

with

$$(12) \qquad (x_j, t_j) \to (x_0, t_0)$$

Owing to (10),(11),

$$(13) \qquad \phi_t \le \left(\delta_{ij} - \frac{\phi_{x_i}\phi_{x_j}}{|D\psi|^2 + \varepsilon^2} \right) \phi_{x_i x_j} \quad \text{at } (x_j, t_j) \ ;$$

since (11) implies

$$D\phi = Du^\varepsilon \ , \quad \phi_t = u_t^{\varepsilon_j} \ , \quad D^2 u^{\varepsilon_j} \le D^2\phi \quad \text{at } (x_j, t_j) \ .$$

Now if $D\phi(x_0, t_0) \ne 0$, then our sending $x_j \to x_0$, $t_j \to t_0$ in (13) gives (8). If $D\phi(x_0, t_0) = 0$, we likewise derive (9) for

$$\eta = \lim_{j \to \infty} \frac{D\phi(x_j, t_j)}{(|D\phi(x_j, t_j)|^2 + \varepsilon_j^2)^{\frac{1}{2}}} \ ,$$

where we pass to a subsequence for which this limit exists. □

The main assertion is then that

Theorem 2. *A viscosity solution of (6) is unique.*

The proof (cf [E-S I]) uses the technology described in Crandall's lectures, suitably modified to handle the case (9). The basic technical trick is, roughly speaking, to take two viscosity solutions, call them u and v, and maximize

$$u(x,t) - v(y,s) - \frac{1}{\varepsilon}[|x - y|^4 + (t - s)^2] \ .$$

The point is that the term "$|x - y|^4$", rather than "$|x - y|^2$" allows us to ignore the right-hand side of (9) should this case arise. See [E-S I] for details.

Theorems 1 and 2 assure us that the PDE (6) has a unique solution, but our definition (7) of Γ_t is still worrisome. The problem is this. Suppose we select a different function $\hat{g} : \mathbb{R}^n \to \mathbb{R}$ such that

$$\Gamma_0 = \{x \mid \hat{g}(x) = 0\} \ ,$$

and we solve

$$\begin{cases} \hat{u}_t = \left(\delta_{ij} - \frac{\hat{u}_{x_i}\hat{u}_{x_j}}{|D\hat{u}|^2} \right) \hat{u}_{x_i x_j} & \text{in } \mathbb{R}^n \times (0, \infty) \\ \hat{g} = \hat{y} & \text{on } \mathbb{R}^n \times (t = 0) \ . \end{cases}$$

How do we know

$$(14) \qquad \Gamma_t = \{x \mid u(x,t) = 0\} = \{x \mid \hat{u}(x,t) = 0\} \ ?$$

If this is false, then Γ_t is not well-defined by Steps 1–3 above.

It turns out that (14) is in fact true. The proof ([E-S I; Theorem 5.1]) depends upon

Theorem 3. *Let u be the unique viscosity solution of (6). Assume* $\Phi : \mathbb{R} \to \mathbb{R}$ *is continuous. Then*

(15)
$$\tilde{u} = \Phi(u)$$

is the unique viscosity solution of

$$
\begin{cases}
\tilde{u}_t = \left(\delta_{ij} - \dfrac{\tilde{u}_{x_i} \tilde{u}_{x_j}}{|D\tilde{u}|^2} \right) \tilde{u}_{x_i x_j}, & in \ \mathbb{R}^n \times (0, \infty) \\
\tilde{u} = \tilde{g} & on \ \mathbb{R}^n \times \{t = 0\} \ ,
\end{cases}
$$

where $\tilde{g} = \Phi(g)$.

In other words an arbitrary nonlinear function of a solution to our mean curvature PDE is still a solution. This is because (15) amounts merely to a relabeling of the level sets, each of which moves via the mean curvature flow. A formal analytic proof is this:

$$
\left(\delta_{ij} - \frac{\tilde{u}_{x_i} \tilde{u}_{x_j}}{|D\tilde{u}|^2} \right) \tilde{u}_{x_i x_j} = \left(\delta_{ij} - \frac{u_{x_i} u_{x_j}}{|Du|^2} \right) [u_{x_i x_j} \Phi' + u_{x_i} u_{x_j} \Phi'']
$$

$$
= \Phi' u_t
$$

$$
= \tilde{u}_t \ .
$$

See [E-S I] for a proof in the viscosity sense, and for a proof of (14).

4. Consistency

Finally we must verify that the level set motion agrees with the classical evolution.

Theorem 4. *Assume* $\Gamma_0 \subset \mathbb{R}^n$ *is a smooth, embedded* $(n-1)$*-dimensional surface. Then the generalized flow* $\{\Gamma_t\}_{t \geq 0}$ *agrees with the unique classical flow starting from* Γ_0, *so long as the latter exists.*

IDEA OF PROOF. Let $\{\Sigma_t\}_{0 \leq t < t_0}$ denote the smooth flow starting with Γ_0. Let $d(x, t)$ be the (signed) distance function to Σ_t. Then, as in Lecture D,

$$
d_t = \sum_{i=1}^{n} \frac{\lambda_i}{1 - \lambda_i d} \qquad (\lambda = \lambda(D^2 d) = \text{ eigenvalues of } D^2 d) \ .
$$

Thus

$$
d_t - \Delta d = \sum_{i=1}^{n} \frac{\lambda_i}{1 - \lambda_i d} - \lambda_i
$$

$$
= \left(\sum_{i=1}^{n} \frac{\lambda_i^2}{1 - \lambda_i d} \right) d \qquad \text{near } \Sigma_t \ .
$$

Consequently,

$$\underline{d} = \alpha \, e^{-\lambda t} d$$

satisfies

$$\underline{d}_t - \Delta \underline{d} \le 0 \quad \text{for appropriate } \alpha, \lambda$$

Since $|D\underline{d}|^2 = \alpha^2 e^{-2\lambda t}$, we have as well

$$\underline{d}_t - \left(\delta_{ij} - \frac{\underline{d}_{x_i} \underline{d}_{x_j}}{|D\underline{d}|^2} \right) \underline{d}_{x_i x_j} \le 0 \; .$$

We now employ the comparison principle: see [E-S I]/ □

5. Questions

Having rigorously justified Steps 1–3 above, and so unambiguously defined the generalized level set flow, the real question is to study $\{\Gamma_t\}_{t \ge 0}$. We ask:

Given various geometric properties of Γ_0, what can be said about the geometric properties of the flow $\Gamma_0 \mapsto \Gamma_t$ $(t \ge 0)$?

F. Geometric properties

We discuss next some partial answers to the question posed at the end of the preceding lecture. We are especially interested in recovering from the level set method various classical differential geometric assertions.

1. Decrease of surface area

Hereafter let $\mathcal{H}^{n-1}$ denote $(n-1)$-dimensional Hausdorff measure. Let us temporarily assume $\{\Gamma_t\}_{t \ge 0}$ is a smooth flow of hypersurfaces moving by mean curvature flow. Then differential geometry tells us that

$$\begin{aligned} \frac{d}{dt} \mathcal{H}^{n-1}(\Gamma_t) &= - \int_{\Gamma_t} \mathbf{v} \cdot \mathbf{H} \, d\mathcal{H}^{n-1} \\ &= - \int_{\Gamma_t} H^2 d\mathcal{H}^{n-1} \end{aligned}$$

(1)

In particular therefore, $t \mapsto \mathcal{H}^{n-1}(\Gamma_t)$ is nondecreasing.

How can we mimic this calculation in our level set approach? The point is that all the information must somehow be contained in the PDE

(2) $$u_t = |Du| \text{div} \left(\frac{Du}{|Du|} \right) = \left(\delta_{ij} - \frac{u_{x_i} u_{x_j}}{|Du|^2} \right) u_{x_i x_j} \; .$$

What we need is a way to recover from the evolution equation (2) information about the evolution of the level sets of u. The technical tools that allow for this are:

(a) Theorem 3 from Lecture E, which asserts that $\tilde{u} = \Phi(u)$ solves (2) for each continuous Φ,

and

(b) the **coarea formula**, which asserts

$$(3) \qquad \int_{\mathbf{R}^n} f|Du|dx = \int_{-\infty}^{\infty} \int_{\{u=s\}} f \; d\mathcal{H}^{n-1} ds$$

for each Lipschitz $u : \mathbf{R}^n \to \mathbf{R}$ and continuous $f : \mathbf{R}^n \to \mathbf{R}$.

Suppose for the moment u is a smooth solution of (2). We now compute in analogy with (1) that

$$
\begin{aligned}
(4) \qquad \frac{d}{dt} \int_{\mathbf{R}^n} |Du|dx &= \int_{\mathbf{R}^n} \frac{Du}{|Du|} \cdot Du_t \; dx \\
&= -\int_{\mathbf{R}^n} \operatorname{div}\left(\frac{Du}{|Du|}\right) u_t \; dx^* \\
&= -\int_{\mathbf{R}^n} H^2|Du|dx
\end{aligned}
$$

where

$$(5) \qquad H = -\operatorname{div}\left(\frac{Du}{|Du|}\right) \qquad \text{(cf. (1) in Lecture D)} .$$

Integrating we find

$$(6) \qquad \int_{\mathbf{R}^n} |Du(x,t_2)|dx \leq \int_{\mathbf{R}^n} |Du(x,t_1)|dx$$

if $0 \leq t_1 \leq t_2$. Now apply (6) with $\tilde{u} = \Phi(u)$ replacing u:

$$(7) \qquad \int_{\mathbf{R}^n} |\Phi'|(u(x,t_2))|Du(x,t_2)|dx \leq \int_{\mathbf{R}^n} |\Phi'|(u(x,t_1))|Du(x,t_1)|dx .$$

By an approximation we see that (7) obtains for

$$\Phi(z) = \begin{cases} 0 & z \leq \gamma \\ \text{linear} & \gamma \leq z \leq \gamma + h \\ 1 & z \geq \gamma + h \end{cases}$$

where $\gamma \in \mathbf{R}$, $h > 0$. Substituting into (3), and using the coarea formula we find

$$(8) \qquad \frac{1}{h}\int_{\gamma}^{\gamma+h} \mathcal{H}^{n-1}(\Gamma_{t_2}^s)ds \leq \frac{1}{h}\int_{\gamma}^{\gamma+h} \mathcal{H}^{n-1}(\Gamma_{t_1}^s)ds .$$

---

[0]I believe (4) contains the only occurrence of integration-by-parts in this entire CIME course.

Then if we let $h \to 0^+$ we expect that

(9) $$\mathcal{H}^{n-1}(\Gamma_{t_2}^\gamma) \leq \mathcal{H}^{n-1}(\Gamma_{t_1}^\gamma)$$

for each level γ at times $0 \leq t_1 \leq t_2$.

This computation suggests that each level set of u has decreasing surface area. This is of course consistent with the classical calculation (1) (applied to each $\{\Gamma_t^\gamma\}$). However we must be careful, as our weak solution of (2) is not smooth and the passage to limits from (8) to (9) need not be true for all levels γ and all times $0 \leq t_1 < t_2$.

Example. Consider in fact the case that $n = 2$ and Γ_0 looks like this:

Γ_0
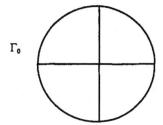

It turns out that **under the generated mean curvature flow the $\{\Gamma_t\}$ develop interiors.**

Γ_t
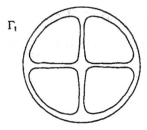

Thus $\mathcal{H}^1(\Gamma_t) = +\infty$ for $t > 0$, whereas $\mathcal{H}^1(\Gamma_0) < \infty$. Consequently (9) is certainly false here. $\qquad\square$

To recover some rigorous deduction from the heuristics above we turn to the approximations u^ε from Lecture E. Recall these solve

(10) $$\begin{cases} u_t^\varepsilon = \left(\delta_{ij} - \dfrac{u_{x_i}^\varepsilon u_{x_j}^\varepsilon}{|Du^\varepsilon|^2}\right) u_{x_i x_j}^\varepsilon & \text{in } \mathbb{R}^n \times (0, \infty) \\ u^\varepsilon = g & \text{on } \mathbb{R}^n \times \{t = 0\} . \end{cases}$$

Let

$$\phi(x) = \phi_\delta(x) = e^{-\delta(1+|x|^2)^{\frac{1}{2}}}$$

and compute

$$
\begin{aligned}
\frac{d}{dt} \int_{\mathbf{R}^n} \phi^2 (|Du^\varepsilon|^2 + \varepsilon^2)^{\frac{1}{2}} \, dx &= \int_{\mathbf{R}^n} \phi^2 \frac{Du^\varepsilon}{(|Du^\varepsilon|^2 + \varepsilon^2)^{\frac{1}{2}}} \cdot Du_t^\varepsilon \, dx \\
&= - \int_{\mathbf{R}^n} \phi^2 \operatorname{div}\left(\frac{Du^\varepsilon}{(|Du^\varepsilon|^2 + \varepsilon^2)^{\frac{1}{2}}} \right) u_t^\varepsilon \, dx \\
&\quad - 2 \int \phi \frac{D\phi \cdot Du^\varepsilon}{(|Du^\varepsilon|^2 + \varepsilon^2)^{\frac{1}{2}}} u_t^\varepsilon \, dx \\
&= - \int_{\mathbf{R}^n} \phi^2 (H^\varepsilon)^2 (|Du^\varepsilon|^2 + \varepsilon^2)^{\frac{1}{2}} \, dx \\
&\quad + 2 \int_{\mathbf{R}^n} \phi \, D\phi \cdot Du^\varepsilon H^\varepsilon \, dx \, ,
\end{aligned}
$$

where

$$H^\varepsilon = -\operatorname{div}\left(\frac{Du^\varepsilon}{(|Du^\varepsilon|^2 + \varepsilon^2)^{\frac{1}{2}}} \right) \, .$$

Consequently

$$\frac{d}{dt} \int_{\mathbf{R}^n} \phi^2 (|Du^\varepsilon|^2 + \varepsilon^2)^{\frac{1}{2}} \, dx \leq C \int_{\mathbf{R}^n} |D\phi|^2 (|Du^\varepsilon|^2 + \varepsilon^2)^{\frac{1}{2}} \, dx \, .$$

Since $|D\phi| \leq \delta\phi$, we may apply Gronwall's inequality to obtain

$$\int_{\mathbf{R}^n} \phi^2 (|Du^\varepsilon(x,t)|^2 + \varepsilon^2)^{\frac{1}{2}} \, dx \leq e^{C\delta^2 t} \int_{\mathbf{R}^n} \phi^2 (|Dg|^2 + \varepsilon^2)^{\frac{1}{2}} \, dx \, .$$

Let $\varepsilon \to 0$ and then $\delta \to 0$ to deduce

$$\int_{\mathbf{R}^n} |Du(x,t)| dx \leq \int_{\mathbf{R}^n} |Dg| dx$$

Then, as before,

$$\frac{1}{h} \int_\gamma^{\gamma+h} \mathcal{H}^{n-1}(\Gamma_t^s) ds \leq \frac{1}{h} \int_\gamma^{\gamma+h} \mathcal{H}^{n-1}(\Gamma_0^s) ds$$

Now for a fixed t and a.e. $\gamma \in \mathbb{R}$, the left-hand side converges to $\mathcal{H}^{n-1}(\Gamma_t^\gamma)$ as $h \to 0$. Thus

$$\mathcal{H}^{n-1}(\Gamma_t^\gamma) \leq \mathcal{H}^{n-1}(\Gamma_0^\gamma) \quad \text{for a.e. } \gamma \in \mathbb{R} \, .$$

A more subtle estimate shows for the particular level set $\Gamma_t = \Gamma_t^0$ that

$$\sup_{t>0} \mathcal{H}^{n-1}(\partial\Gamma_t) \leq C\mathcal{H}^{n-1}(\Gamma_0) \, ,$$

provided Γ_0 is nice enough. See [E-S III] for more details.

2. Estimates of extinction time

We turn now to the question of estimating the time t^* of extinction for a flow $\{\Gamma_t\}_{0 \le t < \infty}$.

If $\Gamma_0 \subset B(0, R)$ for some $R > 0$, then $\Gamma_t \subset B(0, R(t))$ where $R(t) = (R^2 - 2(n-1)t)^{\frac{1}{2}}$. Since $R(t) = 0$ for $t = R^2/2(n-1)$ we see that $t^* \le R^2/2(n-1)$. In particular we have the estimate

$$(11) \qquad t^* \le C \operatorname{diam}(\Gamma_0)^2 .$$

This is quite a crude geometric estimate.

To do better let us proceed again by supposing $\{\Gamma_t\}_{0 \le t < t^*}$ is a smooth flow. Then (1) says

$$(12) \qquad \frac{d}{dt} \mathcal{H}^{n-1}(\Gamma_t) = -\int_{\Gamma_t} H^2 \, d\mathcal{H}^{n-1} .$$

Now a Sobolev-type inequality for manifolds states

$$(13) \qquad \left(\int_{\Gamma_t} |f|^{\frac{n-1}{n-2}} d\mathcal{H}^{n-1} \right)^{\frac{n-2}{n-1}} \le C \int_{\Gamma_t} |Df| + |f||H| d\mathcal{H}^{n-1}$$

for any smooth function f with compact support. As Γ_t is bounded we can take $f \equiv 1$:

$$(\mathcal{H}^{n-1}(\Gamma_t))^{\frac{n-2}{n-1}} \le C \int_{\Gamma_t} |H| d\mathcal{H}^{n-1}$$

$$\le C \left(\int_{\Gamma_t} H^2 \, d\mathcal{H}^{n-1} \right)^{\frac{1}{2}} \mathcal{H}^{n-1}(\Gamma_t)^{\frac{1}{2}} .$$

Thus

$$(\mathcal{H}^{n-1}(\Gamma_t))^{\frac{n-3}{n-1}} \le C \int_{\Gamma_t} H^2 \, d\mathcal{H}^{n-1} ,$$

and so (12) implies

$$\frac{d}{dt} \mathcal{H}^{n-1}(\Gamma_t) \le C \mathcal{H}^{n-1}(\Gamma_t)^{\frac{n-3}{n-1}} .$$

We integrate this differential inequality to deduce

$$(\mathcal{H}^{n-1}(\Gamma_t))^{\frac{2}{n-1}} - (\mathcal{H}(\Gamma_0))^{\frac{2}{n-1}} \le -Ct .$$

Letting $t \to t^*$ we conclude

$$(14) \qquad t^* \le C \mathcal{H}^{n-1}(\Gamma_0)^{\frac{2}{n-1}} .$$

This estimates t^* in terms of the surface area of Γ_0.

To make a rigorous proof for the generalized flow we quote without proof a "clearing out" result proved in [E-S III]:

Theorem. *There exist constants $\alpha, \beta, \eta > 0$ such that if*

$$(15) \qquad \mathcal{H}^{n-1}(\Gamma_{t_0} \cap B(x_0, r)) \leq \eta r^{n-1}$$

for some time $t_0 \geq 0$ and some ball $B(x_0, r) \subset \mathbb{R}^n$, then

$$(16) \qquad \Gamma_t \cap B(x_0, \tfrac{r}{2}) = \emptyset \quad \text{for} \quad \alpha r^2 \leq t - t_0 \leq \beta r^2 .$$

This result, which is based upon some fundamental observations of Brakke, states that if Γ_t has very little surface area within some ball at some time, then Γ_t does not intersect at all the ball with half the radius at certain later times. The proof of the theorem is a kind of "localized" version of the proof of (14), applied to the level surfaces of the approximations u^ε.

As an application, take $t_0 = 0$ and set

$$r = \left[\frac{\mathcal{H}^{n-1}(\Gamma_0)}{\eta}\right]^{\frac{1}{n-1}} .$$

Then

$$\mathcal{H}^{n-1}(\Gamma_0 \cap B(x_0, r)) \leq \mathcal{H}^{n-1}(\Gamma_0) = \eta r^{n-1}$$

for each point x_0. Owing to (16) we have

$$\Gamma_t \cap B(x_0, \tfrac{r}{2}) = \emptyset \quad \text{for} \quad t = \alpha r^2 .$$

As this holds for all x_0, we deduce

$$t^\star \leq \alpha r^2 = C\mathcal{H}^{n-1}(\Gamma_0)^{\frac{2}{n-1}} .$$

This is (14).

Remark. Giga and Yama-uchi [G-Y] have derived the extremely interesting lower bound

$$t^\star \geq 2|U_0|^2 \mathcal{H}^{n-1}(\Gamma_0)^{-2}$$

where $\Gamma_0 = \partial U_0$. $\qquad \square$

3. Geometric structure for a.e. level set

The computations in §1,2 above strongly suggest that

$$\nu = \frac{Du}{|Du|}$$

should be the unit normal vector field to Γ_t^γ and

$$H = \begin{cases} u_t/|Du| & \text{if } |Du| \neq 0 \\ 0 & \text{if } |Du| = 0 \end{cases}$$

should act like the mean curvature. Formally this is so, but as u is not smooth it is difficult to make rigorous computations. The paper [E-S IV] is devoted to showing that such interpretations are valid, at least for a "generic" level set Γ_t^γ.

There are many open questions concerning the regularity of u and its level sets, other geometric properties of the flow $\Gamma_0 \mapsto \Gamma_t$ ($t \geq 0$), etc. It is also extremely important to understand how this notion of generalized mean curvature motion relates to other theories, especially those of Brakke and of De Giorgi.

References

[C-C]   X. Cabre and L. Caffarelli, *Fully Nonlinear Elliptic Equations*, Amer. Math. Soc., to appear.

[C-G-G]   X. Chen, Y. Giga and S. Goto, Uniqueness and existence of viscosity solutions of generalized mean curvature flow equations. *J. Diff. Geom* **33** (1991), 749–786.

[E-S I]   L. C. Evans and J. Spruck, Motion of level sets by mean curvature, I, *J. Diff. Geom.* **33** (1991), 635–681.

[E-S II]   L. C. Evans and J. Spruck, Motion of level sets by mean curvature, II, *Trans. Amer. Math. Soc.* **330** (1992), 321–332.

[E-S III]   L. C. Evans and J. Spruck, Motion of level sets by mean curvature, III, *J. Geom. Analysis* **2** (1992), 121–150.

[E-S IV]   L. C. Evans and J. Spruck, Motion of level sets by mean curvature, IV, *J. Geom. Analysis*, to appear.

[G-T1]   D. Gilbarg and N. Trudinger, *Elliptic Partial Differential Equations of Second Order*, 2nd edition (Springer, 1983).

[G-T2]   D. Gilbarg and N. Trudinger, revised material for the above (1984).

[G-Y]   Y. Giga and K. Yama-uchi, On a lower bound for the extinction time of surfaces moved by mean curvature. *Cal. Var.* **1** (1993), 417–428.

[J]   R. Jensen, The maximum principle for viscosity solutions of second-order fully nonlinear partial differential equations. *Arch. Rat. Mech. Analysis* **101** (1988), 1–27.

[J-L-S]   R. Jensen, P-L. Lions and P. E. Souganidis, A uniqueness result for viscosity solutions of second-order, fully nonlinear PDEs. *Proc. Amer. Math. Soc.* **102** (1988).

Controlled Markov Processes, Viscosity Solutions and Applications to Mathematical Finance

CIME LECTURES

Halil Mete Soner
Department of Mathematics
Carnegie Mellon University
Pittsburgh, PA 15213, U.S.A.

1 Controlled Diffusion Processes

The purpose of this section is to give a concise, nontechnical introduction to stochastic differential equations and to controlled diffusion processes. In the first subsection, we provide a brief description of Brownian motion, stochastic differential equations and their connection to second order, linear prabolic equations. Then we formulate the dynamic programming principle for the optimization problem and derive the resulting nonlinear partial differential equation; in this text refered to as the dynamic programming equation (DPE). In this direction, we provide a proof of the dynamic programming principle when there is a smooth subsolution of (DPE). Using this principle, we prove that the value function is viscosity solution of the (DPE). We also include two examples: the classical linear, stochastic regulator and the Merton's optimal portfolio problem.

1.1 Stochastic Differential Equations.

Here we provide a very brief introduction to stochastic differential equations. For more detailed information, see for instance [23], [21].

Let $(\Omega, \mathcal{F}, P)$ be a probability space, $\{\mathcal{F}_t\}_{t \geq 0}$ be filtration, i.e., Ω is a set, $\mathcal{F}$ is a σ-algebra of its subsets, P is a probability measure on $(\Omega, \mathcal{F})$, $\mathcal{F}_t$ is a non-decreasing family of sub σ-algebras of $\mathcal{F}$. A *stochastic process*

X is a map of $[0, \infty) \times \Omega$ into a metric space. In these notes we employ the standard notation of the probability theory: $X_t := X(t, \cdot)$. (However for a detrerministic function φ, φ_t is the partial derivative of φ.) We say that a stochastic process is

- *adapted*: if for every $t \geq 0$, the map $\omega \to X_t(\omega)$ is $\mathcal{F}_t$ measurable,

- *progresively measurable*: if for every $t \geq 0$, the map

$$(s, \omega) \in [0, t] \times \Omega \to X_s(\omega)$$

  is Borel times $\mathcal{F}_t$ measurable.

A d-dimensional *Brownian motion* B is an adapted stochastic process with values in $\mathcal{R}^d$, $B_0 = 0$ and

- for every $0 \leq s \leq t$, the difference $B_t - B_s$ is independent of $\mathcal{F}_s$ and is normally distributed with mean zero and variance $(t - s)$ times the identity matrix, i.e.,

$$P(\{\omega \,:\, B_t(\omega) - B_s(\omega) \in A\} \mid \mathcal{F}_s) = \frac{1}{(2\pi(t-s))^{d/2}} \int_A \exp\left(-\frac{|x|^2}{2(t-s)}\right) dx,$$

  for every Borel set $A \subset \mathcal{R}^d$.

To avoid technical complications, we assume that the filtration is right continuous.

For given functions,

$$f : \mathcal{R}^n \times [0, \infty) \to \mathcal{R}^n, \qquad \sigma : \mathcal{R}^n \times [0, \infty) \to M^{n \times d} \ (= n \times d \text{ matrices}),$$

we consider the stochastic differential equation

$$(1.1) \qquad dx_s = f(x_s, s)ds + \sigma(x_s, s)dB_s, \quad s > t,$$

with initial data $x_t = x$. By a pathwise solution of this equation, we mean an $\mathcal{F}_t$ adapted stochastic process x_s satisfying

$$x_s = x + \int_t^s f(x_r, r)dr + \int_t^s \sigma(x_r, r)dB_r, \quad s \geq t.$$

In the above expression, the second integral is an Ito-stochastic integral: an adapted, continuous stochastic process with certain properties. In these notes, we will not attempt to define this integral, instead we simply state few fundamental properties of (1.1) and refer to standard textbooks for their proofs: see for instance [23], [21]. First of these properties is the existence and the uniqueness of pathwise solutions of (1.1) under standard growth and Lipschitz assumptions on the data. These solutions are Hölder continuous in the t-variable. For any smooth, compactly supported $\varphi(x, t)$, the stochastic process $\varphi(x_s, s)$ satisfies the <u>Ito's formula</u> for $s \geq t$

$$\varphi(x_s, s) = \varphi(x_t, t) + \int_t^s \left[\frac{\partial}{\partial t}\varphi(x_r, r) + A\varphi(x_r, r) \right] dr + \int_t^s \nabla\varphi(x_r, r) \cdot \sigma(x_r, r) dB_r,$$
(1.2)

where the operator A is defined by

$$A\varphi(x, s) := \nabla\varphi(x, s) \cdot f(x, s) + \frac{1}{2} \, trace\left(D^2\varphi(x, s)\sigma(x, s)\sigma^t(x, s) \right),$$

where σ^t is the adjoint of σ. This formula is also valid when s is replaced by a stopping time $\tau \geq t$ (τ is a stopping time if

$$\{\omega \mid \tau(\omega) \leq s\} \in \mathcal{F}_s, \quad \forall s \geq t.)$$

The classical example of a stopping time is the exit time from an open subset $O \subset \mathcal{R}^n$:

$$\tau_O(\omega) := \inf\{s \geq t \mid x_s(\omega) \notin O\},$$

where x_s is the solution of (1.1). Note that the exit time τ_O depends on the initial conditions x and t, however, in what follows we will supress this dependence.

Feynman-Kac Formula

For $T > 0$, smooth functions $\psi(x)$, $g(x, t)$ and a bounded, open set O, set

$$\phi(x, t) := E\left(\psi(x_T) 1_{T < \tau_O} + g(x_{\tau_O}, \tau_O) 1_{T \geq \tau_O} \mid x_t = x \right), \quad x \in \mathcal{R}^n, s \in [0, T],$$

where $E(\ldots \mid x_t = x)$ is the mathematical expectation conditioned on the initial condition $x_t = x$ and 1_B is the indicator of the set B.

Lemma 1.1 *Suppose that $\phi \in C^{1,2}(O \times (0, T)) \cap C(\bar{O} \times [0, T])$. Then*

(1.3) $$-\frac{\partial}{\partial t}\phi(x, t) - A\phi(x, t) = 0, \quad \text{in } O \times (0, T),$$

(1.4) $$\phi(x, t) = g(x, t), \quad \text{on } \partial O \times (0, T),$$

(1.5) $$\phi(x, T) = \psi, \quad \text{on } \bar{O}.$$

Conversely, if $\varphi \in C^{1,2}(O \times (0, T)) \cap C(\bar{O} \times [0, T])$ is a solution of (1.3), (1.4) and (1.5), then $\varphi = \phi$.

Proof. (1.4) and (1.5) follow easily from the continuity and the definition of ϕ. To derive (1.3), we use (1.2) with ϕ, $s = t + h$, take the expected value and then let $h \downarrow 0$: the result is

$$-\frac{\partial}{\partial t}\phi(x, t) - A\phi(x, t) = \lim_{h \downarrow 0} \frac{1}{h}\left[\phi(x, t) - E(\phi(x_{t+h}, t + h)|x_t = x)\right].$$

Note that when $\tau_O > t + h$, $\phi(x, t) = \phi(x_{t+h}, t + h)$. Hence,

$$|\phi(x, t) - E(\phi(x_{t+h}, t + h)|x_t = x)| \leq 2\|\phi\|_\infty P(\tau_O \leq t + h).$$

For a initial point $x \in O$, using standard estimates from probability theory, we can show that, as $h \downarrow 0$, $P(\tau_O \leq t+h)$ converges to 0 exponentially fast. Thus ϕ satisfies (1.3). The converse is an immediate application of the Ito's rule (1.2). □

As we will see later in these notes, when ϕ is merely continuous, it still satisfies (1.3) in the sense of viscosity solutions.

1.2 Controlled Diffusion Processes.

We now consider problems in which the time evolution of x_s is influenced by another stochastic process u_s, called a *control process*. The control process u_s takes values in a *control space* U which is assumed to be a closed subset of a Euclidean space. Suppose that the functions f and σ in (1.1) depend on the control process. Then (1.1) takes the form

(1.6) $$dx_s = f(x_s, u_s, s)ds + \sigma(x_s, u_s, s)dB_s, \quad s > t,$$

For a given adapted control process, under standard assumptions on f and σ, there is a unique solution of (1.6) satisfying the initial data $x_t = x$.

Moreover, the Ito's formula (1.2) also generalizes:

$$(1.7) \quad \varphi(x_s, s) = \varphi(x_t, t) + \int_t^s \left[\frac{\partial}{\partial s} \varphi(x_r, r) + A^{u_r} \varphi(x_r, r) \right] dr$$

$$+ \int_t^s \nabla \varphi(x_r, r) \cdot \sigma(x_r, u_r, r) dB_r,$$

where the operator A^u is defined by

$$A^u \varphi(x, s) := \nabla \varphi(x, s) \cdot f(x, u, s) + \frac{1}{2} \; trace \left(D^2 \varphi(x, s) \sigma(x, u, s) \sigma^t(x, u, s) \right).$$

Let

$$\nu := (\Omega, \{\mathcal{F}_t\}, P).$$

be the *reference probability space* and $\mathcal{A}_\nu$ be the set of all adapted control processes. Then, for a given (x, t), the finite-horizon, exit-time stochastic optimal control problem is to minimize

$$J(x, t; u.) := E \left(\int_t^{\tau_O} L(x_s, u_s, s) ds + \psi(x_T) 1_{T < \tau_O} + g(x_{\tau_O}, \tau_O) 1_{T \geq \tau_O} \; \middle| \; x_t = x \right)$$

over all $u. \in \mathcal{A}_\nu$. Here L is the *running cost*, ψ is the *terminal cost* and g is the *boundary cost*. We always assume that L, ψ and g are continuous and $\psi(x) = g(x, T)$ for all $x \in \partial O$. For future reference we define the *value function*

$$(1.8) \quad\quad\quad\quad V(x, t) := \inf_{u. \in \mathcal{A}_\nu} J(x, t; u.).$$

Note that the value function, as defined, may depend on the reference probability space ν. However, under reasonable assumptions V is independent of ν and we will give a brief discussion of this in §1.5.

Analogously, the discounted infinite horizon problem is to minimize

$$\hat{J}(x; u.) := E \left(\int_0^{\tau_O} e^{-\beta s} L(x_s, u_s) ds + e^{-\beta \tau_O} g(x_{\tau_O}, \tau_O) 1_{\tau < \infty} \; \middle| \; x_0 = x \right),$$

$$(1.9)$$

where $\beta > 0$ is the discount factor and x_t is the solution of (1.6) with data, f, σ independent of t. The corresponding value function is

$$\hat{V}(x) := \inf_{u. \in \mathcal{A}_\nu} \hat{J}(x, t; u.).$$

We close this subsection with two examples whose solutions are given in §1.4.

Example 1. (Stochastic linear regulator). In this example, $O = \mathcal{R}^n$, $U = \mathcal{R}^m$ and the stochastic differential equation for x_s are linear:

$$dx_s = [A(s)x_s + B(s)u_s]\,ds + \sigma(s)dB_s,$$

with given matrices $A(s)$, $B(s)$ and $\sigma(s)$. The expected total cost is

$$J(x, t; u.) := E\left(\int_t^T [M(s)x_s \cdot x_s + N(s)u_s \cdot u_s]ds + Dx_T \cdot x_T \,\Big|\, x_t = x\right),$$

where $M(s)$, $N(s)$ and D are given symmetric, positive semi-definite matrices.

Example 2. (Merton's portfolio problem). In a simple model of stock portfolio selection, the portfolio consists of two assets, one "risk-free" asset and the other "risky". The price b_s per share for the risk-free asset changes according to $db = abds$ while the price p_s of the risky asset changes according to $dp = p(\alpha ds + \sigma dB_s)$. Here a, α, σ are constants with $a < \alpha$, $\sigma > 0$ and B_s is a one-dimensional Brownian motion. The investor's wealth w_s at time s then changes according to the stochastic differential equation

$$(1.10) \quad dw_s = (1 - \pi_s)w_s ads + \pi_s w_s(\alpha ds + \sigma dB_s) - c_s ds, \quad s > 0,$$

where π_s is the fraction of wealth invested in the risky asset at time s and $c_s \geq 0$ is the consumption rate. We assume that the investor can change the amount invested in the risky asset instantenously and without any transation cost. We will examine the models with proportional transaction cost later in these notes.

The control is the pair (π_s, c_s) taking values in $U = \mathcal{R}^1 \times [0, \infty)$. Note that π_s is allowed to take values outside the interval $[0, 1]$. This corresponds to borrowing or short-selling, which is allowed in our model. The maximization problem is to maximize the expected discounted utility from consumption:

$$\hat{J}(x; u.) := E\left(\int_0^\infty e^{-\beta s}\ell(c_s)ds \,\Big|\, x_0 = x\right),$$

where β is the discount factor and $\ell(c)$ is the utility function. A common example is

$$\ell(c) = \frac{1}{p}(c)^p, \quad c \geq 0,$$

for some $p < 1$ and $p \neq 0$.

1.3 Dynamic Programming: Formal Description.

In this subsection we shall describe, in a purely formal way, the principle of dynamic programming due to Bellman [7], the corresponding dynamic programming equation and a criterion for finding feedback control policies. Our mathematically rigorous statements will take the form of "Verification Theorems" which require the dynamic programming equation to have sufficiently regular classical solutions. It often happens that there is no classical solution. In such cases, we must resort to the theory of viscsoity solutions.

In this section we restrict ourselves to the finite horizon problem with $O = \mathcal{R}^n$. The discussion for other cases are similar. The starting point of the dynamic programming is to consider the value function $V(x,t)$ defined in §1.2:

$$V(x,t) := \inf_{A_\nu} E\left(\int_t^T L(x_s, u_s, s)ds + \psi(x_T) \,\bigg|\, x_t = x\right).$$

The next step (at least conceptually) is to obtain the Bellman's *principle of optimality*. This states that for any $h \in [0, T - t]$

$$(1.11) V(t,x) = \inf_{A_\nu} E\left(\int_t^{t+h} L(x_s, u_s, s)ds + V(x_{t+h}, t + h) \,\bigg|\, x_t = x\right).$$

Speaking intuitively, the expression in brackets represents the sum of the running cost on $[t, t + h]$ and the minimum expected cost obtained proceeding optimaly after $t + h$ with $(x_{t+h}, t + h)$ as the initial data.

Once the dynamic programming principle is established, the dynamic programming equation is obtained from (1.11) by the following heuristic derivation. If we take the constant control $u_s = \bar{u}$ in $s \in [t, t + h]$, then

$$V(t,x) \le E\left(\int_t^{t+h} L(x_s, \bar{u}, s)ds + V(x_{t+h}, t + h) \,\bigg|\, x_t = x\right).$$

We subtract $V(x,t)$ from both sides, divide by h, use the Ito's rule (1.7) and let $h \downarrow 0$:

$$0 \le \lim_{h \downarrow 0} \frac{1}{h} E\left(\int_t^{t+h} L(x_s, \bar{u}, s)ds + [V(x_{t+h}, t + h) - V(x,t)] \,\bigg|\, x_t = x\right)$$

$$= L(x, \bar{u}, t) + \lim_{h \downarrow 0} \frac{1}{h} E\left(\int_t^{t+h} \frac{\partial}{\partial t}V(x_s, s) + A^{\bar{u}}V(x_s, s)ds \,\bigg|\, x_t = x\right)$$

$$= L(x, \bar{u}, t) + \frac{\partial}{\partial t}V(x, t) + A^{\bar{u}}V(x, t).$$

Since this holds for all $\bar{u}$,

$$-\frac{\partial}{\partial t}V(x,t) + \sup_{u\in U} \{ -A^u V(x,t) - L(x,u,t) \} \le 0.$$

Among the various assumptions needed to make this argument rigorous would be an assumption that enables us to use the Ito's formula (1.7).

On the other hand, if $u^* \in \mathcal{A}_\nu$ is an optimal control process, for every $h \in [0, T-t]$ we should have

$$V(t,x) = E\left(\int_t^{t+h} L(x_s^*, u_s^*, s)ds + V(x_{t+h}^*, t+h) \,\Big|\, x_t = x \right)$$

where x^* is the solution of (1.6) with control process u^*. A similar argument gives, under sufficiently strong assumptions (including the continuity of u^* at t),

$$-\frac{\partial}{\partial t}V(x,t) - A^{u_t^*}V(x,t) - L(x,u_t^*,t) = 0.$$

Hence we have obtained the *dynamic programming equation*

$$(1.12)\quad \frac{\partial}{\partial t}V(x,t) + \sup_{u\in U} \{ -A^u V(x,t) - L(x,u,t) \} = 0, \quad x \in O, \ t \in (0,T).$$

For the infinite horizon problem (1.9), the dynamic programming equation has the form

$$(1.13)\qquad \beta V(x) + \sup_{u\in U} \{ -A^u V(x) - L(x,u) \} = 0, \quad x \in O.$$

1.4 Verification Theorems.

In this subsection, we establish a connection between smooth solutions the dynamic programming equation (1.12) and the stochastic optimization problem. These theorems are known as the "Verification Theorems" and they are instructive in providing the necessary intiution how to use the partial differential equations in solving particular optimization problems. After proving a Verification Theorem, we will use it to solve the two examples introduced in §1.2. Our discussion closely follows [18, §4].

We only consider the finite-time horizon problem. Set

$$Q := O \times (0,T),$$

and let $C_p(\bar{Q})$ denote the set of all continuous functions growing at most polynomially.

Theorem 1.2 (Verification) *Let $W \in C^{1,2}(Q) \cap C_p(\bar{Q})$ be a solution of the equation (1.12), the terminal data (1.5) and the boundary condition (1.4). Then:*

1. *$W(x,t) \leq J(x,t;u.)$ for any initial data $(x,t) \in Q$ and control process $u.$.*

2. *If for an initial data $(x,t) \in Q$ and reference probability space ν there exists $u^* \in \mathcal{A}_\nu$ satisfying*

$$u_s^* \in \operatorname{argmin}\left\{ A^{u_s^*} W(x_s^*, s) + L(x_s^*, u_s^*, s) \right\}$$

for Lebesque$\times P$ almost every $(s, \omega) \in [t, T \wedge \tau_O^*] \times \Omega$, then $W(t,x) = J(x,t;u^*)$.
(Here x^* is the solution of (1.6) with control process u^*.)

Proof.

1. Fix $(x,t) \in Q$ and a control process $u.$. Let $x.$ be the solution of (1.6) with control $u.$ and initial data $x_t = x$. Set $\theta := T \wedge \tau_O$. An application of the Ito's formula (1.7) and the polynomial growth of W yields

$$EW(x_\theta, \theta) = W(x,t) + E \int_t^\theta \left[\frac{\partial}{\partial s} W(x_r, r) + A^{u_r} W(x_r, r) \right] dr.$$

Since W solves (1.12),

$$\frac{\partial}{\partial s} W(x_r, r) + A^{u_r} W(x_r, r) \geq -L(x_r, u_r, r).$$

Combining (1.4), (1.5) and the above inequalities, we obtain $W(t,x) \leq J(t,x;u.)$.

2. Observe that we now have

$$\frac{\partial}{\partial s} W(x_r^*, r) + A^{u_r^*} W(x_r^*, r) = -L(x_r^*, u_r^*, r),$$

for almost every (s, ω). Again we use (1.4), (1.5) to obtain $W(t,x) = J(t,x;u^*)$. $\qquad\square$

Remark. In the case of unbounded O, we need to be careful with the growth conditions on W. See [18] for a discsussion of this point for infinite horizon problems.

We continue by using the Verification Theorem to solve the examples 1 and 2 of §1.2.

Example 1. In this example

$$
\begin{aligned}
A^u V(x,t) &= (A(t)x + B(t)u) \cdot \nabla V(x,t) + \frac{1}{2}trace(\, D^2 V(x,t)\sigma(t)\sigma^t(t)\,) \\
L(x,u,t) &= M(t)x \cdot x + N(t)u \cdot u,
\end{aligned}
$$

and (1.12) takes the form,

$$
(1.14)\frac{\partial}{\partial t}V(x,t) - A(t)x \cdot \nabla V(t,x) - \frac{1}{2}trace(\, D^2 V(x,t)\sigma(t)\sigma^t(t)\,)
$$
$$
+ \frac{1}{4}B(t)N^{-1}(t)B^t(t)\nabla V(t,x) \cdot \nabla V(t,x) = M(t)x \cdot x.
$$

As a solution of this equation, we try

$$
W(x,t) = P(t)x \cdot x + q(t),
$$

for some scalar function $q(t)$ and symmetric $n \times n$ matrix $P(t)$. Then (1.5) yields

$$
P(T) = D, \qquad q(T) = 0,
$$

and (1.12) yields

$$
\frac{\partial}{\partial t}P(t) = P(t)B(t)N^{-1}(t)B^t(t) - A(t)P(t) - P(t)A^t(t) - M(t), \quad t \in [0,T],
$$

$$
\frac{d}{dt}q(t) = -trace(\, \sigma(t)\sigma^t(t)P(t)\,), \quad t \in [0,T].
$$

It can be shown that there is a unique solution $P(t)$ of this differential equation known as the *Riccati equation*. To obtain the optimal feedback controls, we observe that

$$
u^*(x,t) = -N^{-1}(t)B^t(t)P(t)x
$$

minimizes $A^u V(x,t) + L(x,u,t)$ over all $u \in U$. Hence we consider the linear equation

$$
dx_s^* = G^*(s)x_s^* ds + \sigma(s)dB_s,
$$

where

$$
G^*(s) := A(s) - B(s)N^{-1}(s)B^t(s)P(s).
$$

Then we use the Verification Theorem with

$$
u_s^* = -N^{-1}(s)B^t(s)P(s)x_s^*.
$$

Example 2. We consider the case

$$\ell(c) = \frac{1}{p}(c)^p, \quad c \geq 0,$$

for some $0 < p < 1$. The (1.13) takes the form (recall that this is maximization problem)

$$\beta V(x) = \max_{\pi} \left\{ \frac{1}{2}\sigma^2\pi^2 x^2 V_{xx}(x) + (\alpha - a)\pi x V_x(x) \right\} + r x V_x(x) + \max_{c>0} \left\{ \ell(c) - c V_x(x) \right\}$$

on $(0, \infty)$ with boundary condition is $V(0) = 0$. Observe that the maximizers are

$$\pi^*(x) = -\frac{(\alpha - a)V_x(x)}{\sigma^2 x V_{xx}(x)}, \qquad c^*(x) = (V_x(x))^{1/(p-1)}.$$

As a solution, we try

$$W(x) = Kx^p, \quad x > 0,$$

For some constant $K > 0$. We substitute W into the equation and obtain a nonlinear equation

$$\left(\beta - \frac{(\alpha - a)^2 p}{2\sigma^2(1 - p)} - rp \right) K = (pK)^{1/(p-1)}$$

for K, which has a unique solution provided

$$\beta > \frac{(\alpha - a)^2 p}{2\sigma^2(1 - p)} + ap.$$

When the above condition is violated, the value function is identically equal to infinity. The optimal feedback controls are

$$\pi^*(x) = \frac{\alpha - a}{(1 - p)\sigma^2}, \qquad c^*(x) = (pK)^{1/(1-p)}x.$$

We obtain the open loop controls to be used in the Verification Theorem, as in the previous example.

1.5 Dynamic Programming Principle.

Problem on $\mathcal{R}^n$. First we study the finite-time horizon problem with $O = \mathcal{R}^n$. Set

$$Q_0 := \mathcal{R}^n \times (0, T).$$

We assume

$$|f(x, u, t) - f(y, u, s)| \leq K (|x - y| + |s - t|), \qquad \forall (x, t), (y, s) \in Q_0, u \in U,$$
$$|\sigma(x, u, t) - \sigma(y, u, s)| \leq K (|x - y| + |s - t|), \qquad \forall (x, t), (y, s) \in Q_0, u \in U,$$
$$|\sigma(x, u, t)| + |f(0, u, t)| \leq K, \qquad \forall (x, t) \in Q_0, u \in U,$$
$$|L(x, u, t)| \leq K \left(1 + |x|^k + |u|^k\right), \qquad \forall (x, t) \in Q_0, u \in U,$$
$$|\psi(x)| \leq K \left(1 + |x|^k\right), \qquad \forall x \in \mathcal{R}^n,$$

for suitable constants K and k. We refer to these as the *standard assumptions* and use them in the rest of these notes. We have not sought for utmost generality, under the weakest possible assumptions.

The following version of the dynamic programming principle is proved in [18, Theorem 7.1, §4, page185].

Theorem 1.3 *Suppose that $O = \mathcal{R}^n$, U is compact and the standard assumptions are satisfied. Then the value function V defined by (1.8) is continuous on Q_0 and it is independent of the reference probability space ν. Moreover, the dynamic programming principle holds:*

- *For every initial data $(x, t) \in Q_0$, reference probability space ν, control process $u. \in \mathcal{A}_\nu$ and a stopping time $\theta \geq t$*

$$V(x, t) \leq E \left(\int_t^\theta L(x_s, u_s, s) ds + V(x_\theta, \theta) \,\bigg|\, x_t = x \right).$$

- *For every $\delta > 0$ and reference probability space ν, there exists a control process $u. \in \mathcal{A}_\nu$ such that*

$$V(x, t) + \delta \geq E \left(\int_t^\theta L(x_s, u_s, s) ds + V(x_\theta, \theta) \,\bigg|\, x_t = x \right),$$

for every stopping time $\theta \geq t$.

The above statement of the dynamic programming principle is slightly stronger that

$$V(x, t) = \inf_{u. \in \mathcal{A}_\nu} E \left(\int_t^\theta L(x_s, u_s, s) ds + V(x_\theta, \theta) \,\bigg|\, x_t = x \right).$$

The compactness assumption on U is relaxed at the end of this subsection.

The proof of this theorem consists of two steps. In the first step, we assume that the data f, σ, L and ψ satisfy strong regularity and boundedness conditions and we also assume that

$$\sigma(x,t)\sigma^t(x,t) \geq \alpha I, \qquad \forall (x,t) \in Q_0,$$

for some suitable constant $\alpha > 0$. Under these assumptions there is a unique smooth solution of the dynamic programming equation satisfying the terminal condition. Then an attendant modification of the proof of the deterministic dynamic programming principle yields the result. The second step in the proof is to approximate the problem at hand by a sequence of problems satisfying the assumptions used in step one. Once the uniform convergence, on bounded set, of the approximating value functions is obtained, the result follows from step one.

Exit time problem. Let us assume that there is a smooth subsolution $G(x,t)$ of (1.12), (1.4) and (1.5), i.e., G satisfies

$$G(x,t) = g(x,t), \quad \text{on } \partial O \times (0,T),$$

$$G(x,T) \leq \psi, \quad \text{on } \bar{O},$$

$$-\frac{\partial}{\partial t}G(x,t) + \sup_{u \in U}\{-A^u G(x,t) - L(x,u,t)\} \leq 0, \quad x \in O, \ t \in (0,T).$$

Following Lions [25], we define

$$\tilde{L}(x,u,t) := L(x,u,t) + A^u G(x,t) + G_t(x,t), \qquad \tilde{\psi}(x) := \psi(x) - G(x,T).$$

Then the value function, $\tilde{V}$, for the exit-time problem with running cost $\tilde{L}$, terminal cost $\tilde{\psi}$ and boundary cost $\tilde{g} \equiv 0$ and previous dynamics, is equal to $V - G$. Hence it suffices to prove the dynamic programming for $\tilde{V}$. This is done by approximating this exit-time problem by a sequence of problems defined on the entire space Q_0: for $\epsilon > 0$ define

$$V^\epsilon(x,t) := \inf_{u.} E\left(\left. \int_t^T \Gamma(s,\epsilon)\tilde{L}(x_s,u_s,s)ds + \Gamma(T,\epsilon)\tilde{\psi}(x_T) \right| x_t = x\right),$$

where

$$\Gamma(s,\epsilon) = \exp\left(\frac{1}{\epsilon}\int_t^s d(x_r)dr\right),$$

and $d(x)$ is the distance between x and $\bar{O}$. Since, by assumption, both $\tilde{L} \geq 0$ and $\tilde{\psi} \geq 0$, V^ϵ converge to $\tilde{V}$ monotonically and hence uniformly on compact subsets of $\bar{Q}$. Hence we have the following result: see [18, §5.2, pages 214-219].

Theorem 1.4 *Suppose that there is smooth subsolution G of (1.12), (1.4) and (1.5), U is compact and the standard assumptions are satisfied. Then the value function V defined by (1.8) is continuous on $\bar{Q}$ and it is independent of the reference probability space ν. Moreover, the dynamic programming principle holds:*

- *For every initial data $(x,t) \in Q_0$, reference probability space ν, control process $u. \in \mathcal{A}_\nu$ and a stopping time $\theta \geq t$*

$$V(x,t) \leq E \left(\int_t^{\theta \wedge \tau_O} L(x_s, u_s, s)ds + V(x_{\theta \wedge \tau_O}, \theta \wedge \tau_O) \,\Big|\, x_t = x \right).$$

- *For every $\delta > 0$ and reference probability space ν, there exists a control process $u. \in \mathcal{A}_\nu$ such that*

$$V(x,t)+\delta \geq E \left(\int_t^{\theta \wedge \tau_O} L(x_s, u_s, s)ds + V(x_{\theta \wedge \tau_O}, \theta \wedge \tau_O) \,\Big|\, x_t = x \right),$$

for every stopping time $\theta \geq t$.

Noncompact control set U. In both of the above theorems we assumed that the control set is compact. In some applications, such as the linear regulator and etc., this is restrictive. However, there is an easy way to circurvent this problem. For a positive integer m, set

$$U_m := \{\, u \in U \,,\, |u| \leq m \,\},$$

and let V_m be the value function of the corresponding stochastic control problem. In all relevant applications, V_m is a good approximation of V. So we may assume that

$$(1.15) \quad \lim_{m \to \infty} \sup \{|V(t,x) - V_m(T,x)| \mid (x,t) \in \bar{Q}, |x| \leq R\} = 0, \quad \forall R > 0.$$

Sufficient conditions for (1.15), such as the coercivity of L, are discussed in [18].

Since U_m is compact, Theorem 1.3 or 1.4 apply to V_m. Then an approximation argument shows that Theorem 1.3 or 1.4 remain valid after replacing the compactness assumption on U with (1.15).

1.6   Viscosity Solutions.

In this section we show that the value function of the finite horizon problem is a viscosity solution of the dynamic programming equation (1.12). Analysis of other problems is similar: see [18]. Boundary conditions are discussed in the next section.

First we recall the definition of viscosity solutions of (1.12), as defined by Crandall and Lions in [14], also see [13]. Viscosity solutions of the stochastic dynamic programming equations were first considered by Lions in [25].

Definition. Let W be a continuous function of $\bar{Q}$.

<u>Subsolution</u> We say that W is a *viscosity subsolution* of (1.12) in Q if for each $\varphi \in C^\infty(Q)$,

$$(1.16) \quad -\frac{\partial}{\partial t}\varphi(x_0, t_0) + \sup_{u \in U} \{ -A^u\varphi(x_0, t_0) - L(x_0, u, t_0) \} \le 0,$$

at every $(x_0, t_0) \in Q$ which is local maximum of the difference $W - \varphi$ in Q.

<u>Supersolution</u> We say that W is a *viscosity supersolution* of (1.12) in Q if for each $\varphi \in C^\infty(Q)$,

$$(1.17) \quad -\frac{\partial}{\partial t}\varphi(x_0, t_0) + \sup_{u \in U} \{ -A^u\varphi(x_0, t_0) - L(x_0, u, t_0) \} \ge 0,$$

at every $(x_0, t_0) \in Q$ which is local minimum of the difference $W - \varphi$ in Q.

<u>Solution</u> We say that W is a *viscosity solution* of (1.12) in Q if it is both a viscosity subsolution and a viscosity supersolution in Q.

It is known that in the above definition, we only need to consider the strict, global extrema.

We are now in a position to prove the main connection between the optimization problem and the theory of viscosity solutions.

Theorem 1.5 *Suppose that the standard assumptions are satisfied and that either U is compact or, if it is not compact, (1.15) holds. Then the value function defined by (1.8) is a viscosity solution of (1.12) in Q.*

To keep the presentation simple we will prove this result only for $O = \mathcal{R}^n$. However, an attendant modification of the following proof yields the same result with any O; see [18, Corrolary 3.1,§5,page 223].

Proof. The continuity of the value function is proved in the previuos section. Observe that, by an approximation argument, it suffices to prove the result for a compact control set U.

<u>Subsolution.</u> Let φ be a smooth test function and $(x_0, t_0) \in Q_0$ be a maximizer of the difference $V - \varphi$. Set

$$a := \varphi(x_0, t_0) - V(x_0, t_0).$$

Then $V \leq \varphi - a$ and the dynamic programming principle, Theorem 1.3, with the constant control $u_s \equiv u$ yields

$$
\begin{aligned}
\varphi(x_0, t_0) \;=\; & V(x_0, t_0) + a \\
\leq\; & E_{x_0} \left(\int_{t_0}^{t_0+h} L(x_s, u, s)ds + V(x_{t_0+h}, t_0 + h) + a \right) \\
\leq\; & E_{x_0} \left(\int_{t_0}^{t_0+h} L(x_s, u, s)ds + \varphi(x_{t_0+h}, t_0 + h) \right),
\end{aligned}
$$

for every $h \in [0, T - t]$ and u.. Here E_{x_0} stands for $E(\cdots | x_{t_0} = x_0)$. Next we subtract $\varphi(x_0, t_0)$ from both sides, use the Ito's rule (1.7), divide by h and let $h \downarrow 0$:

$$
\begin{aligned}
0 \;\leq\; & \lim_{h \downarrow 0} \frac{1}{h} E_{x_0} \left\{ \int_{t_0}^{t_0+h} L(x_s, u, s)ds + [\varphi(x_{t_0+h}, t_0 + h) - \varphi(x_0, t_0)] \right\} \\
=\; & L(x_0, u, t_0) + \lim_{h \downarrow 0} \frac{1}{h} E_{x_0} \left\{ \int_{t_0}^{t_0+h} \frac{\partial}{\partial t} V(x_s, s) + A^u \varphi(x_s, s)ds \right\} \\
=\; & L(x_0, u, t_0) + \frac{\partial}{\partial t} \varphi(x_0, t_0) + A^u \varphi(x_0, t_0).
\end{aligned}
$$

Since this holds for all u, (1.16) follows this.

<u>Supersolution.</u> Let φ be a smooth test function and $(x_0, t_0) \in Q_0$ be a minimizer of the difference $V - \varphi$. For each positive integer m, by the second part of the Theorem 1.3, with

$$h = \frac{1}{m^2} \qquad \theta = t_m := t_0 + \frac{1}{m},$$

there exists a control process u^m satisfying

$$V(x_0, t_0) + \frac{1}{m^2} \geq E_{x_0} \left(\int_{t_0}^{t_m} L(x_s^m, u_s^m, s)ds + V(x_{t_m}^m, t_m) \right),$$

where x^m is the solution of (1.6) with control u^m. Set

$$a := \varphi(x_0, t_0) - V(x_0, t_0).$$

Then $V \geq \varphi - a$ and

$$\varphi(x_0, t_0) = V(x_0, t_0) + a$$

$$\geq E_{x_0} \left(\int_{t_0}^{t_m} L(x_s^m, u_s^m, s)ds + V(x_{t_m}^m, t_m) + a \right) - \frac{1}{m^2}$$

$$\geq E_{x_0} \left(\int_{t_0}^{t_m} L(x_s^m, u_s^m, s)ds + \varphi(x_{t_m}^m, t_m) \right) - \frac{1}{m^2}.$$

Next we subtract $\varphi(x_0, t_0)$ from both sides, use the Ito's rule (1.7), multiply by m and let $m \to \infty$:

$$0 \geq \lim_{m \to \infty} m E_{x_0} \left\{ \int_{t_0}^{t_m} L(x_s^m, u_s^m, s)ds + [\varphi(x_{t_m}^m, t_m) - \varphi(x_0, t_0)] \right\}$$

$$= \lim_{m \to \infty} m E_{x_0} \int_{t_0}^{t_m} \left[L(x_s^m, u_s^m, s) + \frac{\partial}{\partial t}\varphi(x_s^m, s) + A^{u_s^m}\varphi(x_s^m, s) \right] ds$$

$$= \lim_{m \to \infty} m E_{x_0} \int_{t_0}^{t_m} \left[L(x_0, u_s^m, t_0) + \frac{\partial}{\partial t}\varphi(x_0, t_0) + A^{u_s^m}\varphi(x_0, t_0) \right] ds.$$

Here we used the fact that, as $m \to \infty$, the random variable

$$\sup \{ |x_s^m - x_0| \mid s \in [t_0, t_m] \}$$

converges to zero in $L^p(\Omega; dP)$ with any $p < \infty$.

Following Lions [25], we set

$$U(x_0, t_0) := \{ (L, f, a) \in \mathcal{R}^1 \times \mathcal{R}^n \times S^{n \times n} :$$

$$(L, f, a) = (L(x_0, u, t_0), f(x_0, u, t_0), a(x_0, u, t_0)) \text{ for some } u \in U \},$$

where $S^{n \times n}$ is the set of all symmetric $n \times n$ matrices. We write the previous inequality as

$$-\frac{\partial}{\partial t}\varphi(x_0, t_0) - \lim_{m \to \infty} m E \int_{t_0}^{t_m} (L(x_0, u_s^m, t_0), f(x_0, u_s^m, t_0), a(x_0, u_s^m, t_0))$$

$$\cdot \left(1, \nabla\varphi(x_0, t_0), \frac{1}{2}D^2\varphi(x_0, t_0) \right) ds \geq 0,$$

where for (L, f, a), $(l, p, A) \in \mathcal{R}^1 \times \mathcal{R}^n \times S^{n \times n}$

$$(L, f, a) \cdot (l, p, A) := lL + f \cdot p + trace(aA).$$

Since $m \int_{t_0}^{t_m} ds = 1$,

$$m E \int_{t_0}^{t_m} (L(x_0, u_s^m, t_0), f(x_0, u_s^m, t_0), a(x_0, u_s^m, t_0)) \, ds \in \bar{co} CU(x_0, t_0),$$

where $\bar{co} U(x_0, t_0)$ is the closed convex hull of $U(x_0, t_0)$. Hence

$$0 \leq -\frac{\partial}{\partial t}\varphi(x_0, t_0) + \sup \left\{ -(L, f, a) \cdot (1, \nabla\varphi, \frac{1}{2}D^2\varphi) : (L, f, a) \in \bar{co} U(x_0, t_0) \right\}$$

$$= -\frac{\partial}{\partial t}\varphi(x_0, t_0) + \sup \left\{ -(L, f, a) \cdot (1, \nabla\varphi, \frac{1}{2}D^2\varphi) : (L, f, a) \in U(x_0, t_0) \right\}$$

$$= -\frac{\partial}{\partial t}\varphi(x_0, t_0) + \sup_{u \in U} \left\{ -A^u\varphi(x_0, t_0) - L(x_0, u, t_0) \right\}.$$

Hence (1.17) holds. $\qquad\qquad\qquad\qquad\qquad\qquad\qquad\qquad\qquad$ □

In these notes, we do not discuss the uniqueness of viscosity solutions. The User's Guide [13] provides an excellent survey of this topic.

1.7 Boundary Conditions.

It is well known that the value function of a deterministic or degenerate stochastic optimal control problem may not satisfy the boundary data (1.4) pointwise. This may still be the case even if the value function is continuous up to the boundary of O. The simplest way to see this is to consider the state contraint problem: we minimize over all controls u. which keep the trajectories x_t inside the set $\bar{O}$. An alternate way of formulating the state constraint problem is to set the boundary cost $g \equiv \infty$. Then, in many cases, the value function V has a finite value at the boundary; whence it does not satisfy (1.4) pointwise.

Although the boundary condition is not satisfied in the usual sense, to characterize the value function as the unique solution of a certain partial differential eqution we need boundary conditions. It turns out that the boundary condition, although not satisfied in a pointwise or usual manner, is satisfied in a viscosity sense.

We first proceed in a purely formal way and derive the viscosity formulation of the boundary condition (1.4), then we will formalize this weak formulation in the form of a definition and a theorem. We only consider the finite time, exit problem (1.8). Since this is a minimization problem, $V(x, t) \le g(x, t)$ for all $x \in \partial O$. Suppose that at some $x_0 \in \partial O$ and $t_0 \in [0, T)$, $V(x_0, t_0) < g(x_0, t_0)$. So if u^* is the optimal control process starting from (x_0, t_0), then the exit time τ_O^* of the process x^* is strictly positive. Now suppose that for some smooth test function φ, $V - \varphi$ is minimized at (x_0, t_0). Then, by the supersolution part of the Theorem 1.5 with minor modifications, (1.17) holds at (x_0, t_0). In conclusion, at a boundary point (x_0, t_0): we have either $V(x_0, t_0) = g(x_0, t_0)$ or V satisfies the supersolution property at (x_0, t_0).

Following Barles & Perthame [5], Ishii [22] and Soner [29], we define the viscosity solutions of (1.12) and (1.4).

Definition. Let W be a continuous function of $\bar{Q}$.

Subsolution We say that W is a *viscosity subsolution* of (1.12) in Q and the boundary condition (1.4): if it is a viscosity subsolution of (1.12) in Q and if for each $\varphi \in C^\infty(Q)$,

(thir8)$V(x_0, t_0) - g(x_0, t_0)$,

$$-\frac{\partial}{\partial t}\varphi(x_0, t_0) + \sup_{u \in U} \{ -A^u \varphi(x_0, t_0) - L(x_0, u, t_0) \} \} \le 0,$$

at every $(x_0, t_0) \in \partial O \times [0, T)$ which is local maximum of the difference $W - \varphi$ in $\bar{Q}$, if such a maximizer exists.

Supersolution We say that W is a *viscosity supersolution* of (1.12) in Q and the boundary condition (1.4): if it is a viscosity supersolution of (1.12) in Q and if for each $\varphi \in C^\infty(Q)$,

(thir9)$V(x_0, t_0) - g(x_0, t_0)$,

$$-\frac{\partial}{\partial t}\varphi(x_0, t_0) + \sup_{u \in U} \{ -A^u \varphi(x_0, t_0) - L(x_0, u, t_0) \} \} \ge 0,$$

at every $(x_0, t_0) \in \partial O \times [0, T)$ which is local minimum of the difference $W - \varphi$ in $\bar{Q}$, if such a minimizer exists.

Solution We say that W is a *viscosity solution* of (1.12) in Q and the boundary condition if it is both a viscosity subsolution and a viscosity supersolution.

Remarks.

1. The subsolution property is equivalent to following: Let φ be a smooth

test function and $(x_0, t_0) \in \bar{O} \times [0, T)$ be a minimizer of $W - \varphi$ in $\bar{Q}$. Then W is a viscosity subsolution if the following hold:

- (1.16) holds, when $x_0 \in O$,
- (1.18) holds, when $x_0 \in \partial O$.

The supersolution property has a similar reformulation.

2. If W is the value function of the stochastic minimization problem, then (1.18) always hold. We have stated the subsolution property in this manner so as to keep the generality. For a maximization problem, for example, the above formulation of the subsolution is necessary.

3. For the state constraint problem $g \equiv \infty$ and (1.19) is equivalent to (1.17).

Theorem 1.6 *Suppose that the standard assumptions are satisfied and that either U is compact or, if it is not compact, (1.15) holds. Then the value function defined by (1.8) is a viscosity solution of (1.12) in Q and the boundary condition (1.4).*

The proof of this theorem is very similar to the proof of Theorem 1.5; see [18, §2.13].

Remarks.

1. A recent paper by Barles and Burdeau [3] contains several uniqueness results for Dirichlet problems.

2. Other type of boundary conditions, such as Neumann or nonlinear Neumann problems, are studied by Barles and Lions [4] for the deterministic control problems. A general introduction to these type of problems can be found in a recent book by Barles [2]

2 Singular Stochastic Control

In contrast to classical control problems, in which the displacement of the state due to control effort is differentiable in time, the singular control models we consider allow this displacement to be discontinuous. Bather-Chernoff were the first to formulate such a problem in their study of a simplified model of spacecraft control. Since then singular control has found many other applications in diverse areas of communications, queueing systems and mathematical finance.

We start our analysis in with a formal derivation of the dynamic programming equation. This discussion leads us to a formulation of singular control problems with controls which are processes of bounded variation. The related verification theorem is then proved in §2.3. We discuss the viscosity property of the value function in §2.4. Since we want to emphasize only the new issues arising in singular control, in Sections 2-5 we restrict our attention to a simple infinite horizon problem with no absolutely continuous control component. However the theory is not limited only to these problems.

Indeed in the next section, we will apply this general theory to a portfolio selection problem with transaction costs. The portfolio selection problem has a finite horizon and "mixed" type controls; consumption rate and transactions between the bond and the stock. Here the consumption rate is an absolutely continuous control and the transactions between the bond and the stock is a singular type control.

The dynamic programming equation (2.3) below, is a pair of differential inequalities. Also at every point of the state space either one of the inequalities is satisfied by an equality. So the state space splits into two regions, the "no-action" region and the "push region", corresponding to the active inequality in (2.3). Starting from the push region, the optimal state process moves immediately into the no-action region, where its exit is prevented by reflection at the boundary in an appropriate direction. In §2.4 we exhibit this qualitative character of the optimal control process in several examples.

2.1 Formal discussion.

In this section, we consider a special case of the infinite horizon (1.9). We let $U \subset \mathcal{R}^n$ be a closed cone in $\mathcal{R}^n$, i.e.,

$$u \in U, \lambda \geq 0 \Rightarrow \lambda u \in U.$$

We also assume that there are $\hat{f}, \hat{\sigma} \in C^1(\mathcal{R}^n)$ with bounded first order partial derivatives and $\hat{c}, \hat{L} \in C(\mathcal{R}^n)$ satisfying

$$f(x, u) = u + \hat{f}(x), \qquad \sigma(x, u) = \hat{\sigma}(x),$$

$$L(x, u) = \hat{L}(x) + \hat{c}(u), \qquad \hat{c}(\lambda u) = \lambda \hat{c}(u), \quad \forall \lambda \geq 0, \ u \in U,$$

for all $x \in \mathcal{R}^n$, $u \in U$. For simplicity we take the lateral boundary condition $g \equiv 0$ and $\hat{L}, \hat{c} \geq 0$.

Note that the control set U is not bounded and moreover the classical coercivity condition is not satisfied by f and L.

For a smooth function $\varphi \in C^\infty(\mathcal{R}^n)$, let us compute the right hand side of (1.13):

$$\sup_{u \in U} \{-A^u \varphi(x) - L(x, u)\} = -\frac{1}{2} trace(\hat{a}(x)D^2\varphi(x)) - \hat{f}(x) \cdot \nabla\varphi(x) - \hat{L}(x) + \hat{\mathcal{H}}(\nabla\varphi(x)),$$

where $\hat{a}(x) = \hat{\sigma}(x)\hat{\sigma}^t(x)$ and for $p \in \mathcal{R}^n$

(2.1) $$\hat{\mathcal{H}}(p) = \sup_{u \in U}\{-p \cdot u - \hat{c}(u)\}.$$

Observe that if $-p \cdot u - \hat{c}(u) > 0$ for some $u \in U$, then, by the homogeneity of U and $\hat{c}$, $\hat{\mathcal{H}}(p) = +\infty$. Therefore

$$\hat{\mathcal{H}}(p) = \begin{cases} +\infty & , \text{ if } H(p) > 0, \\ \\ 0 & , \text{ if } H(p) \leq 0, \end{cases}$$

where

$$H(p) = \sup_{v \in \hat{K}}\{-p \cdot v - \hat{c}(v)\},$$

$$\hat{K} = \{u \in U , \ |u| = 1\}.$$

One can think of $\hat{K}$ as the set of allowable directions in which control may act.

The above calculation indicates that the dynamic programming equation (1.13) has to be interpreted carefully. However, we formally expect that the value function V satisfies

(2.2) $$H(DV(x)) \leq 0, \qquad x \in O,$$

$$\mathcal{L}V(x) := \beta V(x) - \frac{1}{2}tr \ \hat{a}(x)D^2V(x) - \hat{f}(x) \cdot DV(x) \leq \hat{L}(x), \quad x \in O.$$

Now suppose $H(DV(x)) < 0$ for some $x \in O$. Then in a neighborhood of x, the unique maximizer in (2.1) is zero. Hence at least formally, the optimal feedback control should be equal to zero in a neighborhood of x.

Since the uncontrolled diffusion processes are related to linear equations, we expect

$$\mathcal{L}V(x) = \hat{L}(x), \text{ whenever } H(DV(x)) < 0.$$

We now rewrite the above inequalities in the following more compact form,

(2.3) $$\max\{\mathcal{L}V(x) - \hat{L}(x), H(DV(x))\} = 0, \quad x \in O.$$

Since $g \equiv 0$, the lateral boundary condition is,

(2.4) $$V(x) = 0, \quad x \in \partial O.$$

In Section §2.4, we will prove a Verification Theorem, Theorem 2.3, for the dynamic programming equation (2.3) and the boundary condition (2.4). This provides a rigorous basis to our formal derivation. However due to the linear dependence of L on v, in general there are no optimal controls and nearly-optimal controls take arbitrarily large values. For this reason it is convenient to reformulate the above problem by using the integral of u_s as our control process. This reformulation will be the subject of the next section.

2.2 Reformulation.

As in §1, let $\nu = (\Omega, \{\mathcal{F}_s\}, P, w)$ be a probability reference system with a right continuous filtration $\mathcal{F}_s$. Let us rewrite the state dynamics (1.6) as follows, taking into account the special form of f: Let

$$\hat{u}_s = \begin{cases} |u_s|^{-1}u_s, & \text{if } u_s \neq 0 \\ \\ 0, & \text{if } u_s = 0, \end{cases}$$

$$\xi_t = \int_0^t |u_s| ds.$$

Then (1.6) becomes

$$dx_s = \hat{\sigma}(x_s)dB_s + \hat{f}(x_s)ds + \hat{u}_s d\xi_s, \quad s > 0.$$

We now regard

(2.5) $$z_t := \int_{[0,t)} \hat{u}_s d\xi_s$$

as the control variable at time t. However in order to obtain optimal controls, we must enlarge the class of controls to admit z. which may not be absolutely continuous function of s. But we assume that each component of $z.$ is a function of bounded variation on every finite interval $[0, t]$, namely; each component of $z.$ is the difference of two monotone functions of s. Let $\mu(\cdot)$ be the total variation measure of $z.$ and set

$$\xi_t := \int_{[0,t)} d\mu(s).$$

Then

(2.6) ξ_t nondecreasing, real $-$ valued, left continuous with $\xi_0 = 0$.

Moreover by Radon-Nikodynm theorem, for a given z_t, there exists $\hat{u}_s \in \mathcal{R}^n$ satisfying (2.5) and $|\hat{u}_s| \leq 1$. We identify the process $z.$ by the pair $(\xi., \hat{u}.)$. Let $\hat{\mathcal{A}}_\nu$ the set of all progressively measurable $z. = (\xi., \hat{u}\cdot)$ satisfying (2.6),

$$\hat{u}_s \in U, \text{ for } \mu - \text{almost every } s \geq 0,$$

and

$$E|z(t)|^m < \infty, \ m = 1, 2, \cdots.$$

Then for a given $x \in O$, the usual Picard iteration yields a unique, left continuous solution to

(2.7) $\qquad x_t = x + \int_0^t \hat{\sigma}(x_s) dB_s + \int_0^t \hat{f}(x_s) ds + z_t, \ t \geq 0,$

with

$$x_{t+} - x_t = z_{t+} - z_t.$$

Observe that x_t is not in general continuous. Let $\tau := \tau_O$ be the exit time of $x.$ from $\overline{O}$. Since $x.$ is left continuous and $\mathcal{F}_t$ is right continuous, τ is a $\mathcal{F}_t$ stopping time. We now wish to minimize

$$J(x; \xi, \hat{u}) = E \int_{[0,\tau]} e^{-\beta s} [\hat{L}(x_s) ds + \hat{c}(\hat{u}_s) d\xi_s]$$

over $\hat{\mathcal{A}}_\nu$. Finally, let

$$V(x) = \inf_{\hat{\mathcal{A}}_\nu} J(x; \xi, \hat{u}),$$

We close this section by proving some elementary properties of V.

Lemma 2.1 Let $x \in O$, $u \in U$ and $h > 0$. If $x + hu \in O$, then

$$V(x) - V(x + hu) \leq h\hat{c}(u).$$

In particular if V is differentiable at x, then (2.2) holds at x.

Proof. In view of the homogeneity of $\hat{c}$, we may assume that $|u| \leq 1$. For $(\xi_., \hat{u}_.) \in \hat{A}_\nu$, let

$$\xi_s^h = \begin{cases} 0, & s = 0 \\ \xi_s + h, & s > 0, \end{cases} \qquad \hat{u}_s^h = \begin{cases} u, & s = 0 \\ \hat{u}_s, & s > 0. \end{cases}$$

Then $(\xi^h, \hat{u}^h) \in \hat{A}_\nu$. Let x_s^h be the solution of (2.7) with control $(\xi_.^h, \hat{u}_.^h)$ and initial condition $x_0^h = x$. Then, $x_s^h = x_s + hu$ for all $s > 0$. Moreover,

$$V(x) \leq J(x; \xi_.^h, \hat{u}_.^h) = J(x + hu; \xi_., \hat{u}_.) + h\hat{c}(u).$$

We complete the proof of the lemma after minimizing the right hand side over $\xi_.$ and $\hat{u}_.$. □

The following is a standard convexity result, that will be used later in these notes. For a proof, see [18, Lemma 3.2].

Lemma 2.2 *Suppose that $O = \mathcal{R}^n$, U is convex, $\hat{f}$ and $\hat{\sigma}$ are affine functions and $\hat{L}, \hat{c}$ are convex. Then V is also convex.*

The convexity assumption on U and $\hat{c}$ are satisfied by the following examples:

$$\begin{aligned} \hat{K} &= S^{n-1} = \{u \in \mathcal{R}^n , |u| = 1\}, & \hat{c}(u) = |u|, \\ \hat{K} &= \{\nu_0\}, \\ \hat{K} &= \{\nu_0, -\nu_0\}, \end{aligned}$$

where $\nu_0 \in S^{n-1}$.

2.3 Verification Theorem.

We start with the definition of classical solutions of (2.3). Let $W^{1,\infty}(O; \mathcal{R}^n)$ be the set of all $\mathcal{R}^n$ - valued, bounded, Lipschitz continuous functions of O.

Definition. Let $W \in C_p(\bar{O}) \cap C^1(\bar{O})$ with $DW \in W^{1,\infty}(O; \mathcal{R}^n)$ be given. Set
$$\mathcal{P} = \{x \in \mathcal{R}^n \ : \ H(DW(x)) < 0\}.$$
We say that W is a (classical) solution of (2.3) if $W \in C^2(\mathcal{P})$,

$$\mathcal{L}W(x) = \hat{L}(x), \ \forall x \in \mathcal{P}, \qquad H(DW(x)) \leq 0, \ \forall x \in \bar{O},$$

and
$$\mathcal{L}W(x) \leq \hat{L}(x),$$

for almost every $x \in \mathcal{R}^n$.

The following verification theorem is very similar to Theorem 1.2. See [18, Theorem 4.1., page 322] for its proof.

Theorem 2.3 (Verification) *Let W be a classical solution of (2.3) with the boundary condition (2.4). Then:*
(i). $W(x) \leq J(x; \xi_., \hat{u}.)$ *for any $x \in O$ and control process $(\xi_., \hat{u}.)$ satisfying*
$$\liminf_{t \to \infty} e^{-\beta t} E_x[W(x_t)1_{\tau = \infty}] = 0.$$

(ii). *Assume that $W \geq 0$ and at $x \in O$ there exists $(\xi_.^*, u_.^*)$ such that with probability one:*

$$x_t^* \in \mathcal{P}, \text{ Lebesgue almost every } t \leq \tau^*,$$

$$\int_{[0,t)} [u_s^* \cdot DW(x_s^*) + \hat{c}(u_s^*)]d\xi_s^* = 0, \ \forall t \leq \tau^*,$$

$$W(x_t^*) - W(x_{t+}^*) = \hat{c}(u_t^*)[\xi_{t+}^* - \xi_t], \ \forall t \leq \tau^*,$$

$$\lim_{t \to \infty} E_x[e^{-\beta(t \wedge \tau^*)} W(x_{t \wedge \tau^*}^*)1_{\tau^* = \infty}] = 0.$$

Then
$$J(x; \xi_.^*, u_.^*) = W(x).$$

We continue with two one-dimensional simple examples.

Example 3. Consider a one dimensional problem with $O = (-\infty, \infty), \hat{f} \equiv 0, \hat{\sigma} \equiv \sqrt{2}, \hat{c}(u) = |u|, \hat{K} = \{-1\}$ and $\hat{L}$ is convex. Then $U = (-\infty, 0]$ and the hypotheses of Lemma 2.2 are satisfied. Hence the value function V is convex and the equation (2.3) takes the form

$$(2.8) \max\{\beta V(x) - V_{xx}(x) - \hat{L}(x), \ V_x(x) - 1\} = 0, \ x \in (-\infty, \infty).$$

We will first construct a convex, polynomially growing solution W of (2.8) and then using the Verification Theorem we will show that $W = V$. Now suppose that W is indeed a convex solution of (2.8). Set

$$a = \sup\{x: \ W_x(x) < 1\}.$$

Here a may be equal to $+\infty$. The convexity of W yields

$$W_x(x) < 1, \ \forall x < a.$$

Then using (2.8) we conclude that

$$\beta W(x) - W_{xx}(x) = \hat{L}(x), \ \forall x < a, \quad W_x(x) = 1, \ \forall x \geq a.$$

The value of a is not *a priori* given to us. So we will solve the above equation for every real number a, and then determine the value of a by using (2.8) again. Let $W_a(x)$ be the polynomially growing solution of this equation. Now we assume that $\beta = 1$, $\hat{L}(x) = \alpha x^2$ for some $\alpha > 0$. Then, $W_a(x)$ is given by

$$W_a(x) = \begin{cases} (1 - 2\alpha a)e^{x-a} + \alpha x^2 + 2\alpha, & x \leq a, \\ \\ W_a(a) + x - a, & x > a. \end{cases}$$

Since $W_{a,x}(x) \leq 1$ for every x, W_a solves (2.8) provided

$$W_a(x) - W_{a,xx}(x) - \alpha x^2 \leq 0, \ \forall x > a.$$

An elementary argument shows that W_a is a solution of (2.8) if and only if

$$\lim_{x \uparrow a} W_{a,xx}(x) = 0 \Rightarrow a = \frac{1}{2\alpha} + 1.$$

Let W be equal W_a with the above choice of a, then

$$W(x) \leq V(x).$$

To prove that $W(x) = V(x)$, we need to construct a control process (ξ^*, u^*) satisfying the hypotheses of the Verification Theorem. Since $\hat{K} = \{-1\}$, we take $u_s^* \equiv -1$. Then for a given initial point x, we look for a process ξ^*

$$x_t^* = x + \sqrt{2}B_t - \xi_t^* \leq a, \ \text{for a. e. } t \geq 0,$$

$$\int_{[0,t)} [-W_x(x_s^*) + 1]d\xi_s^* = 0, \ \forall t \geq 0.$$

Since $W_x(x) - 1 < 0$ unless $x \geq a$, the last condition is equivalent to

$$\int_{[0,t)} 1_{\{x_s^* \geq a\}} d\xi_s^* = \xi_t^*, \ \forall t \geq 0.$$

The problem of finding $\xi^*(\cdot)$ these conditions is known as the *Skorokhod problem* and its solution is given by

$$\xi^*_{t+} = \sup\{(x + \sqrt{2}B_s - a) \vee 0 : s \le t\}, \ t \ge 0.$$

The optimal process constructed above satifies the condition of the Verfication Theorem, hence $W = V$. The interpretation of the optimal process is this: If $x \le a$, the process x^* is a *Brownian motion reflected at a* and ξ^* is *the local time at a*. (See [23, (3.8) in §6].) If $x > a$, $\xi^*_{0+} = x - a$, and $x^*_{0+} = a$.

It is well known that the local time is not absolutely continuous. This is why we need to admit controls ξ_s which are not necessarily absolutely continuous functions of s. $\square$

Example 4. Consider the same problem as in the previous example but with $\hat{K} = \{1, -1\}$. Then $U = (-\infty, \infty)$ and (2.3) takes the form,

$$\max\{V(x) - V_{xx}(x) - \alpha x^2, |V_x(x)| - 1\} = 0, \ \forall x \in (-\infty, \infty).$$

Following the procedure devised in the previous example, we look for a convex, polynomially growing function W and constants $-\infty \le a < b \le +\infty$ satisfying

$$\begin{aligned}
W(x) - W_{xx}(x) &= \alpha x^2, & \forall a < x < b, \\
W_x(x) &= -1, & \forall x \le a, \\
W_x(x) &= 1, & \forall x \ge b,
\end{aligned}$$

$$\lim_{x \downarrow a} W_{xx}(x) = \lim_{x \uparrow b} W_{xx}(x) = 0.$$

The last condition follows from the dynamic programming equation and the convexity of W. Now an elementary computation yields

$$W(x) = \begin{cases} \alpha x^2 + 2\alpha + \dfrac{(1 - 2\alpha b)}{\sinh(b)}\cosh(x), & |x| < b, \\ \alpha b^2 + x - b, & x \ge b, \\ \alpha b^2 - x + b, & x \le -b, \end{cases}$$

and $b = -a$ is the unique positive solution of

$$\tanh(b) = b - \frac{1}{2\alpha}.$$

Using the explicit form of W, we deduce that the optimal control process has to satisfy

$$x_t^* \in [-b, b], \quad \text{a.e.}, \tau \geq 0,$$
$$\hat{u}_t^* = 1 \text{ if } x_t^* \leq 0, \quad \hat{u}_t^* = -1 \text{ if } x_t^* > 0,$$
$$\int_{[0,t)} 1_{\{|x_s^*| \geq b\}} d\xi_s^* = \xi_t^*, \quad \forall t \geq 0.$$

Then the solution x_t^* of the above equations is the Brownian motion reflected at the boundary of $(-b, b)$. Also ξ^* is the sum of the local times of x^* at b and $-b$ (see [23, Definition 7.3] and [21, §4].)   □

When $U = \mathcal{R}^n, \hat{c}(u) = |u|, \hat{\sigma} \equiv \sqrt{2}$ times the identity, (2.3) reduces to

$$(2.9) \quad \max\{\beta V(x) - \Delta V(x) - \hat{L}(x), |DV(x)| - 1\} = 0, \quad \forall x \in O.$$

Then by analytical arguments Evans [17] proved the following.

Theorem 2.4 *Suppose that $\hat{L} \in C^2(\bar{O})$ and O is bounded. Then the value function V is the unique classical solution of (2.9) and (2.4).*

The $W^{2,\infty}$ regularity proved by Evans is the best possible general result. However for convex problems (i.e., those satisfying the hypotheses of Lemma 2.2) we expect the value function V to be twice continuously differentiable. Indeed C^2 regularity of the value function is proved by Soner & Shreve [30, 31] under either one of the following assumptions

$$O = \mathcal{R}^n, \hat{f} \equiv 0, \hat{\sigma} = \text{identity}, \hat{c}(u) = |u|, \hat{L} \text{ strictly convex},$$

or

$$U = \{\lambda v_0 : \lambda \geq 0\}, \text{ and hypotheses of Lemma 2.2,}$$

where $v_0 \in \mathcal{R}^n$ is any nonzero vector.

Now consider the first case. The following change of variables is introduced in [30]. Let x^* be the minimizer of V and $\delta > 0$ be sufficiently small. For $t \geq 0$ and $\theta \in S^{n-1}$ let $z(t; \theta) \in \mathcal{R}^n$ be the unique solution of

$$\frac{d}{dt} z(t; \theta) = DV(z(t; \theta)), t > 0, \theta \in S^{n-1},$$

$$z(0; \theta) = x^* + \delta\theta.$$

In [30] it is shown that the map

$$(t, \theta) \to z(t, \theta)$$

is a diffeomorphism between $[0, \infty) \times S^{n-1}$ onto $\{x \in \mathcal{R}^n : |x - x^*| \geq \delta\}$. Also for $t \geq 0$.

$$\frac{d}{dt}[|DV(z(t; \theta))|^2] = 2D^2V(z(t; \theta))DV(z(t; \theta)) \cdot DV(z(t, \theta)).$$

Since V is convex, $|DV(z(t; \theta))|^2$ is nondecreasing in t. Hence we may characterize the region $\mathcal{P}$ by

$$\mathcal{P} = B_\delta(x^*) \cup \{z(t; \theta) : \theta \in S^{n-1}, t < T(\theta)\},$$

where

$$T(\theta) = \inf\{t \geq 0 : |DV(z(t; \theta))|^2 \geq 1\}.$$

This calculation proves that $\mathcal{P}$ is a connected subset of $\mathcal{R}^n$. Also the above characterization of $\mathcal{P}$ is the first step in studying the regularity of $\partial \mathcal{P}$ and V.

For a general problem with a convex value function the appropriate change of variables is

$$\frac{d}{dt}z(t; \theta) \in \partial H(DV(z(t; \theta))),$$

where $\partial H(p)$ is the subdifferential of H in the sense of convex analysis. However the properties of this change of variables have not yet been studied.

We finally remark that the region $\mathcal{P}$ does not have a special geometric property. In particular, $\mathcal{P}$ in general is not a convex set.

2.4 Viscosity solutions.

Following §1.6, we can develop a viscosity theory for the dynamic programming equation (2.3). Since there are no essential new difficulties, here we only state the main theorem, details are given in [18, §8.5, pages 333-338]. We assume that $V \in C_p(\overline{O})$ and it satisfies the dynamic programming, i.e., for every $x \in O$ and stopping time $\theta > 0$,

$$V(x) = \inf_{\mathcal{A}_\nu} \left\{ E \int_0^{\tau \wedge \theta} e^{-\beta s}[\hat{L}(x_s)ds + \hat{c}(\hat{u}_s)d\xi_s] + Ee^{-\beta(\tau \wedge \theta)}V(x(\tau \wedge \theta)) \right\}$$

where τ is the exit time of x. from $\overline{O}$. Then V is a viscosity solution of (2.3) in O.

3 Portfolio Selection with Transaction Costs.

The purpose of this section is to formulate a portfolio selection problem with transaction costs and consumption. We will then formally reduce the two-dimensional problem to the pair of differential inequalities on a one-dimensional interval with the endpoint conditions. Optimal consumption and transaction policies are constructed in the final subsection. Constantinides [11] was first to formulate and formally solve this problem. His formulation was put on a mathematically rigorous basis by Davis and Norman [15] who also solved the problem under some restrictive techinical assumptions. Later, using viscosity solutions, Shreve and Soner [27] gave a solution with minimal assumptions. Here we outline the proof given in [27].

Related problems are studied by Fleming & Zariphopoulou [19], Zariphopoulou [33, 34] and Shreve, Soner and Xu [28]: see §3.4.

3.1 Problem.

We consider a consumption and investment problem of a single agent which is very similar to the one described in Example 2 §1.2. Recall that the portfolio consists of two assets, one "risk free" low-yield asset (which we call a "bond") and the other "risky" but higher yield asset (which we call a "stock"). The agent may consume from his investment in the bond, but simultaneously he may transfer his stock holdings to the bond. However, this results in a transaction cost, which is linearly proportional in the size of the transaction. Let $\mu \in (0,1)$ be the cost of transaction from stock to bond. In this model, we also allow transactions from bond to stock with a linear transaction cost. Let $\lambda \in (0,1)$ be the cost of these type of transactions.

Following the Merton's portfolio problem, we assume that the price b_s per share for the bond changes according to $db = abds$ while the price p_s of the stock follows a stochastic equation $dp = p(\alpha ds + \sigma dB_s)$. Let x_s and y_s be the dollars invested in the bond and the stock, respectively. Then x_s and y_s change according to

$$(3.1) \qquad dx_s = [ax_s - c_s]ds + (1 - \mu)dM_s - dN_s$$

$$(3.2) \qquad dy_s = \alpha y_s ds - dM_s + (1 - \lambda)dN_s + \sigma y_s dB_s,$$

where $c_s \geq 0$ is the consumption rate and M_s and N_s are the total transactions up to time $s \geq 0$, from stock to bond and bond to stock, respectively.

Let $\ell(c)$ be the utility of consuming at rate $c > 0$. The agent's goal is to maximize his discounted total utility of consumption

$$J(x, y; c, M, N) = E \left(\int_0^\infty e^{-\beta t} \ell(c_t) dt \;\middle|\; x_s = x, y_s = y \right),$$

over all progressively measurable $c., M., N.$ satisfying,

(3.3) M, N are left continuous, nondecreasing, $M_0 = N_0 = 0$,

(3.4) $c_t \geq 0, \; \forall t \geq 0$

(3.5) $x_t + (1 - \mu)y_t \geq 0, \; (1 - \lambda)x_t + y_t \geq 0, \; \forall t \geq 0.$

Set

$$O = \{(x, y) \in \mathcal{R}^n \; : \; x + (1 - \mu)y > 0, (1 - \lambda)x + y > 0\}.$$

We can now restate (3.5) as a state constraint:

$$(x_t, y_t) \in \bar{O}, \; \forall t \geq 0.$$

We claim that for any $(x, y) \in \bar{O}$ there are $c., M., N.$ satisfying (3.3), (3.4) and (3.5). Indeed let $(x, y) \in \bar{O}$ with $x + (1 - \mu)y = 0$ be given. Then $y \geq 0$ and

$$c_t \equiv N_t \equiv 0, \; M_t \equiv y, \; \forall t > 0,$$

satisfies (3.3) and (3.4). Also $x_t \equiv y_t \equiv 0$ for all $t > 0$, and therefore $c., M., N.$ also satisfy the state constraint. Similarly for $(x, y) \in \bar{O}$ with $(1 - \lambda)x + y = 0$,

$$c_t \equiv M_t \equiv 0, \; N_t \equiv x, \; \forall t > 0$$

satisfies the constraints. Since there are admissible controls for each boundary point $(x, y) \in \partial O$, there are admissible controls for every $(x, y) \in \bar{O}$. Let $\mathcal{A}(x, y)$ be the set of all admissible controls. When $\sigma > 0$, O is the largest open set such that each point $(x, y) \in O$ can be driven to the origin with an appropriate choice of control. For this reason Davis and Norman [15] call O the "solvency region".

If consumption were not allowed in our model, it would be a special case of the problem formulated in §2 with a state constraint and $\hat{c} \equiv 0$,

$$U = \{m(1 - \mu, -1) : m \geq 0\} \cup \{n(-1, 1 - \lambda) : n \geq 0\},$$

$$\hat{f}(x,y) = (ax, \alpha y), \qquad \hat{\sigma}(x,y) = \begin{bmatrix} 0 & 0 \\ 0 & \sigma y \end{bmatrix}.$$

The model with consumption is still very similar to the one discussed in §2. Indeed one can prove an entirely similar Verification Theorem for the dynamic programming equation,

$$(3.6) \min\{\mathcal{L}V(x,y) + F(V_x(x,y)), -(1-\mu)V_x(x,y) + V_y(x,y),$$

$$V_x(x,y) - (1-\lambda)V_y(x,y)\} = 0, \quad (x,y) \in O,$$

where

$$F(\xi) = \inf_{c \geq 0}\{c\xi - \ell(c)\},$$

$$\mathcal{L}V(x,y) = \beta V(x,y) - ax V_x(x,y) - \alpha y V_y(x,y) - \frac{1}{2}\sigma^2 y^2 V_{yy}(x,y).$$

Let us now assume that $\ell(c)$, the utility of consuming at rate $c \geq 0$, is of HARA type, i.e., for some $p \in (0,1)$,

$$\ell(c) = \frac{1}{p}c^p, \quad c \geq 0.$$

In these notes we will not consider the cases $p < 0$ and the logarithmic utility function. However, a similar analysis can be carried out for those cases: see Shreve & Soner [27]. We now compute that

$$F(\xi) = \frac{p-1}{p}(\xi)^{\frac{p}{p-1}}, \quad \xi > 0.$$

Also observe that if $(c., M., N.) \in \mathcal{A}(x,y)$ and $\rho > 0$, then $\rho(c., M., N.) \in \mathcal{A}(\rho x, \rho y)$ with

$$J(\rho x, \rho y; \rho c, \rho M, \rho N) = \rho^p J(x, y; c, M, N).$$

Hence
$$(3.7) \qquad V(\rho x, \rho y) = \rho^p V(x,y), \quad \forall \rho \geq 0.$$

Let
$$\hat{W}(r) = V(r, 1-r), \quad r \in (-\frac{1-\mu}{\mu}, \frac{1}{\lambda}).$$

Then for $(x,y) \in O$

$$x + y > 0, \quad \frac{x}{x+y} \in (-\frac{1-\mu}{\mu}, \frac{1}{\lambda}),$$

$$V(x,y) = (x+y)^p \hat{W}(\frac{x}{x+y}).$$

By formal differentiation, we obtain a formal reduction of (3.6):

$$(3.8) \min\{\hat{\mathcal{L}}\hat{W}(r) + F(p\hat{W}(r) + (1-r)\hat{W}_r(r)), \hat{W}(r) - d_4(r)\hat{W}_r(r),$$

$$\hat{W}(r) + d_5(r)\hat{W}_r(r)\} = 0, \quad \forall r \in (-\frac{1-\mu}{\mu}, \frac{1}{\lambda}),$$

where

$$\hat{\mathcal{L}}\hat{W}(r) = d_1(r)\hat{W}(r) + d_2(r)\hat{W}_r(r) - d_3(r)\hat{W}_{rr}(r),$$

$$d_1(r) = \beta - p[\alpha - (\alpha - a)r + \frac{\sigma^2}{2}(p-1)(1-r)^2],$$

$$d_2(r) = (\alpha - a)r(1-r) - \sigma^2(1-p)(1-r)^2 r,$$

$$d_3(r) = \tfrac{1}{2}\sigma^2(1-r)^2 r^2,$$

$$d_4(r) = (1 + \mu(r-1))/p\mu,$$

$$d_5(r) = (1 - \lambda r)/p\lambda.$$

Also at the boundary ∂O the only admissible consumption policy is $c. \equiv 0$. Hence

$$(3.9) \qquad \hat{W}(-\frac{1-\mu}{\mu}) = \hat{W}(\frac{1}{\lambda}) = 0.$$

Studying an equivalent one dimensional problem, Davis and Norman [15] constructed a solution $\hat{W} \in C^2$ of (3.8) and (3.9), under a restrictive, implicit set of technical assumptions on α, a, σ, β and γ. Later this problem was analyzed under minimal assumptions in [27].

3.2   Regularity.

In this section we study the regularity of the value function. We only assume that $\sigma > 0$ and the problem is well-posed, i.e.,

$$V(x,y) < \infty, \qquad \forall (x,y) \in O.$$

We start with several properties of the value function V. The following is straightforward to verify.

Lemma 3.1 *The value function V is continuous and concave on $\overline{O}$.*

Using the dynamic programming and the continuity of V, we can easily show that V is a viscosity solution of (3.6) in O. Let $\partial V(x, y)$ be the set of subdifferentials of V in the sense of convex analysis, i.e., for $(x, y) \in O$

$$\partial V(x, y) = \{p = (p_x, p_y) : V(x, y) + p_x(\overline{x} - x) + p_y(\overline{y} - y) \geq V(\overline{x}, \overline{y}), \forall (\overline{x}, \overline{y}) \in \overline{O}\}.$$

The above set is nonempty, bounded and convex. By (3.7) for any $(x, y) \in O, (p_x, p_y) \in \partial V(x, y)$ and $\rho > 0$, we have

$$\begin{aligned}
(\rho^p - 1)V(x, y) &= V(\rho x, \rho y) - V(x, y) \\[2mm]
&\leq p_x(\rho x - x) + p_y(\rho y - y) \\[2mm]
&= (\rho - 1)[x p_x + y p_y]
\end{aligned}$$

Let ρ go to 1 from above and below: the result is

$$x p_x + y p_y = pV(x, y), \quad \forall (x, y) \in O, (p_x, p_y) \in \partial V(x, y).$$

Also as in Lemma 2.1, we can prove that

$$V(x, y) \geq V(x + (1 - \mu)m, y - m)$$
$$V(x, y) \geq V(x - n, y + (1 - \lambda)n)$$

for every $n, m \geq 0$ and $(x, y) \in O$. These yield

(3.10) $$-(1 - \mu)p_x + p_y \geq 0, \quad p_x - (1 - \lambda)p_y \geq 0.$$

Moreover, since V is nondecreasing in the x-variable, $p_x \geq 0$. We claim that both p_x and p_y are both strictly positive. Indeed, suppose that $(0, \hat{p}_y) \in \partial V(x, y)$ for some $(x, y) \in O$. Since $p_x \geq 0$ at every point and V is concave, we obtain

$$(p_x, p_y) \in \partial V(x - h, y) \Rightarrow p_x = 0,$$

for every $h \geq 0$ and $(x - h, y) \in O$. Therefore

$$0 \leq V(x, y) \leq V(x - h, y), \quad \forall h > 0, (x - h, y) \in O.$$

Since $V = 0$ on the boundary, the above inequality yields that $V(x, y) = 0$. But it is clear that $V > 0$ in O. Hence $p_x > 0$ and by (3.10) we obtain

$$p_x > 0, p_y > 0, \quad \forall (p_x, p_y) \in \partial V(x, y), (x, y) \in O.$$

By (3.7),

$$\rho^{p-1}(p_x, p_y) \in \partial V(\rho x, \rho y).$$

Now define

$$\mathcal{P} = \{(x, y) \in O : -(1-\mu)p_x + p_y > 0, p_x - (1-\lambda)p_y > 0, \forall (p_x, p_y) \in \partial V(x, y)\}.$$

Then $\mathcal{P}$ is a cone, i.e., there are $r_1, r_2 \in [-\frac{1-\mu}{\mu}, \frac{1}{\lambda}]$ such that

$$\mathcal{P} = \{(x, y) \in O : r_1 < \frac{x}{x+y} < r_2\}.$$

We call $\mathcal{P}$ the *no transaction* region and define *sell bond* region SB and *sell stock* region SSt by

$$SB = \{(x, y) \in O : \frac{x}{x+y} > r_2\},$$

$$SSt = \{(x, y) \in O : \frac{x}{x+y} < r_1\}.$$

We will show that in SB the optimal strategy is to replace bonds by stocks until we reach the no transaction region $\mathcal{P}$ and analogously in SSt we replace stocks by bonds.

The following is proved in [18, Lemma 7.2, page 346].

Lemma 3.2 *Let* $(p_x, p_y) \in \partial V(x, y)$.

$$-(1-\mu)p_x + p_y = 0, \quad \forall (x, y) \in SSt,$$

$$p_x - (1-\lambda)p_y = 0, \quad \forall (x, y) \in SB.$$

Until now we have not used the fact that the value function is a viscosity solution. In [18] and [27] the viscosity property of V is used to prove its regularity. In particular, it is shown that V is regular enough to consider the dynamic programming equation (3.6) classically.

3.3 Optimal investment-transaction policy.

Recall that for $r > 0$,

$$F(r) = \inf_{c>0}\{cr - \frac{1}{p}(c)^p\},$$

and the minimizer is $c^* = r^{1/p-1}$. Therefore the candidate for an optimal Markov consumption policy is,

$$c^*(x, y) = (V_x(x, y))^{\frac{1}{p-1}}.$$

In view of the regularity of V, $c^*(x, y)$ is continuous on O. Also since $p < 1$ and V is concave, we have

$$c^*(x, y) > c^*(\bar{x}, y), \text{ if } x > \bar{x}.$$

Due to this monotonicity, for every progressively measurable nondecreasing processes $M.$ and $N.$ there exists a unique solution $(x., y.)$ of (3.1), (3.2) with $c_s = c^*(x_s, y_s)$. See [27] for details.

We continue by constructing the optimal controls $(c^*_\cdot, M^*_\cdot, N^*_\cdot)$. Later we will show that this control is indeed optimal.

1. $(x, y) \in \overline{\mathcal{P}}$. By a result of Lions and Sznitman [26] on reflected diffusion processes, there are unique processes $x^*_\cdot, y^*_\cdot$ and nondecreasing processes $M^*_\cdot, N^*_\cdot$ such that for $t < \tau^* = \inf\{s : (x^*_s, y^*_s) \in \partial O\}$,

$$M^*_t = \int_0^t 1_{\{(x^*_s, y^*_s) \in \partial_1 \mathcal{P}\}} dM^*_s,$$

$$N^*_t = \int_0^t 1_{\{(x^*_s, y^*_s) \in \partial_2 \mathcal{P}\}} dN^*_s,$$

$$(x^*_0, y^*_0) = (x, y), \quad (x^*_t, y^*_t) \in \overline{\mathcal{P}}, \quad \forall t \geq 0,$$

and $(x^*_\cdot, y^*_\cdot)$ solves (3.1), (3.2) with $c_s = c^*(x^*_s, y^*_s)$ where

$$\partial_i \mathcal{P} = \{(x, y) \in O : \frac{x}{x + y} = r_i\}, \text{ for } i = 1, 2.$$

When

$$-\frac{1 - \mu}{\mu} < r_1 < r_2 < \frac{1}{\lambda},$$

$(x^*_\cdot, y^*_\cdot)$ is a diffusion process reflected at $\partial \mathcal{P}$. The reflection angle is $(1 - \mu, -1)$ on $\partial_1 \mathcal{P}$ and $(-1, 1 - \lambda)$ on $\partial_2 \mathcal{P}$. Moreover M^*_t and N^*_t are the local times of $(x^*_\cdot, y^*_\cdot)$ on the lines $\partial_1 \mathcal{P}$ and $\partial_2 \mathcal{P}$ at time t, respectively. If $r_1 = -\frac{1-\mu}{\mu}$, then the processes is stopped at $\partial_1 O$. Similarly if $r_2 = 1/\lambda$, the process is stopped at $\partial_2 O$. Since the region $\mathcal{P}$ has a corner at the origin, the result of Lions and Sznitman is not directly applicable to the above situation. However the process (x^*_s, y^*_s) is stopped if it reaches the origin.

2. Suppose that SSt is not empty and $(x, y) \in SSt$. Then

$$\frac{x}{x + y} \in (-\frac{1 - \mu}{\mu}, r_1),$$

and in particular $r_1 > -(1 - \mu)/\mu$. Set

$$M_{0+}^* = [r_1 y + (r_1 - 1)x]\frac{1}{1 - \mu + r_1\mu},$$

and $N_{0+}^* = 0$. Since $(x, y) \in SSt$, $M_{0+}^* > 0$. Also

$$x_{0+}^* = x + (1 - \mu)M_{0+}^*,$$

$$y_{0+}^* = y - M_{0+}^*,$$

and we calculate that

$$\frac{x_{0+}^*}{x_{0+}^* + y_{0+}^*} = r_1.$$

Therefore $(x_{0+}^*, y_{0+}^*) \in \overline{\mathcal{P}}$. We now construct x^*, y^*, c^*, M^*, N^* as in case 1 starting from (x_{0+}^*, y_{0+}^*) and M_{0+}^*.

3. Suppose that SB is not empty and $(x, y) \in SB$. Then

$$\frac{x}{x + y} \in (r_2, \frac{1}{\lambda}),$$

and in particular $r_2 < 1/\lambda$. Set

$$N_{0+}^* = \frac{(1 - r_2)x - r_2 y}{1 - r_2\lambda},$$

and $M_{0+}^* = 0$. As in the previous case, $N_{0+}^* > 0$ since $(x, y) \in SB$. Also

$$x_{0+}^* = x - N_{0+}^*,$$

$$y_{0+}^* = y + (1 - \lambda)N_{0+}^*.$$

Then $(x_{0+}^*, y_{0+}^*) \in \partial_2\mathcal{P}$ and we construct x^*, y^*, c^*, M^*, N^* as in case 1 starting from (x_{0+}^*, y_{0+}^*) and N_{0+}^*.

The following is proved in [18, Theorem 7.1, page 354].

Theorem 3.3 *For any* $(x, y) \in O$,

$$J(x, y; c^*, M^*, N^*) = V(x, y).$$

3.4 Other Related Problems.

Proportional transaction costs is one way of introducing market friction into our model. However, there are several other ways. For instance, in Merton's problem borrowing is allowed with the same interest rate as the non-risky is allowed. This is studied by Zariphopulou in [34] and Fleming & Zariphopoulou [19]. A model with Markov chain is considered again by Zariphopoulou in [33]. The case $\sigma = 0$ is studied by Shreve, Soner and Xu [28] to gain insight for the problems with non-zero but small volatiity σ.

Once the regularity and uniqueness of the value function is established, there several natural questions concerning the qualitative behavior of the free boundaries. For instance, consider the case with no transaction costs. Then $r_1 = r_2 := r^*$ and the optimal policy is to keep the ratio $x_s^*/(x_s^* + y_s^*)$ equal to r^*. Then one question is to analyze the behavior of r_1 and r_2 for small but positive λ and μ. Other question is whether we always have $r_1 < r^* < r_2$. This was conjectured by Davis & Norman [15]. However it is not always true. Another natural question is whether r_1 is positive or r_2 is less than one. Interestingly, r_1 may become negative. These are investigated in [27]

4 Pricing a European Option.

The purpose of this section is to discsuss how to price contingent claims with or without transaction costs. To keep the discussion simple, we consider only the European options and derive the celebrated Black & Scholes formula by almost-sure hedging [8]. Similar analysis is also possible for more complicated options. A general modern treatment of the contingent claims via almost-sure hedging can be found [23, §5]. An alternate approach to option pricing based on utility maximization, first introduced by Hodges & Neuberger [20] and further developed by Davis, Panas & Zariphopoulou [16], again yields the Black & Scholes formula for European options when there are no transaction costs. However, in the presence of transaction costs, this new approach is the only currently avaliable method for option pricing.

In §4.1, we will describe the problem and derive the Black & Scholes formula. Then we introduce the utility maximization approach and redrive the Black & Scholes formula when there are no transaction costs. In §4.3, we show that, in the presence of transaction sosts, almost sure hedging approach yields a trivial result. Then we describe the utility maximization approach and the formal results of Leland [24].

4.1 Black & Scholes Formula.

Consider the financial market with one risky and one non-risky asset described in Example 2 of §1.2 (Merton's problem) and in §3. Let us briefly recall that the price b_s per share for the risk-free asset changes according to $db = abds$ while the price p_s of the risky asset changes according to $dp = p(\alpha ds + \sigma dB_s)$. Here a, α, σ are constants with $a < \alpha$, $\sigma > 0$ and B_s is a one-dimensional Brownian motion. Suppose that the investor is not allowed to consume ($c_s \equiv 0$). Then the investor's wealth w_s (or the total amount invested by the invested at time $s \geq 0$) evolves in time according to the stochastic differential equation

$$(4.1) \qquad dw_s = (1 - \pi_s)w_s ads + \pi_s w_s(\alpha ds + \sigma dB_s), \quad s > 0,$$

where, as before, π_s is the fraction of wealth invested in the risky asset at time s and the investor can change this fraction instantenously and without any transation cost. We will examine the models with proportional transaction cost later.

The *European option*, with *maturity T* and the *exercise price q*, is a contract that we sign at time $t = 0$, which gives us the option to buy, at time T, one share of the stock (the risky asset) at the price q. At maturity, if the price of the stock p_T is below the exercise price q, the contract is wortless to us; on the other hand, if $p_T > q$, we can exercise our option that is: buy one stock for the preassigned price q. Then immediately sell it for p_T. Hence the European option is equivalent to a payment of $(p_T - q)^+$ at maturity.

The option pricing problem is this: What is the fair price to pay at time $t = 0$ for the European option? Suppose that for a given $x^* > 0$ there is an adapted hedging policy π^* such that w^*, the solution of (4.1) with control π^* and initial data $w_0^* = x^*$, satisfies

$$(4.2) \qquad\qquad w_T^* = (p_T - q)^+.$$

Then the investor who buys the option for the price x^* could have instead invested this amount, x^*, in the market and use the investment policy π^*. By (4.2), this investment policy would dublicate the payoff of the option.

Consequently, the price of the option should not be greater than x^*. Now suppose that the price of the option x is strictly less that x^*. An investor may take a short position in the hedging portfolio; thus acquiring $x^* = w_0^*$. She uses x of this to buy the option and invests the rest, $x^* - x$, in the bond. At the maturity, the option and the investment in the bond is worth

$$(p_T - q)^+ + (x^* - x)e^{rT},$$

while her short position, by (4.2), is equal to $w_T^* = (p_T - q)^+$. So her net worth is $(x^* - x)e^{rT}$; thus an arbitrage opportunity, a way of making money from nothing, exists. Since such arbitrage opportunities can not exist, the fair price of the option can not be less than x^* and hence it has to be equal to it.

We have argued that the *fair price* of the European option is equal to the number x^* which allows the construction of a hedging policy π_t^* with initial wealth x^*. Now the task is to compute this price. Let e_t be the value of the European option, with exercise price q and maturity T, at time $t \in [0, T]$. Clearly, e_t may depend on the values of the stock up to time t or equivalently it is a $\mathcal{F}_t$ measurable random variable. In particular, $e_T = (p_T - q)^+$, since at time T the option is matured instantenously. The idea of Black and Scholes is continuous hedging. They assume that there is a deterministic function

$$G : [0, \infty) \times [0, T] \to [0, \infty),$$

such that $e_t = G(p_t, t)$ for every $t \in [0, T]$. Since

$$dp_t = \alpha p_t dt + \sigma p_t dB_t,$$

by Ito's formula, (1.7),

$$(4.3) = [G_t(p_t, t) + p_t \alpha G_p(p_t, t) + \frac{1}{2}\sigma^2 p_t^2 G_{pp}(p_t, t)]dt + \sigma p_t G_p(p_t, t)dB_t.$$

(Here, although in conflict with our previous notation, a subscript of a deterministic function indicates differentiation with respect to that variable.) By our foregoing discussion on the fair price, there is π_t^* such that

(4.4) $$de_t = (1 - \pi_t^*)e_t \alpha dt + \pi_t^* e_t(\alpha dt + \sigma dB_t)$$

By (4.3) and (4.4),

$$\pi_t^* e_t \sigma = \sigma p_t G_p(p_t, t) \Rightarrow \pi_t^* = G_p(p_t, t)\frac{p_t}{e_t}$$

$$e_t a + (\alpha - a)\pi_t^* e_t = G_t(p_t, t) + \alpha p_t G_p(p_t, t) + \frac{1}{2}\sigma^2 p_t^2 G_{pp}(p_t, t).$$

Use the expession for π_t^* in the second equation and note that $e_t = G(p_t, t)$: the result is

$$G_t(p_t, t) + a p_t G_p(p_t, t) + \frac{1}{2}\sigma^2 p_t^2 G_{pp}(p_t, t) - aG(p_t, t) = 0.$$

Consequently, G is the unique solution of the linear parabolic equation

$$(4.5) \quad G_t(p, t) + a p G_p(p, t) + \frac{1}{2}\sigma^2 p^2 G_{pp}(p, t) - aG(p, t) = 0, \quad \text{on } (0, \infty) \times (0, T)$$

with terminal data

(4.6) $$G(p, T) = (p - q)^+, \quad p \in [0, \infty).$$

By the Feynman-Kac formula, we may easily express G probabilisticaly.

Note that the option price G is *independent* of the mean return rate α of the stock!

4.2 Option Pricing via Utility Maximization.

Following [20] and [15, §2], we consider two utility maximization problems and define an option price by comparing these two problems.

A utility function is a concave increasing function $\ell : \mathcal{R}^1 \to \mathcal{R}^1$ with $\ell(0) = 0$. We wish to price the European option, with maturity T and exercise price q. To simplify the computations, we take

$$a = 0.$$

Suppose that an agent (who is to sell an European option) has an initial endowment of $w_t = x$ at some time $t \in [0, T]$ and that she follows an investment policy of $\pi_.$. Then her wealth process changes according to (4.1). If she has not written an option, her final wealth is equal to w_T and her utility maximization problem is to maximize

$$E\left(\ell(w_T) \mid w_t = x\right)$$

over all adapted, integrable investment policies. The value function

$$V^f(x, t) := \sup_{\pi_.} E\left(\ell(w_T) \mid w_t = x\right)$$

solves the dynamic programming equation

$$(4.7) \quad -\frac{\partial}{\partial t}V^f(x,t) + \inf_{\pi}\left\{-A^\pi V^f(x,t)\right\} = 0, \qquad \text{on } \mathcal{R}^1 \times (0,T),$$

where

$$A^\pi V^f(x,t) := \pi\alpha x V_x^f(x,t) + \frac{1}{2}\sigma^2\pi^2 x^2 V_{xx}^f(x,t).$$

In the case she has written an option, her wealth still changes according to (4.1) for $s < T$. However at the maturity she has a liability: she has to pay the value of the option, $(p_T - q)^+$. Therefore, at time T, her total wealth is $w_T - (p_T - q)^+$ and the utility maximization problem is to maximize

$$E\left(\ell(w_T - (p_T - q)^+) \mid w_t = x\right).$$

over all adapted, integrable investment policies. Now the value function depends both on the initial endowment and the initial value of the stock:

$$V(x,p,t) := \sup_{\pi.} E\left(\ell(w_T - (p_T - q)^+) \mid w_t = x, \; p_t = p\right).$$

The dynamic programming equation is

$$-\frac{\partial}{\partial t}V(x,p,t) + \inf_{\pi}\left\{-A^\pi V(x,p,t) - \sigma^2 p x \pi V_{px}(x,p,t)\right\} - \mathcal{B}V(x,p,t) = 0,$$
$$(4.8)$$

where

$$\mathcal{B}V(x,p,t) := \alpha p V_p(x,p,t) + \frac{1}{2}\sigma^2 p^2 V_{pp}(x,p,t).$$

Now suppose that the initial endowment of the agent is $w_t = x$ and the stock price is $p_t = p$. Hodges & Neuberger [20] postulate that the agent would write an option for a price e only if

$$V(x+e,p,t) \geq V^f(x,t).$$

So the price of an option is defined as

$$e(x,p,t) := \inf\left\{e \; : \; V(x+e,p,t) \geq V^f(x,t)\right\}.$$

Note that this price may depend on (x,p,t).

Lemma 4.1 $e(x,p,t)$ *is equal to the Black & Scholes price.*

Proof. Let G be the solution of (4.5) and (4.6). Set

$$W(x, p, t) := V^f(x - G(p, t), t).$$

We claim that W solves (4.8). Indeed, by (4.5),

$$
\begin{aligned}
\mathcal{B}W(x, p, t) &= -[\alpha p G_p(p, t) + \frac{1}{2}\sigma^2 p^2 G_{pp}(p, t)]V_x^f(\cdots) + \frac{1}{2}\sigma^2 p^2 (G_p(p, t))^2 V_{xx}^f(\cdots) \\
&= (-\alpha p G_p(p, t) + G_t(p, t))V_x^f(\cdots) + \frac{1}{2}\sigma^2 p^2 (G_p(p, t))^2 V_{xx}^f(\cdots),
\end{aligned}
$$

where $(\cdots)$ stands for $(x - G(p, t), t)$. Using this, we calculate

$$
-\frac{\partial}{\partial t}W(x, p, t) + \inf_\pi \left\{ -A^\pi W(x, p, t) - \sigma^2 p x \pi W_{px}(x, p, t) \right\} - \mathcal{B}W(x, p, t)
$$

$$
= -V_t^f(\cdots) + \inf_\pi \left\{ -\alpha[\pi x - p G_p]V_x^f(\cdots) - \frac{1}{2}\sigma^2(\pi x - p G_p)^2 V_{xx}^f(\cdots) \right\}
$$

$$
= -V_t^f(\cdots) + \inf_{\bar{\pi}} \left\{ -\alpha\bar{\pi}(x - G(p, t))V_x^f(\cdots) - \frac{1}{2}\sigma^2\bar{\pi}^2(x - G(p, t))^2 V_{xx}^f(\cdots) \right\}.
$$

By (4.7) at $(\cdots)$, the last term is equal to zero. Hence W is a solution of (4.8) and, since

$$V^f(x, T) = \ell(x), \qquad V(x, p, T) = \ell(x - (p - q)^+) \qquad G(p, T) = (p - q)^+,$$

$$W(x, p, T) = V^f(x - G(p, T), T) = V(x, p, T).$$

Therefore, by a uniqueness theorem for (4.8), $W = V$ and consequently $e(x, p, t) = G(p, t)$.

In the foregoing argument, we relied on a uniqueness theorem. There is an easier proof: see [16, Theorem 1]. $\qquad \square$

4.3 Transaction Costs.

In this subsection, we summarize the results of Soner, Shreve & Cvitanic [32]. Briefly, [32] proves that, in the presence of transaction costs, there is nontrivial way of perfectly hedging a European option. This is expected, since the total transactions of the Black & Scholes hedging portfolio is infinite.

Our model is the same as the one of §3 without consumption: As before x_s and y_s are the dollars invested in the bond and the stock, respectively and x_s, y_s change according to

$$(4.9) \qquad dx_s = ax_s ds + (1 - \mu)dM_s - dN_s$$

$$(4.10) \qquad dy_s = \alpha y_s ds - dM_s + (1 - \lambda)dN_s + \sigma y_s dB_s,$$

where M_s and N_s are the total transactions up to time $s \geq 0$, from stock to bond and bond to stock, respectively. Instead of perfectly hedging an option, as it is done §1.1 (see (4.2)), we look for dominating investment policies. To make this statement precise we need to make two definitions:

Definitions.
1. We say that $(x, y) \in \mathcal{R}^n$ dominates $(a, b) \in \mathcal{R}^n$ and, we write $(x, y) \succ (a, b)$, if there are $m, n \geq 0$ satisfying

$$a = x - n + (1 - \mu)m, \qquad b = y - m + (1 - \lambda)n$$

2. We say that, at (x, y), a pair $(M., N.)$ satisfying (3.3), (3.5) is a *dominating policy*, if

$$(4.11) \qquad (x_T, y_T) \succ (-q, p_T) 1_{\{p_T \geq (1-\lambda)q\}},$$

where (x_s, y_s) is the solution of (4.9), (4.10) with investment policy $(M., N.)$ and initial data $(x_t, y_t) = (x, y)$. (We assume that the option is exercised if $p_T \geq (1 - \lambda)q$.)

The problem we wish to solve is this: given the price of one share of the stock at time t, $p_t = p$, find the fair price of the European option using the approach described in §4.1. Mathematically, the problem is to characterize the set of all pairs (x, y) at which there is at least one dominating policy. Let $H(p, t) \subset \mathcal{R}^2$ be this set. The following follows immediately from the definitions.

Lemma 4.2 *If $(a, b) \in H(p, t)$ and $(x, y) \succ (a, b)$, then $(x, y) \in H(p, t)$. Moreover, for any $t \in [0, T]$, $p \geq 0$, $(0, p) \in H(p, t))$ In particular, $H(p, t) \supset \{(x, y) : (x, y) \succ (0, p)\}$.*

Proof. We only prove the second statement. Set $N_s = M_s \equiv 0$. Then the unique solution of (4.9), (4.10), with this investment policy and initial data $(x_t, y_t) = (0, p)$, is equal to $(x_s, y_s) = (0, p_s)$ and

$$(x_T, y_T) = (0, p_T) \succ (-q, p_T) 1_{\{p_t \geq (1-\lambda)q\}}.$$

Therefore $N_s = M_s \equiv 0$ is dominating and $(0, p) \in H(p, t)$. $\qquad\square$

The following is proved in [32].

Theorem 4.3 *For any* $t \in [0, T)$, $p \geq 0$, $H(p,t) = \{ (x,y) \, : \, (x,y) \succ (0,p) \}$.

Since in the region $\{ (x,y) \, : \, (x,y) \succ (0,p) \}$ the trivial policy of holding one share of stock is optimal, this theorem simply says that in the presence of transaction costs there is no nontrivial hedging policy.

Note that, trivially,

$$H(p,T) = \left\{ (x,y) \, : \, (x,y) \succ (-q,p) 1_{\{p \geq (1-\lambda)q\}} \right\},$$

and $H(p,t)$ is discontinuous at T.

Remark. Consider the problem of maximizing

$$P\left((x_T, y_T) \succ (-q, p_T) \, 1_{\{p_T \geq (1-\lambda)q\}} \; \middle| \; x_t = x, y_t = y, p_t = p \right),$$

over all transaction policies M, N. Then the value function $V(x,y,p,t)$ solves

$$\min\{-V_t - \mathcal{A}V, -(1-\mu)V_x(x,y) + V_y(x,y), V_x(x,y) - (1-\lambda)V_y(x,y)\} = 0,$$
(4.12)

where

$$\mathcal{A}V = \alpha x V_x + \alpha y V_y + \frac{1}{2}\sigma^2 \left(y^2 V_{yy} + p^2 V_{pp} + 2yp V_{yp} \right).$$

The terminal data is

$$(4.13) \quad V(x,y,p,T) = \begin{cases} 1, & \text{on } (x,y) \succ (-q,p) 1_{\{p \geq (1-\lambda)q\}}, \\ 0, & \text{otherwise.} \end{cases}$$

In view of Lemma 4.2,

$$V(x,y,p,t) = 1, \qquad (x,y) \succ (0,p), \; t < T.$$

An alternate proof of Theorem 4.3 is to construct a supersolution W of (4.12), (4.13) such that

$$W(x,y,p,t) < 1, \qquad \forall (x,y) \not\succ (0,p), \; t < T.$$

Such a proof has the advantage of providing a quantitative measure of hedging probability. However, the difficulty is in the discontinuity of $H(p,t)$. □

Remark. Although the minimal dominating portfolio appraoch is not useful in the presence of transaction costs, it has been proved to be successful in problems with constraints [10]. □

Utility function approach

We showed that the only policy that assures probability-one hedging is the trivial policy of holding one share of the stock. Clearly this policy is too conservative to be off any practical use. An alternate approach is to use the utility maximization method of Hodges and Neuberger [20] which was discussed in §4.2 when there are no transaction costs. The definition of an option is very similar to the one given in that section. This approach is fully developed by Davis, Panas and Zariphopoulou [16] and we refer to that paper for more detail. The exponential utility function

$$\ell(c) := 1 - e^{-\gamma c}, \qquad c \in \mathcal{R}^1,$$

with the parameter $\gamma > 0$, provides an interesting example and it was studied in [16]. For the exponential utility function, the dimension of the corresponding dynamic programming equation is reduced by one thus simplifing the numerical computations.

Option price obtained by this approach depends oth on the utility fuction and on the initial wealth of the writer of the option. In a more recent paper, Constantinides and Zariphopoulou [12] modified the approach of [16] and obtained universal, independent of utility, upper and lower bounds for the option price. Based on their work, Barles and Soner [6] used asymptotic analysis to derive a nonlinear Black & Scholes equation with a nonlinear volatility depending on the Gamma of the option. This equation is somehow related to Leland's approach which is described in the next subsection.

4.4 Leland's approach to transaction costs.

Proceeding in a completely formal way, Leland [24] proposed an option pricing formula. His approach is very similar to that of §4.1 and the final result states that the price of an option is still determined by the Black & Scholes differential equation, (4.5), (4.6), with an increased volatility σ. Formal computations of Leland contains one very crucial approximation which causes his policy to hedge the option with very high probability but not almost surely and the analysis of the hedge probability is still not avaliable. One exception is the work of Boyle & Vorst [9] which considers a bionomial tree model with discrete time steps Δt. They obtain a perfectly hedging investment policy and a corresponding option price for the discrete time problem. They then study the asymptotic limit, as $\Delta t \downarrow 0$. They show that if the proportional transaction costs, μ and λ in (4.9) and (4.10), behave like $\sqrt{\Delta t}$, then in the limit option price is again determined by the Black & Scholes differential equation, (4.5), (4.6), with an increased volatility σ. But their correction of the volatility differs from Leland's correction by a constant multiple. In their model, Boyle & Vorst allow transactions at every time step. If, instead, we restrict the transactions to every other time step or to some other frequency, we obtain a different correction to volatility again differing by a constant multiple. Equivalently, we may consider models in which the stock price has more than two possible value at the end of each time step. This also yields a different correction. Therefore further study of this very desirable approach is still necessary. Recently, Avallaneda & Paras [1] started such a study.

We continue with Leland's derivation. Our presentation follows [1]. To simplify we take $a = 0$ and

$$\lambda = \mu = k.$$

As in §4.1, let e_t be the value of the option at time t and assume that there is a deterministic function

$$G : [0, \infty) \times [0, T] \to [0, \infty), \qquad e_t = G(p_t, t).$$

We assume that G is convex and refer to [1] for the non-convex case. For a stochastic process ξ_t and $\Delta t > 0$, set

$$\Delta \xi_t := \xi_{t+\Delta t} - \xi_t.$$

Then

$$\Delta p_t \simeq p_t \left(\alpha \Delta t + \sigma \Delta B_t \right).$$

Suppose that, as in Black-Scholes model, there is a replicating policy. Let Z_t be the number of stocks used in the replicating policy. Then e_t is equal to $Z_t p_t$ plus the money invested in the bond. Since $a = 0$,

$$\Delta e_t \simeq Z_t \Delta p_t - k p_t |\Delta Z_t|$$

$$\simeq p_t (\alpha Z_t \Delta t - k |\Delta Z_t|) + \sigma p_t Z_t \Delta B_t.$$

Since $e_t = G(p_t, t)$, Ito's formula yields

$$\Delta e_t \simeq \left(G_t + \alpha p_t G_p + \frac{1}{2} \sigma^2 p_t^2 G_{pp} \right) \Delta t + \sigma p_t G_p \Delta B_t.$$

These two equations imply that

$$\sigma p_t Z_t = \sigma p_t G_p \quad \Rightarrow Z_t = G_p(p_t, t),$$

and

(4.14) $$\qquad p_t \alpha Z_t - k p_t \frac{|\Delta Z_t|}{\Delta t} = G_t + \alpha p_t G_p + \frac{1}{2} \sigma^2 p_t^2 G_{pp}.$$

Since $Z_t = G_p(p_t, t)$,

$$|\Delta Z_t| \simeq |\sigma p_t G_{pp} \Delta B_t| + h.o.t.,$$

where $h.o.t.$ are terms of order Δt or higher. Leland's crucial approximation is this:

$$|\Delta B_t| \simeq \sqrt{\frac{2}{\pi}} \sqrt{\Delta t}.$$

Clearly, as $\Delta t \downarrow 0$, $\Delta B_t = B_{t+\Delta t} - B_t$ converges to zero like $\sqrt{\Delta t}$, but the justification for the constant is not very clear. So instead we use the approximation

$$|\Delta B_t| \simeq \hat{a} \sqrt{\Delta t},$$

for some constant $\hat{a} > 0$, and analyze the dependence of our formula on this parameter. Recall that G is assumed to be convex. So

$$|\Delta Z_t| \simeq a \sigma p_t G_{pp} \sqrt{\Delta t}.$$

Use this in (4.14)

$$p_t \alpha Z_t - \frac{\hat{a} \sigma p_t^2}{\sqrt{\Delta t}} G_p p = G_t + \alpha p_t G_p + \frac{1}{2} \sigma^2 p_t^2 G_{pp}.$$

Hence the price G solves

$$G_t + \frac{1}{2} \sigma^2 p_t^2 \left(1 + \frac{2k\hat{a}}{\sigma \sqrt{\Delta t}} \right) G_{pp} = 0.$$

Note that this is the Black & Scholes equation (4.5) with increased volatility

$$\hat{\sigma} := \sigma \left(1 + \frac{2k\hat{a}}{\sigma \sqrt{\Delta t}} \right)^{\frac{1}{2}}.$$

Leland obtained this result with $\hat{a} = \sqrt{2/\pi}$ while Boyle & Vorst result is with $\hat{a} = 1$.

References

[1] M. Avellaneda and A. Paras. Optimal hedging portfolios for derivative securities in the presence of large transaction costs. preprint, 1994.

[2] G. Barles. *Solutions de viscosite des equations de Hamilton-Jacobi.* Springer-Verlag, 1994.

[3] G. Barles and J. Burdeau. The Dirichilet problem for semilinear second-order degenerate elliptic equations and applications to stochastic exit time control problems. preprint, 1995.

[4] G. Barles and P.-L. Lions. Fuly nonlinear neumann type boundary conditions for first-order Hamilton-Jacobi equations. *Nonlinear Anal. Th. Mt. Appl.*, 16:143–153, 1991.

[5] G. Barles and B. Perthame. Exit time problems in control and vanishing viscosity solutions of Hamilton-Jacobi-Bellman equations. *SIAM J. Cont. Opt.*, 26:1113–1148, 1988.

[6] G. Barles and H.M. Soner. Option pricing with transaction costs and a nonlinear Blacck-Scholes equation. preprint, 1996.

[7] R. Bellman. *Dynamic Programming.* Princeton University Press, Princeton, 1957.

[8] F. Black and M. Scholes. The pricing of options and corporate liabilities. *J. Political Economy*, 81:637–659, 1973.

[9] P. P. Boyle and T. Vorst. Option replication in discrete time with transaction costs. *The Journal of Finance*, 47:271–293, 1973.

[10] M. Broadie, J. Cvitanic and H.M. Soner. On the cost of super-replication under portfolio constraints. preprint, 1996.

[11] G.M. Constantinides. Capital market equilibrium with transaction costs. *Journal Political Economy*, 94:842–862, 1986.

[12] G.M. Constantinides and T. Zariphopoulou. Universal bounds on option prices with proportional transaction costs. preprint, 1995.

[13] M.G. Crandall, H. Ishii, and P.-L. Lions. User's guide to viscosity solutions of second order partial differential equations. *Bull. AMS*, 27/1:1–67, 1992.

[14] M.G. Crandall and P.-L. Lions. Viscosity solutions of Hamilton-Jacobi equations. *Trans. AMS*, 277:1–43, 1983.

[15] M. Davis and A. Norman. Portfolio selection with transaction costs. *Math. O.R.*, 15:676–713, 1990.

[16] M. Davis, V.G. Panas, and T. Zariphopoulou. European option pricing with transaction fees. *SIAM J. Cont. Opt.*, 31:470–493, 1993.

[17] L.C. Evans. A second order elliptic equation with a gradient constraint. *Comm. P.D.E.*, 4:4552–572, 1979. Erratum: 4 (1979), 1199.

[18] W.H. Fleming and H.M. Soner. *Controlled Markov Processes and Viscosity Solutions*. Springer-Verlag, New-York, 1993.

[19] W.H. Fleming and T. Zariphopoulou. An optimal investment-consumption models with borrowing. *Math. O.R.*, 16:802–822, 1991.

[20] S.D. Hodges and A. Neuberger. Optimal replication of contingent cliams under transaction costs. *Review of Future Markets*, 8:222–239, 1989.

[21] N. Ikeda and S. Watanabe. *Stochastic Differential Equations and Diffusion Processes*. North Holland-Kodansha, Amsterdam and Tokyo, 1989.

[22] H. Ishii. A boundary value problem of the Dirichlet type Hamilton-Jacobi equations. *Ann. Scuola Nor. Sup. Pisa Cl. Sci.*, 16:105–135, 1989.

[23] I. Karatzas and S.E. Shreve. *Brownian Motion and Stochastic Calculus*. Springer-Verlag, New York, 1988.

[24] H.E. Leland. Option pricing and replication with transaction costs. *J. Finance*, 40:1283–1301, 1985.

[25] P.-L. Lions. Optimal control of diffusion processes and Hamiton-Jacobi-Bellman equations, Part II. *Comm. PDE*, 8:1229–1276, 1983.

[26] P.-L. Lions and A.S. Sznitman. Stochastic differential equations with reflecting boundary conditions. *Comm. Pure Appl. Math.*, 37:511–537, 1984.

[27] S.E. Shreve and H.M. Soner. Optimal investment and consumption with transaction costs. *Ann. Appl. Prob.*, 4:609–692, 1994.

[28] S.E. Shreve, H.M. Soner, and G.-L. Xu. Optimal investment and consumption with two bonds and transaction costs. *Math. Finance*, 1:53–84, 1991.

[29] H.M. Soner. Optimal control with state space constraint I. *SIAM J. Cont. Opt.*, 24:552–562, 1986.

[30] H.M. Soner and S.E. Shreve. Regularity of the value function of a two-dimensional singular stochastic control problem. *SIAM J. Cont. Opt*, 27:876–907, 1989.

[31] H.M. Soner and S.E. Shreve. A free boundary problem related to singular stochastic control; parabolic case. *Comm. PDE*, 16:373–424, 1991.

[32] H.M. Soner, S.E. Shreve, and J. Cvitanic. There is no nontrivial hedging portfolio for option pricing with transaction costs. *Ann. Appl. Prob.*, 5:327–355, 1995. to appear.

[33] T. Zariphopoulou. Investment-consumption models with transaction fees and Makrov chain parameters. *SIAM Cont. Opt.*, 30:613–636, 1992.

[34] T. Zariphopoulou. Investment and consumption models with constraints. *SIAM Cont. Opt.*, 32:50–85, 1994. to appear.

Front Propagation: Theory and Applications

by

Panagiotis E. Souganidis
Department of Mathematics
University of Wisconsin–Madison
Madison, WI 53706

0. Introduction

In this note I discuss in some detail, but without all the occasionally cumbersome technicalities, a general theory about propagating fronts, i.e. "hypersurfaces" moving with prescribed normal velocity, depending on the location of the front in space and time, its normal and its principal curvatures. I also discuss, in the same spirit, the way this theory allows for the rigorous study of the detailed asymptotics of a wide class of problems, like reaction-diffusion equations, interacting particle systems (stochastic Ising models), turbulent flame propagation and premixed combustion, etc., which model phase transitions.

The mathematical problem to study the evolution of a hypersurface with presecribed normal velocity is a classical one in geometry. Its connection with the applications comes up either by direct modeling, as in image processing and the phenomenological theory of sharp-interfaces in continuum mechanics, or in the analysis of the asymptotic behavior of certain systems. This last claim can be perhaps understood by the following vague argument: An evolving, with respect to time t, order parameter that, depending on the specific context, describes the different phases of a material or the total (averaged) magnetization of a stochastic system or the temperature of a reacting-diffusing system, etc., approaches, typically as $t \to \infty$, the equilibrium states of certain systems. Depending on the values of a threshold parameter, such systems either have a unique or more than one equilibrium state. The existence of multiple equilibrium states can be associated to phase transitions. When more than one equilibrium states exist, the evolving order parameter develops, for $t \gg 1$, interfaces, which are the boundaries of the regions where it (the order parameter) converges to the different equilibria. The problem is then to justify the appearance of these interfaces and to understand in a qualitative way their dynamics, geometry, regularity, etc..

The main mathematical characteristic of the evolution of hypersurfaces with prescribed normal velocity is the development of singularities in finite time, independently of the smoothness of the initial surface. A great deal of work has been done recently to interpret the evolution past the singularities. The outcome has been the development of a weak notion of evolving fronts called generalized front propagation. The generalized evolution with prescribed normal velocity starting at a given surface is defined for all $t \geq 0$, although it may become extinct in finite time. Moreover, it agrees with the classical differential-geometric flow, as long as the latter exists. The generalized motion may, on the other hand, develop singularities, change topological type and exhibit various other pathologies.

In spite of these peculiarities, the generalized motion has been proven to be the right way to extend the classical motion past singularities. Some of the most definitive results in this direction are about the fact that the generalized evolution governs the asymptotic behavior of solutions of semilinear reaction-diffusion equations and systems. Such equations are often used in continuum mechanics to describe the time evolution of an order parameter determining the phases of a material (phase field theory). Similar equations but with different reaction terms are also used to model turbulent combustion in random environments.

Another recent striking application of the generalized front propagation is the fact that it governs the macroscopic (hydrodynamic) behavior, for large times and in the context of grain coarsening, of interacting particle systems like the stochastic Ising model with long-range interactions and general spin flip dynamics. Such systems are standard Gibbsian models used in statistical mechanics to describe phase transitions. It turns out that the generalized front propagation not only describes the limiting behavior of such systems but also provides, in this context, a theoretical justification, from the microscopic point of view, of several phenomenological sharp interface models in phase transitions.

The notes are organized as follows: Section 1 is about the generalized front propagation. It contains most of the definitions and results which are relevant for the study of the asymptotic problem. In Section 2 I present a general abstract theorem, which allows for the rigorous justification of the appearance of interfaces of a large class of asymptotic problems by checking a number of basic conditions and by justifying formal asymptotics under arbitrary smoothness assumptions. Section 3 is about the asymptotics of reaction-diffusion equations with bistable nonlinearities. In Section 4 I discuss again reaction-diffusion equations but with KPP-type nonlinearities and I briefly present some facts about turbulent flame propagation and combustion. Section 5 discusses the asymptotics of stochastic Ising models.

Since these are supposed to be lecture notes and not a book it will be impossible to state and prove everything with the outmost mathematical rigor. Indeed most of the theorems are stated without most of the routine hypotheses. Along the same lines, the proofs which I present here will contain most of the ideas but may occasionally be a bit formal. I will refer, however, to specific references for the exact statements and details. I will also give as many references as possible with the understanding, of course, that there are some page limitations.

The notion of viscosity solutions and their properties are, of course, fundamental for everything I am presenting here. The notes are written under the assumption that the reader is familiar with viscosity solutions. For a number, if not most, of the technical issues and statements, one can look at the notes of Crandall [Cr] in this volume, the "User's Guide" by Crandall, Ishii and Lions [CIL] and the references therein. Finally in the text I will often say solution instead of viscosity solution.

1. Generalized Front Propagation.

This section is about the evolution of interfaces (fronts, surfaces) $\Gamma_t = \partial\Omega_t \subset \mathbb{R}^N$, where Ω_t is an open subset of $\mathbb{R}^N$, with normal velocity

$$(1.1) \qquad\qquad V = v(Dn, n, x, t),$$

where n and Dn are the exterior normal vector to the front and its gradient respectively, starting at $\Gamma_0 = \partial\Omega_0$.

Typical examples of motions governed by (1.1), which are of great interest in geometry but also appear in the kind of applications referred to in the Introduction, are, among others, the motion by mean curvature

$$(1.2) \qquad\qquad V = -\text{tr } Dn,$$

the motion by the Gaussian curvature

$$(1.3) \qquad\qquad V = \kappa_1 \cdots \kappa_{N-1},$$

where $\kappa_1, \ldots, \kappa_{N-1}$ are the principal curvatures of the front and the anisotropic law

$$(1.4) \qquad\qquad V = -\text{tr}[\theta(n, x, t)Dn] + c(n, x, t),$$

where θ is a nonnegative matrix and c is a scalar.

As mentioned above the main mathematical characteristic of such evolutions is the development of singularities in finite time, independently of the smoothness of the initial surface. Classical examples in this direction are the evolution by mean curvature of "figure-eights", "crosses", "bar-bells" and "tori" in $\mathbb{R}^N$. For more discussion regarding this issue I refer to the notes by Evans [E2] in this volume as well as to the papers by Evans and Spruck [ESp], Ilmanen [Il1], Soner [Son1], Soner and Souganidis [SonS], etc..

A great deal of work has been done recently to interpret the evolution (1.1) past the singularities. A number of a priori different approaches have been put forward leading to different definitions (e.g. level set, distance function, phase field, etc..) The task is then to show that all these approaches yield the same generalized front propagation, which can be accomplished assuming that there is "no-fattening", as is explained later in this section. For the purposes of these notes I will use as the building block the level set approach and I will derive the other definitions as corollaries. In doing so I will also remark about the intrinsic merits of the other approaches and/or definitions. In what follows I discuss motions governed by (1.1) and not only mean curvature, for which one has, due to some additional geometric structure, more information and yet more ways to go about studying it. I refer to [E2] for a detailed discussion of the mean curvature motion and its regularity.

The level set approach, which represents the evolving surface as a level set of an auxiliary partial differential equation, was introduced for numerical calculations by Osher and Sethian [OsS]; I also refer to Ohta, Jasnow and Kawasaki [OJK] in the physics literature and to Barles [Ba1] for a first-order model for flame propagation. The mathematical theory of this approach, which is based on viscosity solutions, was extensively developed independently by Evans and Spruck [ESp] for the motion by mean curvature and by Chen, Giga and Goto [CGG] for more general geometric motions. This work was later extended by Ishii and Souganidis [IS] and Goto [Go] for more general motions which include (1.3) and more general initial surfaces – see also Barles, Soner and Souganidis [BSS], Giga, Goto, Ishii and Sato [GGIS], Gurtin, Soner and Souganidis [GSS], Ohnuma and Sato [OhS], Ishii [Is2], Giga and Giga [GGi], etc..

The idea of the level set approach can be described as follows: Let $\Gamma_t = \partial\Omega_t$, with Ω_t open subset of $\mathbb{R}^N$, be a smooth front at time $t > 0$ moving with normal velocity V given by (1.1). Furthermore assume that there exists a smooth function $u : \mathbb{R}^N \times (0,\infty) \to \mathbb{R}$ such that $\Omega_t = \{x \in \mathbb{R}^N : u(x,t) > 0\}$, $\Gamma_t = \{x \in \mathbb{R}^N : u(x,t) = 0\}$ and $|Du| \neq 0$ on Γ_t. A classical calculation yields that

$$V = \frac{u_t}{|Du|}, \quad n = -\frac{Du}{|Du|} \quad \text{and} \quad Dn = -\frac{1}{|Du|}\Big(I - \frac{Du \otimes Du}{|Du|^2}\Big)D^2u.$$

Inserting these formulae in (1.1) and assuming that all level sets of u are moving with the same velocity, one gets the expression

$$(1.5) \qquad\qquad u_t = F(D^2u, Du, x, t),$$

where F is related to v in (1.1) by

$$(1.6) \qquad F(X, p, x, t) = |p|v\Big(-\frac{1}{|p|}(I - \frac{p \otimes p}{|p|^2})X, -\frac{p}{|p|}, x, t\Big),$$

for all $(x,t) \in \mathbb{R}^N \times (0,\infty)$, $p \in \mathbb{R}^N$ and $X \in \mathcal{S}^N$, the space of $N \times N$ symmetric matrices.

An immediate consequence of (1.6) is that F is as smooth as v with possible discontinuity at $p = 0$ and that F is geometric, i.e. it satisfies, for all $(x,t) \in \mathbb{R}^N \times (0,\infty)$, $p \in \mathbb{R}^N$ and $X \in \mathcal{S}^N$,

$$(1.7) \qquad F(\lambda X + \mu(p \otimes p), x, t) = \lambda F(X, p, x, t) \quad \text{for all } \lambda > 0 \text{ and } \mu \in \mathbb{R}.$$

In order for (1.1) to be well-posed it is also necessary to assume that it is parabolic, i.e. that v is nonincreasing in the Dn argument. In view of (1.6) this translates to F been degenerate elliptic, i.e.

$$(1.8) \qquad F(X, p, x, t) \geq F(Y, p, x, t) \quad \text{if } X \geq Y,$$

for all $X, Y \in \mathcal{S}^N$, $p \in \mathbb{R}^N$ and $(x,t) \in \mathbb{R}^N \times (0,\infty)$. Note that (1.7) yields that F is degenerate at least in the $p \otimes p$-direction.

The level set approach to front propagation can now be described as follows: Given $\Gamma_0 = \partial\Omega_0 \subset \mathbb{R}^N$, the front at time $t = 0$, choose $u_0 : \mathbb{R}^N \to \mathbb{R}$ such that

$$\Gamma_0 = \{x \in \mathbb{R}^N : u_0(x) = 0\} \text{ and } \Omega_0 = \{x \in \mathbb{R}^N : u_0(x) > 0\},$$

solve (in the appropriate way) the pde

$$(1.9) \quad \begin{cases} \text{(i) } u_t = F(D^2u, Du, x, t) & \text{in } \mathbb{R}^N \times (0, \infty), \\ \\ \text{(ii) } u = u_0 & \text{on } \mathbb{R}^N \times \{0\}, \end{cases}$$

and define for $t > 0$ the front Γ_t by

$$\Gamma_t = \{x \in \mathbb{R}^N : u(x, t) = 0\}.$$

The main issues associated with such a program are whether (1.9) is well-posed globally in time and whether the so defined Γ_t depends only on Γ_0 and not the form of u_0 outside Γ_0.

The first issue was settled in [ESp] and [CGG] for motion by mean curvature and general geometric motions with bounded dependence on the curvatures and regular (continuous) v respectively. Motions with unbounded dependence on the curvatures were considered in [Go] and [IS]. Motions with irregular, i.e. discontinuous v were considered by Gurtin, Soner and Souganidis [GSS], Ohnuma and Sato [OhS] and Ishii [Is2] and, more recently, Giga and Giga [GGi].

In what follows I concentrate on motions in $\mathbb{R}^N$ (see Giga and Sato [GS] for motions with boundary conditions) and on the case where v depends in bounded way on n (see the references mentioned above for the other cases), and hence, in view of (1.6), one has, for $(x, t) \in \mathbb{R}^N \times (0, \infty)$,

$$(1.10) \qquad\qquad F^*(0, 0, x, t) = F_*(0, 0, x, t).$$

Recall [Cr] that f^* and f_* denote the upper- and lower-semicontinuous envelope of f respectively.

Next I state a comparison theorem about discontinuous solutions of (1.9) which follows after some easy modifications from the proof of the analogous result for continuous solutions in [CGG] – see [BSS] for a more detailed discussion of this fact and [Cr] for some of the key ingredients of such a proof. For the precise statement it is necessary to make some additional assumptions on the dependence of F on (X, p, x, t), which, however, for the sake of brevity I omit.

Here and below $UC(\mathcal{O})$ denotes the space of uniformly continuous functions $u : \mathcal{O} \to \mathbb{R}$ and $BUC(\mathcal{O})$ denotes the space of bounded uniformly continuous functions $u : \mathcal{O} \to \mathbb{R}$. Finally, viscosity volution (unless otherwise specified) will always be referred to as solutions.

Theorem 1.1: *Assume (1.7), (1.8) and (1.10). If $u \in UC(\mathbb{R}^N \times [0, \infty))$ is a subsolution of (1.9)(i) and $v : \mathbb{R}^N \times [0, \infty) \to \mathbb{R}$ is a discontinuous supersolution and $u(\cdot, 0) \le v(\cdot, 0)$ on $\mathbb{R}^N$, then $u(\cdot, t) \le v(\cdot, t)$ on $\mathbb{R}^N$ for all $t > 0$. A similar statement holds if u is a discontinuous subsolution and $v \in UC(\mathbb{R}^N \times [0, \infty))$ is supersolution.*

An immediate consequence of Theorem 1.1 is the following proposition, which asserts the well-posedness of (1.9).

Proposition 1.2: *Assume (1.7), (1.8) and (1.10). Then, for any $u_0 \in UC(\mathbb{R}^N)$, there exists a unique solution $u \in UC(\mathbb{R}^N \times [0, +\infty))$ of (1.9). Moreover, if u and v are respectively sub- and super-solutions of (1.9) (in $UC(\mathbb{R}^N \times [0, \infty))$), then*

$$u(\cdot, 0) \le v(\cdot, 0) \text{ in } \mathbb{R}^N \Rightarrow u \le v \text{ in } \mathbb{R}^N \times [0, +\infty).$$

□

Next I discuss the issue of whether Γ_t depends only on Γ_0. This follows from the fact that F is geometric which implies that (1.9) is invariant by nondecreasing changes $u \mapsto \phi(u)$, which is stated in the following lemma:

Lemma 1.3: *Assume (1.7) and let $\phi : \mathbb{R} \to \mathbb{R}$ be nondecreasing. If $u : \mathbb{R}^N \times [0, \infty) \to \mathbb{R}$ is a solution of (1.9)(i), then $\phi(u)$ is also a solution.*

Proof: 1. Assume that ϕ is differentiable and that u is smooth. The stability properties and the definition of viscosity solution allow for the removal of these conditions by considering suitable approximations.

2. It is now a calculus exercise to conclude, using (1.7). □

The next theorem now follows:

Theorem 1.4: *Assume the hypotheses of Theorem 1.1 and let $u, v \in UC(\mathbb{R}^N \times [0, \infty))$ be solutions of (1.9)(i) such that $\{x : u(x, 0) > 0\} = \{x : v(x, 0) > 0\}$, $\{x : u(x, 0) < 0\} = \{x : v(x, 0) < 0\}$ (and hence $\{x : u(x, 0) = 0\} = \{x : v(x, 0) = 0\}$). Then for all $t > 0$, $\{x : u(x, t) > 0\} = \{x : v(x, t) > 0\}$, $\{x : u(x, t) < 0\} = \{x : v(x, t) < 0\}$ and hence $\{x : u(x, t) = 0\} = \{x : v(x, t) = 0\}$.* □

Theorem 1.4 justifies the term equation of geometric type for (1.9), since it yields that the evolution of the level set $\Gamma_0 \to \Gamma_t$ depends only on Γ_0 and on the "sign" of the initial datum in the different regions, which in turn gives a sense to the expressions "inside Γ_0" and "outside Γ_0", and not in the choice of the initial datum. Such a result was first obtained by Evans and Souganidis [ESo1] in the case where F is independent of $D^2 u$ using arguments from the theory of differential games. Then [ESp] and [CGG] for mean curvature and general geometric motions respectively obtained this theorem under the additional assumption that Γ_0 be compact. In the generality stated here the theorem has proved in [IS].

Proof of Theorem 1.4: 1. To simplify the argument assume in addition that

$$\left(\lim_{|x| \to \infty} |u(x, 0)| \right) \left(\lim_{|x| \to \infty} |v(x, 0)| \right) > 0;$$

for the general case, which needs additional approximations see [IS].

2. Define the functions $\phi, \psi : \mathbb{R} \to \mathbb{R}$ given by $\phi(t) = \inf\{v(y, 0) \mid u(y, 0) \ge t\}$ and $\psi(t) = \sup\{v(y, 0) \mid u(y, 0) \le t\}$. It is immediate that ϕ and ψ are nondecreasing, lower- and supper-semicontinuous respectively and that

$$\phi(u(\cdot, 0)) \le v(\cdot, 0) \le \psi(u(\cdot, 0)) \quad \text{on } \mathbb{R}^N.$$

Moreover, the assumptions on $u(\cdot,0)$ and $v(\cdot,0)$ yield that ϕ and ψ are actually continuous at 0 with $\phi(0) = \psi(0) = 0$.

3. Arguing as if ϕ and ψ were actually continuous, i.e. by considering appropriate regularizations as well as the stability properties of viscosity solutions, and using Lemma 1.3 and Theorem 1.1 one concludes that

$$\phi(u) \le v \le \psi(u) \text{ in } \mathbb{R}^N \times [0, +\infty),$$

which in turn yields the desired conclusion, since the assumptions on $u(\cdot,0)$ and $v(\cdot,0)$ yield that $\phi(t) > 0$ if $t > 0$ and $\psi(t) < 0$ if $t < 0$. $\qquad\square$

Let $\mathcal{E}$ be the collection of all triples (A, B, C) where A, B and C are mutually disjoint subsets of $\mathbb{R}^N$ such that $A \cup B \cup C = \mathbb{R}^N$. In view of the previous discussion the level-set approach can be summarized as follows:

Definition 1.5: *Given* $(\Gamma_0, \Omega_0^+, \Omega_0^-) \in \mathcal{E}$, *solve (1.9)(i) with initial datum* $u = d_0$, *where* d_0 *is the signed distance to* Γ_0 *with the convention that* $d_0 > 0$ *in* Ω_0^+ *and* $d_0 < 0$ *in* Ω^-. *The level set evolution of* $(\Gamma_0, \Omega_0^+, \Omega_0^-)$ *with normal velocity* $V = v$ *is defined to be the triple*

$$(\Gamma_t, \Omega_t^+, \Omega_t^-) = (\{x : u(x,t) = 0\}, \{x : u(x,t) > 0\}, \{x : u(x,t) < 0\}),$$

where F *and* v *are related by (1.6).*

Next I discuss the geometric pde which correspond to the examples of motions stated at the beginning of the section.

A first example is about hypersurfaces propagating with normal velocity $v(n, x, t)$, which arise in the theory of (turbulent) flame propagation. The geometric equation in this case is

$$(1.11) \qquad u_t = v\left(-\frac{Du}{|Du|}, x, t\right)|Du| \text{ in } \mathbb{R}^N \times (0, \infty).$$

Another example is, of course, the motion by mean curvature, with a geometric pde of the form

$$u_t = \Delta u - \frac{(D^2 u Du, Du)}{|Du|^2} \text{ in } \mathbb{R}^N \times (0, \infty),$$

where $(\cdot, \cdot)$ denotes the usual inner product in $\mathbb{R}^N$. The mean curvature pde has an additional scaling property which yields that it is invariant under any change $u \mapsto \phi(u)$ and not only nondecreasing ones! This particular equation and the properties of its solution has been the topic of extensive study – I refer to [E2] for discussion about it.

A simple generalization of the mean curvature motion is obtained by setting

$$v = g(\text{tr}\,(-Dn))$$

in (1.1), where $g \in C(\mathbb{R})$. Then

$$F(X, p) = |p|g\left(-\frac{1}{|p|}\,\text{tr}\left[\left(I - \frac{p \otimes p}{|p|^2}\right)X\right]\right).$$

If g is nonincreasing in $\mathbb{R}$, then F satisfies (1.7) and (1.8). On the other hand, if for instance, $g(r) = (-r)^\alpha$ and $\alpha > 1$ is an odd number then

$$F^*(0,0) = \infty \quad \text{and} \quad F_*(0,0) = -\infty,$$

i.e. (1.10) fails. In this situation one has to use the arguments of [Go] and [IS]. Replacing mean curvature by Gaussian curvature, leads to

$$F(X,p) = |p| \det \left(-\frac{1}{|p|}(I - \frac{p \otimes p}{|p|^2})X \right)$$

and its generalizations – see [IS] and [GGo] for a discussion of a large number of examples of this nature. Incidentally a geometric pde with a certain power of the Gaussian curvature is the pde derived by Lions [Lio] and Alvarez, Guichard, Lions and Morel [AGLM] in their studies of image processing.

Another example of geometric motion which arises in the theory of phase transitions either as a phenomenological model (see Gurtin [Gu]) or as the asymptotic limit of stochastic Ising models (see Section 5 below) or cellular automata models (Ishii, Pires and Souganidis [IPS]) is the case of anisotropic motion, where (1.9)(i) is of the form

$$u_t = |Du| \text{div}\left(H(\frac{Du}{|Du|}) \right) + |Du|\beta\left(\frac{Du}{|Du|} \right) \quad \text{in } \mathbb{R}^N \times (0,\infty).$$

When H is smooth and convex the pde is covered by Theorem 1.1. There are, however, some very interesting models of phase transitions which yield equations like the one above but with H not convex, in which case (1.8) is not satisfied. Following a relaxation process, these problems give rise to (1.9)(i) with F's satisfying (1.8) but additional to discontinuities in the gradient. This is the object of the study of [GSS], [OhS] and [Is2]. Finally there are other examples which lead to non smooth H's. This includes, for example, the case of crystalline energy. Such problems in one dimension are studied in [GGi]

I now return to the discussion of the level set propagation. This approach seems to avoid all the geometrical difficulties related to the onset of singularities, etc., since the evolution is well defined and unique. On the other hand, as it is the case for the theory of viscosity solutions, it provides very little information about the possible regularity of the evolving front. I refer to [E2] and the references therein about a discussion of this topic for the case of the mean curvature evolution.

A more basic question is whether Γ_t has an empty interior for $t > 0$. In principle, one expects Γ_t to be a hypersurface in $\mathbb{R}^N$ provided that Γ_0 is a nice surface, something which makes the possibility of the creation of interior unreasonable. Before I continue this discussion, I give a more precise definition:

Definition 1.6: *Let $(\Gamma_t, \Omega_t^+, \Omega_t^-)$ be the evolution of $(\Gamma_0, \Omega_0^+, \Omega_0^-)$ with normal velocity V by the level set approach. One says that there is no interior for $t > 0$ iff*

$$\overline{\Omega_t^+} = \Omega_t^+ \cup \Gamma_t \quad \text{and} \quad int(\Omega_t^+ \cup \Gamma_t) = \Omega_t^+.$$

If u is a solution of the level set pde such that $\Omega_t = \{x : u(x,t) > 0\}$ and $\Gamma_t = \{x : u(x,t) = 0\}$, the conditions of Definition 1.6 can be written as

$$\overline{\{x : u(x,t) > 0\}} = \{x : u(x,t) \geq 0\} \quad \text{and} \quad \text{int}\{x : u(x,t) \geq 0\} = \{x : u(x,t) > 0\}.$$

In view of this observation, when the level set evolution $(\Gamma_t, \Omega^+, \Omega_t^-)$ of $(\Gamma_0, \Omega_0^+, \Omega_0^-)$ with $\Gamma_0 = \partial\Omega_0^+ = \partial\Omega_0^-$ does not develop interior for $t > 0$, it follows that $\Omega_t^- = \mathbb{R}^N\setminus\overline{\Omega_t^+}$. In most examples it can be easily shown that Γ_t has no interior in $\mathbb{R}^N$ for all $t > 0$ iff $\cup_{t>0}(\Gamma_t \times \{t\})$ has no interior in $\mathbb{R}^N \times (0,\infty)$. For motion with constant normal velocity this follows from the finite speed of propagation. For motion by mean curvature it can be shown using explicit solutions of the form $\psi((N-1)t - |x|^2)$ as barriers. Such kind of argument can be generalized to more complicated motions.

In view of this remark, I now present a new formulation of the "no-empty interior" question in terms of whether (1.9) has unique discontinuous solutions with initial datum $\mathbb{I}_{\Omega_0} - \mathbb{I}_{\mathbb{R}^N\setminus\overline{\Omega}_0}$, where $\mathbb{I}_A$ denotes the characteristic function of the set A. This formulation is of interest by itself since it provides a criterion which can be checked in some cases, for example, for first-order motions.

Theorem 1.7: *Assume that $\Gamma_0 = \partial\Omega_0 = \partial(\mathbb{R}^N\setminus\overline{\Omega}_0)$ and consider the level set evolution $(\Gamma_t, \Omega_t^+, \Omega_t^-)$ of $(\Gamma_0, \Omega_0, \mathbb{R}^N\setminus\overline{\Omega}_0)$. The set $\cup_{t>0}(\Gamma_t \times \{t\})$ has empty interior in $\mathbb{R}^N \times (0,\infty)$ if and only if there exists a unique solution of (1.9)(i) with initial datum $\mathbb{I}_{\Omega_0} - \mathbb{I}_{\mathbb{R}^N\setminus\overline{\Omega}_0}$.*

This result was obtained in [BSS]. I next present the main steps in the proof of this theorem, which is based very strongly on the notion of discontinuous viscosity solutions (see, for example [Ba2], [BJ], [BP], [CIL], [Is1] and [Cr]), their stability properties and the geometric character of of (1.9)(i), and, in particular, Lemma 1.3.

Proof of Theorem 1.7: 1. Let $u \in UC(\mathbb{R}^N \times [0,\infty))$ be a solution of (1.9)(i) such that $u = d_0$ on $\mathbb{R}^N \times \{0\}$, where d_0 is the signed distance to Γ_0. Lemma 1.3 yields that for any $\epsilon > 0$ the function $u^\epsilon = \tanh(\epsilon^{-1}u)$ is also a solution of (1.9)(i).

2. The stability properties of discontinuous viscosity solutions imply that the function $u_\infty = \lim_{\epsilon\to 0} u^\epsilon$ is also a solution of (1.9)(i). Moreover, the properties of tanh yield

$$u_\infty(x,t) = \begin{cases} 1 & \text{if } u(x,t) > 0, \\ -1 & \text{if } u(x,t) < 0, \\ 0 & \text{if } (x,t) \in \text{int}\{u = 0\}. \end{cases}$$

For the rest of the points (x,t), the value of $u_\infty(x,t)$ depends on the lsc or usc envelope one considers in the definition of the discontinuous solutions.

3. Pick $\alpha \in (0,1)$ and set

$$\overline{u}_\infty = \lim_{\epsilon\to 0} \tanh(\epsilon^{-1}(u_\infty + a)) \quad \text{and} \quad \underline{u}_\infty = \lim_{\epsilon\to 0} \tanh(\epsilon^{-1}(u_\infty - a)).$$

The functions $\overline{u}_\infty$ and $\underline{u}_\infty$ are again solutions of (1.9)(i). Moreover,

$$\overline{u}_\infty = \begin{cases} 1 & \text{in } \{u \geq 0\}, \\ -1 & \text{in } \{u < 0\}, \end{cases} \quad \text{and} \quad \underline{u}_\infty = \begin{cases} 1 & \text{in } \{u > 0\}, \\ -1 & \text{in } \{u \leq 0\}. \end{cases}$$

If $\cup_{t>0}(\Gamma_t \times \{t\})$ has non-empty interior, $\overline{u}_\infty$ and $\underline{u}_\infty$ are two different discontinuous solutions of (1.9)(i) with initial datum $\mathbb{I}_{\Omega_0} - \mathbb{I}_{\mathbb{R}^N \setminus \overline{\Omega}_0}$.

4. Let w be a solution of (1.9)(i) with $w = \mathbb{I}_{\Omega_0} - \mathbb{I}_{\mathbb{R}^N \setminus \overline{\Omega}_0}$ on $\mathbb{R}^N \times \{0\}$ and choose a sequence $(\phi_n)_n$ of smooth functions such that

$$-1 \leq \phi_n \leq 1, \quad \phi_n \equiv 1 \text{ on } [0, +\infty), \quad \inf_n \phi_n = -1 \text{ on } (-\infty, 0] \text{ and } \phi_n' \geq 0.$$

Since $w^* \leq \phi_n(d_0)$ on $\mathbb{R}^N \times \{0\}$, Lemma 1.3 and Theorem 1.1 yields $w^* \leq \phi_n(u)$ in $\mathbb{R}^N \times (0, \infty)$ and, in view of the properties of ϕ_n,

$$w^* \leq -1 = \inf_n \phi_n(u) \quad \text{on } \{u < 0\}.$$

On the other hand (1.10) implies that $+1$ and -1 are respectively super- and sub-solutions of (1.9). Therefore

$$-1 \leq w_* \leq w^* \leq 1 \text{ on } \mathbb{R}^N \times [0, \infty)$$

and, finally, $w^* = -1$ in $\{u < 0\}$, if the set $\cup_{t>0}(\Gamma_t \times \{t\})$ has empty interior,.

5. The same line of argument shows that $w_* = 1$ in $\{u > 0\}$, which in view of the non-empty interior assumption, implies that w is identified uniquely.

$\square$

Theorem 1.7 and some of the steps in its proof allow for an alternative way to define the generalized evolution introduced in Barles, Soner and Souganidis [BSS], which is stated below, in terms of discontinuous solutions of (1.9)(i) which only take the values 1 "inside" and -1 "outside" the front. Incidentally, there is nothing magic about the values ± 1; one can choose instead 0 and 1 or for that matter any two different real numbers. One may argue that this is a more intrinsic definition since, in view of (1.10), it only checks the pde on the front, what is happening away from it being irrelevant. The drawback, however, is that it may define something which is not unique if there is interior.

I continue with a discussion of the properties of the signed distance

(1.12)
$$d(x, t) = \begin{cases} d(x, \Gamma_t) & \text{if } x \in \Omega_t, \\ -d(x, \Gamma_t) & \text{if } x \notin \Omega_t, \end{cases}$$

to a front $\Gamma_t = \partial \Omega_t$ evolving with a given normal velocity. One may use the signed distance to the front to give another formulation for the weak propagation. This was done by Soner [Son1] for the case of (x, t)-independent normal velocities and can be done following [BSS] for the general case. The point of view taken here, however, is not to give another definition but rather to develop some tools which play a fundamental role in the study of the generation of fronts as asymptotic limits of problems like the one's discussed in the Introduction, i.e. phase field theory, stochastic systems, etc.. It should be noted, however, that it is very natural to try to describe the front in terms of the signed distance function, since in the smooth case one, of course, has that

$$V = d_t, \quad n = -Dd \text{ and } Dn = -D^2 d, \quad \text{and} \quad d_t = F(D^2 d, Dd, x, t) \text{ on } \Gamma_t.$$

As usual I begin with an open set $\Omega_0 \subset \mathbb{R}^N$ and consider the level set evolution $(\Gamma_t, \Omega_t^+, \Omega_t^-)$ of $(\Gamma_0, \Omega_0, \mathbb{R}^N \backslash \overline{\Omega}_0)$ and the signed distance function to Γ_t given by (1.12).

To state the main result we define the extinction time $t^* \in (0, +\infty]$ for Γ_t by

$$t^* = \sup\{t > 0 \text{ such that } \Gamma_t \neq \phi\}.$$

Notice that depending on the evolution under consideration it is possible to have $t^* < \infty$. This is for example the case of a bounded set Γ_0 moving by mean curvature. On the other hand, if one is considering the motion governed by (1.11) with $v > 0$, then $t_* = \infty$.

Theorem 1.8: *Assume that Γ_t has empty interior for all $t > 0$. Then the functions $\underline{d} = d \wedge 0$ and $\overline{d} = d \vee 0$ satisfy respectively*

$$(1.13) \qquad \underline{d}_t \leq F(x - \underline{d}D\underline{d}, t, D\underline{d}, D^2\underline{d}) \text{ in } \mathbb{R}^N \times (0, t^*)$$

and

$$(1.14) \qquad \overline{d}_t \geq F(x - \overline{d}D\overline{d}, t, D\overline{d}, D^2\overline{d}) \text{ in } \mathbb{R}^N \times (0, t^*).$$

Moreover,

$$(1.15) \qquad -(D^2\underline{d}D\underline{d} \mid D\underline{d}) \leq 0 \text{ in } \{\underline{d} < 0\}$$

and

$$(1.16) \qquad -(D^2\overline{d}D\overline{d} \mid D\overline{d}) \geq 0 \text{ in } \{\overline{d} > 0\}.$$

The assumption that Γ_t has empty interior was made only to simplify the presentation. In fact one can show that (1.13)–(1.16) still hold when Γ_t has empty interior but for different solutions. Indeed let $\overline{\Gamma}_t = \partial\{x : \overline{u}_\infty(x, t) = 1\}$ and $\underline{\Gamma}_t = \partial\{x : \underline{u}_\infty(x, t) = 1\}$ where $\underline{u}_\infty$ and $\overline{u}_\infty$ are defined as in the proof of Theorem 1.7. Then (1.13), (1.15) and (1.14), (1.16) hold true but for $d(x, \overline{\Gamma}_t)$ and $d(x, \underline{\Gamma}_t)$. This again is related to the connections between the non-empty difficulty and the possible nonuniqueness in the geometric formulation of the motion in terms of discontinuous solutions to (1.9). For a detailed discussion I refer to [Son1] and [BSS].

Another important point is that the assertions of Theorem 1.8 are sharp independently of the smoothness of the distance function, which in the context of the weak evolution is only Lipschitz continuous with respect to x and discontinuous with respect to t. To see this assume for the moment that F is (x, t)-independent. It is then a fact that even when the signed distance function is smooth it does not satisfy $d_t = F(D^2d, Dd)$ away from the front as it can be seen, for example, for the motion by mean curvature by an explicit computation. Indeed following, for example, Gilbarg and Trüdinger [GT], one sees that

$$d_t - \text{tr}\left(I - \frac{Dd \otimes Dd}{|Dd|^2}\right)D^2d \begin{cases} \geq 0 & \text{if } d > 0, \\ \geq 0 & \text{if } d < 0. \end{cases}$$

In the case of (x, t)-dependent F's the meaning of the term $x - dDd$, with the bars at the appropriate place, is that if $x \notin \Gamma_t$ one has to move distance d in the Dd direction to return on the front.

Finally, before I continue with the proof of Theorem 1.8, I state a corollary which follows immediately from the above and which re-emphasizes the point about the sharpness of Theorem 1.8.

Corollary 1.9: *Assume that* $\Gamma_0 = \partial\Omega_0 = \partial(\mathbb{R}^N\backslash\overline{\Omega}_0)$ *and consider the mean curvature evolution* $(\Gamma_t, \Omega_t^+, \Omega_t^-)$ *of* $(\Gamma_0, \Omega_0, \mathbb{R}^N\backslash\overline{\Omega}_0)$. *Then the signed distance function to* Γ_t *satisfies*

$$d_t - \Delta d \begin{cases} \leq 0 & in \ \{d < 0\} \cap \mathbb{R}^N \times (0, t_*), \\[2mm] \geq 0 & in \ \{d > 0\} \cap \mathbb{R}^N \times (0, t_*). \end{cases}$$

Proof of Theorem 1.8: 1. I only prove (1.13) and (1.15), since (1.14) and (1.16) can be obtained by similar arguments.

2. Observe that the function

$$w = \begin{cases} 0 & if \ u_\infty = 1, \\[2mm] -\infty & if \ u_\infty = -1, \end{cases}$$

is a solution of (1.9)(i), where u_∞ is defined in the proof of Theorem 1.7.

3. Let the function $\overline{w} : \mathbb{R}^N \times [0, T] \to \mathbb{R}$ be defined by

$$\overline{w}(x, t) = \sup_{y \in \mathbb{R}^N} \{w(y, t) - |x - y|\}.$$

An easy calculation yields that

$$\overline{w}(x, t) = d \wedge 0.$$

On the other hand, standard arguments from the theory of viscosity solutions (see, for example, Jensen, Lions and Souganidis [JLS], [CIL], [Cr], etc.) imply that $\overline{w}$ is a subsolution of (1.9)(i). The inequalities (1.13) and (1.15) follow then easily.

$\square$

It has become, hopefully, clear by now that understanding the empty interior condition is of great importance, since it may lead to some rather unintuitive situations. Unfortunately, if no conditions are imposed on Γ_0, interior may be created for $t > 0$. See for example [ESp] and Soner [Son1]] for some simple examples in this direction for motion by mean curvature starting with non-smooth surfaces like "crosses", i.e. intersection of planes, and "figure-eights", i.e. two spheres touching at a point. It can, however, be argued that the interior in the examples of [ESp] and [Son1] is due mainly to the fact that the initial data are not smooth which, in turn, yields that the normal direction is somehow not well defined. This, of course, raises the question of finding some necessary and sufficient conditions of

Γ_0 so that no interior is created. I will address this question below for the case of first-order and second-order motions whose geometric pde's are of the form

$$(1.17) \qquad u_t = v(x,t)|Du| \quad \text{in } \mathbb{R}^N \times (0,\infty)$$

and

$$(1.18) \qquad u_t = F(D^2 u, Du) \quad \text{in } \mathbb{R}^N \times (0,\infty).$$

Throughout this discussion I assume that

$$(1.19) \qquad \Gamma_0 = \partial\Omega_0 = \partial(\mathbb{R}^N \backslash \overline{\Omega}_0),$$

with Ω_0 an open subset of $\subset \mathbb{R}^N$, which, in particular, implies that Γ_0 has no interior. The following theorem is proved in [BSS], where I refer for the details.

Theorem 1.10: *Assume* $v \in W^{1,\infty}(\mathbb{R}^N \times (0,T))$ *for all* $T > 0$ *and that either (i)* v *does not change sign in* $\mathbb{R}^N \times (0,+\infty)$ *or, (ii)* v *is independent of* t. *Then* $\Gamma_t = \{x : u(x,t) = 0\}$ *does not develop interior, where* $u \in UC(\mathbb{R}^N \times (0,\infty))$ *is the solution of (1.9) with* F *as in (1.17).*

This theorem does not include the case where $\alpha \equiv \alpha(x, \frac{p}{|p|})$ which was studied by Soravia [Sor]. On the other hand, Theorem 1.10 is sharp in its context because there is an explicit counterexample in the case where α depends on t and changes sign. This example, which is discussed in detail in [BSS], says that there exists an open interval $I \subset \mathbb{R}$ such that the equation

$$\begin{cases} u_t = (t - x)|u_x| & \text{in } \mathbb{R} \times (0,\infty), \\ u = \mathbb{I}_I - \mathbb{I}_{\mathbb{R}\backslash I} & \text{on } \mathbb{R} \times \{0\}, \end{cases}$$

has more than one discontinuous solutions. Theorem 1.7 then yields that the evolution of I with normal velocity $t - x$ develops interior.

I next turn to the case of the motion governed by (1.18), the typical example here being motion by mean curvature. To state the next result one needs the following additional condition on F, which is satisfied in the mean curvature case, that yields that the equation is invariant under rotation and dilation:

$$(1.20) \qquad F(\mu Q^t p, \mu^2 Q^t X Q) = \mu^m F(p, X)$$

for some $m \in \mathbb{R}$ and all $\mu > 0$, $p \in \mathbb{R}^N$, $X \in \mathcal{S}^N$ and $Q \in \mathcal{O}(N)$, where Q^t is the adjoint of Q and $\mathcal{O}(N)$ is the group of $N \times N$ orthogonal matrices, i.e. $Q^t = Q^{-1}$. The following theorem is proved in [BSS].

Theorem 1.11: *Assume that (1.20) and the hypotheses of Theorem 1.1 hold. Let* $(\Gamma_t, \Omega_t^+, \Omega_t^-)$ *be the generalized evolution of* $(\Gamma_0, \Omega^+ \Omega_0^-)$ *governed by (1.18). Assume that* Γ_0 *is of class* C^2, *compact and that there exist nonnegative constants* c_i, $(i = 1, 2, 3)$, *a skew symmetric matrix* H *and* $x_0 \in \mathbb{R}^N$ *such that*

$$(1.21) \quad c_1(x - x_0) \cdot Dd(x) + c_2 H(x - x_0) \cdot Dd(x) + c_3 F(Dd(x), D^2 d(x)) \neq 0 \quad \text{on } \Gamma_0,$$

where d *is the signed distance to* Γ_0. *Then the set* $\cup_{t>0}(\Gamma_t \times \{t\})$ *has empty interior in* $\mathbb{R}^N \times (0, +\infty)$.

The left-hand side of (1.21) is the generator of rotations, dilations and translations in (x, t) evaluated at $t = 0$ on Γ_0. Condition (1.21), which was introduced in [BSS] and includes, as special cases, results of Ilmanen [Il1] and Soner [Son1] for motion by mean curvature, is not necessary. Indeed a work of Soner and Souganidis [SonS] for bodies of rotation moving by mean curvature shows that there exist smooth Γ_0's which do not satisfy (1.21), but their evolution never develops interior. It is proved however in [SonS] that (1.21) holds near the singularities of Γ_t. A related observation is that if (1.21) holds at a later time, this again yields no interior. For the case of mean curvature, [ESp] also showed that, under some assumptions on Γ_0, almost every level set of the solution to the mean curvature pde does not develop interior; I refer to [E2] for more details. More recently, Angenent, Chopp and Ilmanen [ACI] and Angenent, Ilmanen and Velazquez [AIV] constructed specific examples of smooth Γ_0 with the property that their mean curvature evolution develops interior for $t > 0$.

Since the proof of Theorem 1.11 is rather simple and intuitive, I present it below.

Proof of Theorem 1.11: 1. Let $u \in UC(\mathbb{R}^N \times (0, \infty))$ be the unique solution of (1.9) with F as in (1.18) and $u = d_0$ on $\mathbb{R}^N \times \{0\}$. For $h > 0$, define the function

$$u_h(x, t) = \Phi(u((1 + c_1 h)e^{c_2 h H}(x - x_0) + x_0, (1 + c_1 h)t + c_3 h)),$$

where Φ is a nondecreasing smooth function with $\Phi(0) = 0$ to be chosen later. Since $Q = e^{c_2 h H} \in \mathcal{O}(N)$, (1.20) and Lemma 1.3 yield that u_h is also a solution of (1.9). Moreover, if h is small enough, there exists some $\eta > 0$ such that

$$(1.22) \qquad\qquad |u(\cdot, 0) - u_h(\cdot, 0)| \geq \eta h \quad \text{on } \mathbb{R}^N.$$

3. Assuming for the moment the last estimate one obtains from Theorem 1.1 that either $u_h \leq u - \eta h$ or $u_h \geq u + \eta h$ in $\mathbb{R}^N \times (0, \infty)$. If $\cup_{t>0}(\Gamma \times \{t\})$ has interior, either of the above inequalities, however, yields a contradiction. Indeed if $u = 0$ in some neighborhood of a point (x_0, t_0), then so does u_h for h sufficiently small.

4. To prove (1.22) first observe that Φ may be chosen so that one only needs to work in a small neighborhood of Γ_0. Since, for a suitable choice of such a neighborhood, u is smooth, a simple Taylor's expansion yields

$$u((1 + c_1 h)e^{c_2 h H}(x - x_0) + x_0, c_3 h) = u(x, 0) + h(c_1(x - x_0) \cdot Du(x, 0) + c_2 H(x - x_0) \cdot Du(x, 0) + c_3 u_t(x, 0) + o(h).$$

Using (1.21) and the fact that $u = d$ on $\mathbb{R}^N \times \{0\}$ one concludes. $\qquad \square$

To conclude the discussion about the non-empty interior difficulty, I remark that one can not expect to have a general theorem guaranteeing no interior without severe restrictions on the velocity if the latter happens to be dependent on t. The intuitive reason behind this claim is that, in principle, all motions have some "pathological" situations where interior develops. One can take any such motion, perturb its velocity by a time dependent forcing term so that to drive the front to the "pathological" configuration and then simply turn off the forcing term. Examples of such situations were given by Barles, Soner and Souganidis [BSS] for time-depending forcing term and Belletini and Paolini [BP1] for constant forcing term.

As mentioned earlier, one can have a weak formulation of the propagation of a front in terms of whether the signed distance to the front satisfies the inequalities of Theorem 1.8. A natural question to ask is whether these inequalities are enough to identify the distance function uniquely, i.e. if z satisfies (1.13)–(1.16) and $z(x,0) = d(x, \Gamma_0)$, is it true that $z \equiv d$? In addition to being a natural mathematical question to ask, having such information plays an important role on some of the analysis of the asymptotic problems to be discussed later in these notes.

Below I consider, as an example, the initial value problem

$$(1.23) \quad \begin{cases} \text{(i)} \quad u_t = \theta\Big(\Delta u - \dfrac{(D^2 u Du \mid Du)}{|Du|^2}\Big) + a(x,t)|Du| \quad \text{in } \mathbb{R}^N \times (0, \infty), \\[2mm] \text{(ii)} \quad u = d \text{ on } \mathbb{R}^N \times \{0\}, \end{cases}$$

with $\theta \geq 0$ and $\alpha \in W^{1,\infty}(\mathbb{R}^N \times (0, \infty))$. Some of the arguments and the conclusions hold if $\theta = \theta(x,t)$ (under suitable assumptions) as well as for anisotropic motions. This is discussed in some detail in [BSS].

As usual $\Gamma_t = \{x : u(x,t) = 0\}$, where u solves (1.9)(i). Theorem 1.8 and the discussion following it say that the functions $d_1 = d(x, \overline{\Gamma}_t)$ and $d_2 = d(x, \underline{\Gamma}_t)$, where $\overline{\Gamma}_t = \partial\{x : u(x,t) > 0\}$ and $\underline{\Gamma}_t = \partial\{x : u(x,t) \geq 0\}$, satisfy the inequalities

$$(1.24) \qquad z_t - \theta\Delta z - \alpha(x - zDz, t) \leq 0, \quad 1 - |Dz| = 0 \quad \text{in } \{z < 0\},$$

and

$$(1.25) \qquad z_t - \theta\Delta z - \alpha(x - zDz, t) \geq 0, \quad |Dz| - 1 = 0 \quad \text{in } \{z > 0\}.$$

Of course, if the no-interior condition holds for every $t > 0$, (1.24) and (1.25) are satisfied by $z = d(x,t)$.

Next I consider the question whether (1.24) and (1.25) are enought to identify z as the signed distance function. The following theorem which is proved in [BSS] gives a positive answer to this question.

Theorem 1.12: *If the usc (resp. lsc) function z satisfies (1.23)(ii), (1.24), $|Dz| \leq 1$ and $z \wedge 0$ is continuous from below (resp. (1.23)(ii), (1.25), $|Dz| \leq 1$ and $z \vee 0$ is continuous from above), then*

$$z \leq d_2 \quad in \ \{z < 0\} \supset \{d_2 < 0\}, \ (resp. \ z \geq d_1 \ in \ \{z > 0\} \supset \{d_1 > 0\}).$$

If z satisfies (1.23)(ii), (1.24), (1.25), $|Dz| \leq 1$ and $z \wedge 0$ and $z \vee 0$ are continuous from below and above respectively, and Γ_t does not develop interior for all $t > 0$, then $z = d$ in $\mathbb{R}^N \times [0, \infty)$.

Next I present another formulation, which is developed in Barles and Souganidis [BS2], for the generalized front propagation which, as it turns out, is equivalent to the level set one under the no-interior assumption. This formulation, which is undoubtedly rather cumbersome to state and is related to the notion of "barriers" introduced recently by DeGiorgi [D], is rather intuitive geometrically, since it uses smooth (test) surfaces to test the motion in a way similar to the one smooth (test) functions are used to test a solution of (1.9)(i). It applies to evolution of sets which are local an satisfy an avoidance inclusion type geometrical maximum principle. The goal of [BS2] is not to develop yet another definition for the generalized evolution (there are already too many), but rather to develop a general program in order to study in a systematic way the appearance of moving interfaces in asymptotic problem. This is the topic of the next section where I refer to for a more involved discussion of this program and method.

Definition 1.13: *A family $(\Omega_t)_{t \in [0,T]}$ of open subsets in $\mathbb{R}^N$ propagates with normal velocity $V = v$ iff, for any $(x,t) \in \mathbb{R}^N \times (0,T)$ and any smooth function $\varphi : \mathbb{R}^N \to \mathbb{R}$ such that, for some $r > 0$,*

$$\{y : \phi(y) \geq 0\} \subset \Omega_t \cap B_r(x) \quad (resp. \quad \{y : \phi(y) \leq 0\} \subset \mathbb{R}^N \backslash \Omega_t \cap B_r(x),$$

if $|D\phi| \neq 0$ on $\{\phi = 0\}$, then there exists h_0 depending only on r and the C^3-norm ϕ on $B_r(x)$ such that, for any $0 < h < h_0$ and for α sufficiently small

$$(1.26) \qquad \{y : \phi(y) + h[F_*(D^2\phi(y), D\phi(y), y, t) + \alpha] > 0\} \cap \overline{B_r(x)} \subset \Omega_{t+h},$$

(resp.

$$(1.27) \quad \{y : \phi(y) + h[F^*(D^2\phi(y), D\phi(y), y, t) + \alpha] < 0\} \cap \overline{B_r(x_0)} \subset \mathbb{R}^N \backslash \overline{\Omega_{t+h}},$$

where F and v are related by (1.6).

The first key point behind this definition, which of course, can be split into three parts so that one defines propagation for $V \geq v$, $V \leq v$ and $V = v$, is that this type of property characterizes completely the fact that the family $(\Omega_t)_{t \in [0,T]}$ moves with normal velocity $V = v$, provided that one tests it with a sufficiently large collection of families $(\tilde{\Omega}_t)_t$ and for any time interval $[t, t+h]$, where h is sufficiently small. The second key point is that it suffices to consider families of the form $\tilde{\Omega}_t = \{x : \phi(x) > 0\}$ and $\tilde{\Omega}_t = \{x : \phi(x) < 0\}$ flowing smoothly with velocity $V = v$ on the time interval $[t, t+h]$, since after all what is written in

(1.26) and (1.27) is nothing else but the Euler approximation of the level set pde corresponding to F.

I refer to Ilmanen [Il1] where properties of this type were remarked for motion by mean curvature – "avoidance and inclusion properties" – without, however, been put forward as a possible definition of such motions. As mentioned earlier Definition 1.13 is also related to the notion of "barriers" introduced recently by DeGiorgi [D], who uses such geometric maximum principle-type ideas to define the propagation of manifolds in $\mathbb{R}^N$; see also a recent paper by Bellettini and Paolini [BP2].

Definition 1.13 can also be restated and extended in many other ways depending on the context that is needed. For example if motions in bounded sets are considered, then it is necessary to use the evolution of smooth sets instead of the Euler approximation of the geometric pde. I refer to [BS2] for an at-length discussion of these issues.

The fact that the level-set evolution $(\Gamma_t, \Omega_t, \Omega_t^-)$ of $(\Gamma_0, \Omega_0, \mathbb{R}^N \backslash \overline{\Omega}_0)$ with normal velocity v satisfies the conditions of Definition 1.13, is immediate from the comparison properties of the level-set geometric pde. The next theorem establishes the other direction and, hence, in view of Theorem 1.7 the equivalence of this new formulation with the level-set approach under the "empty interior assumption".

Theorem 1.14: *A family $(\Omega_t)_{t \in [0,T]}$ of open subsets of $\mathbb{R}^N$ propagates with normal velocity $V = v$ in the sense of Definition 1.13 iff $\chi = \mathbb{I}_{\Omega_t} - \mathbb{I}_{\mathbb{R}^N \backslash \overline{\Omega}_t}$ is a viscosity solution of (1.9)(i), where v and F are related by (1.6).*

Proof: 1. It is enough to show that χ is a supersolution of (1.9)(i), since the proof of the subsolution property follows along the same lines.

2. Let $(x,t) \in \mathbb{R}^N \times (0,T)$ be a strict local minimum point of $\chi_* - \varphi$, where $\varphi \in C^3(\mathbb{R}^N \times [0,T])$. Changing if necessary φ to $\varphi - \varphi(x,t)$ one may assume without any loss of generality that $\varphi(x,t) = 0$. One needs to show that the following inequality holds

$$\frac{\partial \varphi}{\partial t}(x,t) \geq F_*(D^2 \phi(x,t), D\varphi(x,t), x, t).$$

3. This inequality is, of course, obvious if (x,t) is either in the interior of the set $\{\chi_* = 1\}$ or in the interior of the set $\{\chi_* = -1\}$. Indeed, in either case χ_* is constant in a neighborhood of (x,t). Hence,

$$\frac{\partial \varphi}{\partial t}(x,t) = 0, \quad D\varphi(x,t) = 0, \quad D^2\varphi(x,t) \geq 0.$$

The desired inequality is then just a straightforward consequence of the ellipticity condition and the geometric assumption on F which imply in particular that $F_*(0,0,x,t) \leq 0$.

4. If (x,t) is on the boundary of the sets $\{\chi_* = 1\}$ and $\{\chi_* = -1\}$, then in view of the lower semicontinuity of χ_*,

$$\chi_*(x,t) = -1.$$

Since (x, t) is also a strict local minimum point of $\chi_* - \varphi$, there exists some $r > 0$ such that

$$\chi_*(x, t) - \varphi(x, t) = -1 < \chi_*(y, s) - \varphi(y, s) \quad \text{if } 0 < |y - x| + |t - s| < 2r.$$

Therefore

$$-1 + \varphi(y, s) < \chi_*(y, s) \quad \text{if } \quad 0 < |y - x| + |t - s| < 2r.$$

It follows that $\chi_*(y, s) = 1$ if $\varphi(y, s) \geq 0$ and if $(y, s) \neq (x, t)$, since χ_* takes only the values -1 and 1. This implies, for any $0 < h < r$.

$$\{y : \varphi(y, t - h) \geq 0\} \cap B_r(x) \subset \Omega_{t-h}.$$

5. If $|D\varphi(x, t)| \neq 0$, for $k > 0$ sufficiently large, introduce the functions $\varphi_k : \mathbb{R}^N \times [0, T]$ defined by

$$\varphi_k(y, s) = \varphi(y, s) - k|y - x|^4,$$

and observed that all of the above arguments are still valid for φ_k. Moreover, it follows that there exist $k > 0$ and $\bar{h} > 0$ such that, for $0 < h \leq \bar{h}$,

$$\begin{cases} \text{(i)} & \{y : \varphi_k(y, t - h) \geq 0\} \subset \Omega_{t-h} \cap B_r(x), \\ \text{(ii)} & |D\varphi_k(y, t - h)| \neq 0 \quad \text{on} \quad \{y : \varphi_k(y, t - h) = 0\}. \end{cases}$$

This technical last assertion is proved in [BS2] where I refer to the details.

6. Since the family $(\Omega_t)_{t \in [0, T]}$ propagates with a normal velocity $V = v$ and the C^3-norms of the functions $\varphi_k(\cdot, t - h)$ are uniformly bounded in $B_r(x)$ for h small, for h small enough and $\alpha > 0$, we have

$$\{y : \varphi_k(y, t-h) + h[F_*(y, t-h, D\varphi_k(y, t-h), D^2\varphi_k(y, t-h)) + \alpha] > 0\} \cap \overline{B}_r(x) \subset \Omega_{t+h}.$$

In particular, since $(x, t) \notin \Omega_t$ and $\phi_k(x, t) = \phi(x, t) = 0$,

$$\varphi_k(x, t - h) - \phi_k(x, t) + h[F_*(x, t - h, D\varphi_k(x, t - h), D^2\varphi_k(x, t - h)) + \alpha] \leq 0.$$

Dividing this last inequality by h, letting $h \to 0$ and using the lower-semicontinuity of F_* yields

$$-\frac{\partial \varphi_k}{\partial t}(x, t) + [F_*(D^2\varphi_k(x, t), D\varphi_k(x, t), x, t) + \alpha] \leq 0.$$

But

$$\frac{\partial \varphi_k}{\partial t}(x, t) = \frac{\partial \varphi}{\partial t}(x, t), \quad D\varphi_k(x, t) = D\varphi(x, t), \quad D^2\varphi_k(x, t) = D^2\varphi(x, t).$$

Letting $\alpha \to 0$ yields the desired inequality.

6. If $|D\varphi(x,t)| = 0$, one may assume without any loss of generality that $D^2\varphi(x,t) = 0$ – see for example Barles and Georgelin [BG]. Therefore, since $F_*(x,t,0,0) \leq 0$, it is enough to show

$$\frac{\partial\varphi}{\partial t}(x,t) \geq 0.$$

Using again that (x,t) is a local minimum point of $\chi_* - \varphi$ and the fact that $\varphi(x,t) = D\varphi(x,t) = D^2\varphi(x,t) = 0$, for any $0 < h < r$, one gets

$$-1 + \varphi(x,t) - h\varphi_t(x,t) + O(|y-x|^3) + o(h) \leq \chi_*(y,t-h) \quad \text{if } |y-x| < r.$$

7. If there exists a sequence $(y_\epsilon, t - h_\epsilon) \to (x,t)$ such that

$$\chi_*(y_\epsilon, t - h_\epsilon) = -1,$$

and

$$|x - y_\epsilon|^3 = o(h_\epsilon),$$

then the conclusion follows. Otherwise, for any $C > 0$, there must exist $h_0 > 0$ such that, if $0 < h < h_0$,

$$\chi_*(y, t - h) = 1 \quad \text{if } |y - x|^2 \leq 2Ch.$$

8. Next consider the function $\phi(y) = Ch - |y - x|^2$. It is clear enough that, for $h < h_0$,

$$\{y : \phi(y) \geq 0\} \subset \Omega_{t-h} \cap B_r(x) \quad \text{and} \quad |D\phi(y)| \neq 0 \quad \text{on} \quad \{y : \phi(y) = 0\}.$$

Since the family $(\Omega_t)_{t\in[0,T]}$ propagates with a normal velocity $V \geq v$, one has, for α small enough,

$$\{y : \phi(y) + h[F_*(D^2\phi(y), D\phi(y), y, t) + \alpha] > 0\} \cap \overline{B_r(x)} \subset \Omega_t.$$

But, since F is locally bounded, it is immediate that

$$\phi(x) + h[F_*(x,t,D\phi(x),D^2\phi(x)) + \alpha] > 0,$$

If C is large enough which is, of course, a contradiction. $\square$

2. An Abstract Method

As mentioned in the Introduction one of the main goals of the theory of the generalized front propagation is the rigorous justification of the appearance of moving fronts in the asymptotic limit of several models like reaction-diffusion equations, particle systems, etc., and the identification of the dynamics of the resulting interfaces, their regularity, etc..

In most of the applications one is given a family $(u_\epsilon)_{\epsilon>0}$ of uniformly bounded functions in, for simplicity, $\mathbb{R}^N \times (0,T)$. Typically the u_ϵ's are the solutions of a reaction-diffusion equation with a small parameter ϵ or the averaged magnetization of a system with interactions of range ϵ^{-1}. The desired result is that there exists an evolution $(\Gamma_t, \Omega_t^+, \Omega_t^-)$ with normal velocity v, identifying which being one of the problems, such that as $\epsilon \to 0$,

$$(2.1) \qquad u_\epsilon \to \begin{cases} a & \text{in } \bigcup_{t \in (0,T)} \Omega_t^+ \times \{t\}, \\[2ex] b & \text{in } \bigcup_{t \in (0,T)} \Omega_t^- \times \{t\}, \end{cases}$$

for some $a, b \in \mathbb{R}$ with $a < b$, where the family $(\Omega_t)_{t \in [0,T]}$ propagates with some normal velocity v, identifying v being one of the problems.

In this section I present a new general method, which is developed in Barles and Souganidis [BS2] to study such questions. At a first glance the statements will appear cumbersome and somehow artificial especially since no concrete examples of such asymptotic problems have been discussed in these notes yet. On the other hand, as it will become, hopefully, clear to the reader after going through this as well as the next three sections of these notes, this method is very natural.

The intuitive idea behind this general methodology is to reduce the whole problem to the rigorous justification of (2.1) and the appearance of interfaces under no regularity restrictions, i.e. under the assumption that the resulting geometric motion is smooth. Then using the theory presented in the previous section and especially Definition 1.13 and Theorem 1.14 as well as the general properties of the family $(u_\epsilon)_{\epsilon > 0}$ one may obtain (2.1). Loosely speaking the u_ϵ's can be thought of as defining a "flow" on open sets, which is supposed to approximate the corresponding geometric motion. Checking the convergence (2.1) when everything is smooth amounts to checking the "generator property" for the approximate flow. Presented in this, admittedly informal way, this method resembles the formulation developed by Barles and Souganidis [BS1] to study the convergence of numerical approximations to fully nonlinear second order pde's. The terminology adopted in describing below the main steps of this methodology is inspired from the abstract approach in Image Analysis developed in [AGLM] and [Lio]. Finally, as before, everything is formulated for the case where the problem takes place in $\mathbb{R}^N \times (0, T)$. The method, of course, applies to bounded settings too, after the necessary modifications. Concrete examples are discussed in the following sections.

The family $(u_\epsilon)_{\epsilon > 0}$ of functions $u_\epsilon : \mathbb{R}^N \times [0, T] \to \mathbb{R}$ is assumed to satisfy the following properties, for all $\epsilon > 0$, $h > 0$ and $t \in [0, T]$:

(H1) *Causality*: There exists a family of maps $S_{t,t+h}^\epsilon : L^\infty(\mathbb{R}^N) \to L^\infty(\mathbb{R}^N)$ such that
$$u_\epsilon(\cdot, t + h) = S_{t,t+h}^\epsilon u_\epsilon(\cdot, t) \quad \text{in } \mathbb{R}^N.$$

(H2) *Monotonicity*: For any $u, v \in L^\infty(\mathbb{R}^N)$,
$$\text{if } u \le v \text{ a.e. in } \mathbb{R}^N \text{ then } S_{t,t+h}^\epsilon u \le S_{t,t+h}^\epsilon v \text{ in } \mathbb{R}^N.$$

(H3) *Existence of equilibria:* There exists $a_\epsilon, b_\epsilon \in \mathbb{R}$ such $a_\epsilon < b_\epsilon$ and $S^\epsilon_{t,t+h} a_\epsilon \equiv a_\epsilon$ and $S^\epsilon_{t,t+h} b_\epsilon \equiv b_\epsilon$ in $\mathbb{R}^N$. Moreover,

$$a_\epsilon \leq u_\epsilon \leq b_\epsilon \quad \text{on } \mathbb{R}^N \times \{0\}$$

and there exists $a, b \in \mathbb{R}$ such that $a < b$ and, as $\epsilon \to 0$,

$$a_\epsilon \to a \quad \text{and} \quad b_\epsilon \to b.$$

(H4) *Consistency:* (i) For any $(x,t) \in \mathbb{R}^N \times (0,T)$ and for any smooth function $\phi : \mathbb{R}^N \to \mathbb{R}$ such that $\{y : \phi(y) \geq 0\} \subset B_r(x)$ for some $r > 0$, and $|D\phi| \neq 0$ on $\{y : \phi(y) = 0\}$, there exists $\delta > 0$ and $h_0 > 0$, h_0 depending only on r and the C^3-norm of ϕ in $B_r(x)$, such that

$$\lim \inf_*(S^\epsilon_{t,t+h}[(b_\epsilon - \delta)\mathbb{I}_{\{\phi \geq 0\}} + a_\epsilon \mathbb{I}_{\{\phi < 0\}}])(y) = b,$$

if $y \in B_r(x)$ and $\phi(y) + h[F_*(D^2\phi(y), D\phi(y), y, t) + \alpha] > 0$ for some $\alpha > 0$ and for $h \leq h_0$, and

(ii) For any $(x,t) \in \mathbb{R}^N \times (0,T)$ and for any smooth function $\phi : \mathbb{R}^N \to \mathbb{R}$ such that $\{y : \phi(y) \leq 0\} \subset B_r(x_0)$ for some $r > 0$, and $|D\phi(x)| \neq 0$ on $\{y : \phi(y) = 0\}$, there exists $\delta' > 0$ and $h'_0 > 0$, h'_0 depending only on the C^3-norm of ϕ in $B_r(x)$, such that

$$\lim \sup{}^*(S^\epsilon_{t,t+h}[b_\epsilon \mathbb{I}_{\{\phi > 0\}} + (a_\epsilon + \delta')\mathbb{I}_{\{\phi \leq 0\}}])(y) = a,$$

if $y \in B_r(x)$ and $\phi(y) + h[F^*(D^2\phi(x), D\phi(x), x, t) - \alpha] < 0$ for some $\alpha > 0$ and for $h \leq h'_0$.
Here, as usual,

$$\lim_* v_\epsilon(x,t) = \lim_{\substack{(y,s) \to (x,t) \\ \epsilon \to 0}} \inf v_\epsilon(y,s) \quad \text{and} \quad \lim{}^* v_\epsilon(x,t) = \lim_{\substack{(y,s) \to (x,t) \\ t \to 0}} \sup v_\epsilon(y,s).$$

The result proved in [BS2], stated under, for simplicity, the no interior condition, is then the following:

Theorem 2.1: *Assume that for all $\epsilon > 0$, $t \geq 0$ and $h > 0$ (H1)–(H4) hold. Set, for $t > 0$,*

$$\Omega^1_t = \{x : \lim \inf_* u_\epsilon(x,t) = b\} \quad \text{and} \quad \Omega^2_t = \{x : \lim \sup{}^* u_\epsilon(x,t) = a\},$$

and define Ω^1_0 and Ω^2_0 by

$$\Omega^1_0 = \bigcap_{h > 0} \left(\bigcup_{0 < h \leq t} \Omega^1_t \right) \quad \text{and} \quad \Omega^2_0 = \bigcap_{h > 0} \left(\bigcup_{0 < h \leq t} \Omega^2_t \right).$$

If $\Omega^1_0 = \mathbb{R}^N \backslash \overline{\Omega}^2_0$ and the evolution $(\Gamma_t, \Omega^+_t, \Omega^-_t)$ of $(\partial \Omega^1_0, \Omega^1_0, \Omega^2_0)$ with normal velocity F does not develop interior, then $\Omega^1_t = \Omega^+_t$ and $\Omega^2_t = \Omega^-_t$ $u_\epsilon \to b$ in $\mathcal{O} = \bigcup_{t > 0} \Omega^1_t \times \{t\}$ and $u_\epsilon \to a$ in $\mathcal{O}' = \bigcup_{t > 0} \Omega^2_t \times \{t\}$. □

In most of the examples, checking assumption (H4) presents the only real difficulty. Indeed (H1) is in general obviously satisfies by the definition of the u_ϵ's, while (H2) comes generally for a maximum principle-type property and (H3) from the structure of the problem.

Checking (H4) consists in proving a similar result to the one desired for u_ϵ but only for smooth data (as smooth as necessary), for compact smooth fronts and for small time (again as small as necessary). It is clear enough that these properties are a priori far easier to obtain – although they are not completely trivial – than the general case for u_ϵ and this reduction to an easier case is one of the improvements brought by this new method.

The additional assumption saying that Ω_0^1 is not empty is used to initialize the moving front and is checked, in general, exactly in the same way as (H4), i.e. by studying the small-time behavior of solutions which, locally in time, generate a smooth front. Giving different definitions for Ω_0^1 and Ω_0^2 is necessary in order to take into account the boundary layer which may occur at $t = 0$ and which needs to be studied separately.

Proof of Theorem 2.1: 1. Let $(x,t) \in \mathbb{R}^N \times (0,T)$ and a smooth function $\phi : \mathbb{R}^N \to \mathbb{R}$ be such that, for some $r > 0$, $\{y : \phi(y) \geq 0\} \subset \Omega_t^1 \cap B_r(x)$ and $|D\phi| \neq 0$ on $\{y : \phi(y) = 0\}$. It is necessary to show that there exists $h_0 > 0$ depending only on r and then C^3-norm of ϕ in $B_r(x)$ such that for any $0 < h < h_0$ and for α small enough

$$\{y : \phi(y) + h[F_*(D^2\phi(x), D\phi(x), x, t) + \alpha] > 0\} \cap \overline{B_r(x)} \subset \Omega_{t+h}.$$

3. In view of (H1), (H2) and (H3), one has for all $t \in [0,T]$,

$$a_\epsilon = S_{0,t}^\epsilon a_\epsilon \leq S_{0,t}^\epsilon u_\epsilon(\cdot, 0) = u_\epsilon(\cdot, t) \leq b_\epsilon = S_{0,t}^\epsilon b_\epsilon \quad \text{in } \mathbb{R}^N,$$

and, hence

$$\limsup{}^* u_\epsilon \leq b \quad \text{in } \mathbb{R}^N \times [0,T].$$

Classical arguments in the theory of viscosity solution yield that, as $\epsilon \to 0$, $u_\epsilon \to b$ locally uniformly in $\cup_{t>0} \Omega_t^1 \times \{t\}$. It then follows that for ϵ small enough

$$u_\epsilon(x,t) \geq b_\epsilon - \delta \quad \text{on } \{y : \phi(y) \geq 0\} \subset\subset \Omega_t^1,$$

and, by (H2),

$$S_{t,t+h}^\epsilon u_\epsilon(\cdot, t) \geq S_{t,t+h}^\epsilon [(b_\epsilon - \delta)\mathbb{I}_{\{\phi \geq 0\}} + a_\epsilon \mathbb{I}_{\{\phi < 0\}}](\cdot) \quad \text{in } \mathbb{R}^N.$$

Choose $0 < h \leq h_0$ and apply (H4) to get

$$\liminf{}_* u_\epsilon(y, t+h) = \liminf{}_* S_{t,t+h}^\epsilon u_\epsilon(\cdot, t)(y)$$
$$\geq \liminf{}_* (S_{t,t+h}^\epsilon [(b_\epsilon - \delta)\mathbb{I}_{\{\phi \geq 0\}} + a_\epsilon \mathbb{I}_{\{\phi < 0\}}])(y) = b,$$

if $\phi(y) + h[F_*(D^2\phi(y), D\phi(y), y, t) + \alpha] > 0$ for some $\alpha > 0$.
This implies that

$$\{y : \phi(y) + h[F^*(D^2\phi(y), D\phi(y), y, t) + \alpha] > 0\} \subset \Omega_{t+h}^1,$$

for any $0 < h < h_0$, where h_0 depends only on the C^3-norm of ϕ in $B_r(x)$. The proof is now complete.

2. The argument if $\{y : \phi(y) \leq 0\} \subset\subset \mathbb{R}^N \backslash \Omega_t \cap B_r(x)$ is similar. $\square$

3. Asymptotics of reaction-diffusion equations of bistable type - Phase Theory.

Reaction-diffusion equations of the form

$$(3.1) \qquad u_t - \Delta u + W'(u) = 0,$$

set in, for example, in $\mathbb{R}^N \times (0, \infty)$, where W is a double-well potential as well as their generalizations with a general second-order elliptic operator instead of Δ and W depending also on (x, t), arise in many areas of applications like phase transitions, flame propagation, pattern formation, chemical kinetics, etc.. For example (3.1) was put forward by Allen and Cahn [AC] to describe the evolution of an order parameter identifying the phases of a polycrystalline material.

It is expected that as $t \to \infty$ the solution u will converge to the different equilibrium solutions of (3.1), which turn out to be the local extrema of W. An interface therefore develops as $t \to \infty$, as the boundary of the regions where the solution u of (3.1) converges to the different equilibria. A great deal of work has been done over the last years towards understanding these interfaces. Some of the most striking results in this direction will be discussed in this section.

Formal arguments, some of which are explained below, yield that the front, which develops for $t \gg 1$, moves with normal velocity V which can be expanded in terms of $1/t$ like

$$(3.2) \qquad V = \alpha + \kappa t^{-1} + O(t^{-2}) \qquad (t \gg 1),$$

where here κ denotes the mean curvature.

A convenient and by now classical way to study the fronts and understand (3.2) is to scale (3.1) in space and time, the goal being to reproduce, in finite time and in bounded space regions, the long-time behavior of (3.1). To find the first term in the expansion in (3.2), the correct scaling is, of course, the hyperbolic scaling $(x, t) \mapsto (\epsilon^{-1}x, \epsilon^{-1}t)$. Defining

$$u^\epsilon(x, t) = u(\epsilon^{-1}x, \epsilon^{-1}t),$$

where u solves (3.1), one finds that u^ϵ solves

$$(3.3) \qquad u_t^\epsilon - \epsilon \Delta u^\epsilon + \epsilon^{-1} W'(u^\epsilon) = 0.$$

The problem then becomes the study of the asymptotics of (3.3), as $\epsilon \to 0$, with given initial conditions.

When $\alpha = 0$ in (3.2), the appropriate scaling is the parabolic one $(x, t) \mapsto (\epsilon^{-1}, \epsilon^{-2}t)$. Writing

$$u^\epsilon(x, t) = u(\epsilon^{-1}, \epsilon^{-2}t),$$

one is lead to the study of the asymptotics of

$$(3.4) \qquad u_t^\epsilon - \Delta u^\epsilon + \epsilon^{-2} W'(u^\epsilon) = 0$$

again with given initial datum.

This section is divided into yeo parts as follows: In Part 1 I present rigorous results about (3.3) and (3.4) with some proofs and also discuss some extensions. Part 2 is about new results regarding more complications versions of (3.3) and (3.4) with anisotropies, bounded domains, etc., which are based on the theory presented in Section 2 of these notes.

3.1. Rigorous results.

I begin by recording and recalling some of the main properties of the potential W. They are:

(3.5)
$$\begin{cases} W : \mathbb{R} \to \mathbb{R} \text{ is a double-well potential with minima} \\ \text{at } \pm 1 \text{ and a local maximum at } \mu \in (-1, 1). \end{cases}$$

It follows, see for example Aronson and Weinberger [AW], Fife and McLeod [FM], etc., that

(3.16)
$$\begin{cases} \text{there exist unique } \alpha \in \mathbb{R} \text{ and } q : \mathbb{R} \to \mathbb{R} \text{ such that} \\ \alpha \dot{q} + \ddot{q} = W'(q) \text{ on } \mathbb{R}, \ q(\pm\infty) = \pm 1, \ q(\mu) = 0 \text{ and } \dot{q} > 0 \text{ in } \mathbb{R}. \\ \text{Moreover,} \\ \alpha = \left(\int_{-\infty}^{\infty} \dot{q}^2(\xi) d\xi \right)^{-1} (W(1) - W(-1)). \end{cases}$$

The results regarding the asymptotic behavior of (3.3) and (3.4) are:

Theorem 3.1: *Assume (3.5) and (3.6) and let u^ϵ be the solution of (3.3) with initial datum u_0 on $\mathbb{R}^N \times \{0\}$. Define $\Omega_0 = \{x : u_0(x) > 0\}$, $\Gamma_0 = \{x : u_0(x) = 0\}$ and assume that $\Gamma_0 = \partial\Omega_0 = \partial\mathbb{R}^N\backslash\overline{\Omega}_0$. If $(\Gamma_t, \Omega_t^+, \Omega_t^-)$ is the level set evolution of $(\Gamma_0, \Omega_0, \mathbb{R}^N\backslash\overline{\Omega}_0)$ with normal velocity $-\alpha$, then, as $\epsilon \to 0$,*

$$u^\epsilon \to \begin{cases} 1 & \mathcal{O} = \bigcup_{t>0}(\Omega_t^+ \times \{t\}), \\ & \text{locally uniformly in} \\ -1 & \mathcal{O}' = \bigcup_{t>0}(\Omega_t^- \times \{t\}). \end{cases}$$

$\square$

Theorem 3.2: *Assume (3.5), (3.6) and that $W(-1) = W(+1) = 0$. If u^ϵ is the solution of (3.4) with initial datum u_0 on $\mathbb{R}^N \times \{0\}$, define $\Omega_0 = \{x : u_0(x) > 0\}$, $\Gamma_0 = \{x : u_0(x) = 0\}$, and assume that $\Gamma_0 = \partial\Omega_0 = \partial(\mathbb{R}^N\backslash\overline{\Omega}_0)$. If $(\Gamma_t, \Omega_t^+, \Omega_t^-)$ is the level set evolution by mean curvature of $(\Gamma_0, \Omega_0, \mathbb{R}^N\backslash\overline{\Omega}_0)$, then, as $\epsilon \to 0$,*

$$u^\epsilon \to \begin{cases} +1 & \mathcal{O} = \bigcup_{t>0}(\Omega_t^+ \cup \{t\}), \\ & \text{locally uniformly in} \\ -1 & \mathcal{O}' = \bigcup_{t>0}(\Omega_t^- \times \{t\}). \end{cases}$$

$\square$

Some discussion about the history of Theorems 3.1 and 3.2 is now in order. Theorem 3.1 was proved, without this particular characterization of the sets $\mathcal{O}$ on $\mathcal{O}'$, by Gärtner [Ga] using a combination of probabilistic and pde techniques and by Barles, Bronsard and Souganidis [BBS] based on arguments related to viscosity solutions. In the form stated here and in a more general setting, it was proved in [BSS].

Theorem 3.2 has a considerably longer history. As mentioned earlier, this behavior of (3.4) was conjectured by Allen and Cahn [AC] and was justified formally in [Fi], [Ca1,2] and [RSK]. Rigorous results under the asumption that the mean curvature motion is smooth, i.e. short-time type results, were obtained by De-Mottoni and Schatzman [DM], Chen [Ch] and later, for different but related W's and with sharp rates of convergence, by Nochetto, Paolini and Verdi [NPV]. Also Bronsard and Kohn [BrK] proved a global in time result but for radial solutions. It should also be noted that DeGiorgi [D] formulated an alternative approach to mean curvature evolution using the asymptotics of (3.4) as a definition. The first global in time result for (3.4), under no special assumptions, was obtained by Evans, Soner and Souganidis [ESS]. Nontrivial extensions and generalizations of Theorem 3.2 for more general pde's and motions, some of which will be presented at the end of this part of this section, were obtained in [BSS]. The particular case of the asymptotics of the Allen-Cahn equation in Theorem 3.2 as stated here has been the object of even more study. As mentioned earlier in this note there is another way to study mean curvature evolution, which is, however, very restricted to this kind of motion, due to Brakke [Br] and based on varifolds -- see [E2] for details. Based on this approach to mean curvature, Ilmanen [Il2] gave another proof, after [ESS], for Theorem 3.2 with information about what happens when interior develops, something not addressed by the viscosity theory. Later Soner [Son2] reconciled some of the arguments of [Il2] with the one's from viscosity solutions.

Here I will present three different proofs for Theorem 3.2 and I will refer to [BSS] for the proof of Theorem 3.1. The reason for giving these different proofs is that each one of them contains different ingredients, which may be useful in different situations, and also indicate the scope and strength of the theory of viscosity solutions.

For the first two proofs of Theorem 3.2, I will make the additional assumption that

$$(3.7) \qquad u^\epsilon = q(\epsilon^{-1} d_0) \qquad \text{on } \mathbb{R}^N \times \{0\},$$

where, as usual, d_0 is the signed distance function to Γ_0 and q is the traveling wave, in this case, standing wave solution associated with W' by (3.6). The aforementioned extra condition on u^ϵ at $t = 0$ is essential for the proof I am presenting below. It can, however, be removed at the end by a short time analysis, carried out for (3.4) by [Ch], which implies that in time $t = O(\epsilon^2 |\ell n\ \epsilon|)$ the solution of u^ϵ of (3.4) can be put between two functions which look like $q^\epsilon(\epsilon^{-1} d_0)$ for some approximate q^ϵ's.

First proof of Theorem 3.1: 1. Let $z^\epsilon : \mathbb{R}^N \times [0, \infty) \to \mathbb{R}$ be defined by

$$u^\epsilon = q(\epsilon^{-1} z^\epsilon).$$

A straightforward computation yields that z^ϵ solves the following initial value problem:

(3.8)
$$\begin{cases} z_t^\epsilon - \Delta z^\epsilon + \frac{1}{\epsilon} Q(\epsilon^{-1} z^\epsilon)(|Dz^\epsilon|^2 - 1) = 0 & \text{in } \mathbb{R}^N \times (0, \infty), \\ z^\epsilon = d_0 & \text{on } \mathbb{R}^N \times \{0\}, \end{cases}$$

where

$$Q(\xi) = -(\dot{q}(\xi))^{-1} \ddot{q}(\xi).$$

2. Since, in view of (3.6),

(3.9)
$$\operatorname{sgn} Q(\xi) = \operatorname{sgn}(-\ddot{q}(\xi)) = \operatorname{sgn} q(\xi),$$

it is immediate that

(3.10)
$$z_t^\epsilon - \Delta z^\epsilon \begin{cases} \geq 0 & \text{in } \{z^\epsilon > 0\}, \\ \leq 0 & \text{in } \{z^\epsilon < 0\}. \end{cases}$$

Since $|Dz^\epsilon| \leq 1$ on $\mathbb{R}^N \times \{0\}$, it also follows that

(3.22)
$$|Dz^\epsilon| \leq 1 \quad \text{in } \mathbb{R}^N \times (0, \infty).$$

This last estimate can be either obtained by standard maximum principle type estimates or directly by arguing about what happens at the maximum in $\mathbb{R}^N \times \mathbb{R}^N \times (0, \infty)$ of the function

$$(x, t) \mapsto z^\epsilon(x, t) - z^\epsilon(y, t) - |x - y|.$$

3. Let t_* be the extinction time of the mean curvature evolution of $\{d_0 > 0\}$ and assume that

$$\sup_{\epsilon > 0} \|z^\epsilon\|_{L^\infty(K)} < \infty$$

for any compact subset K of $\mathbb{R}^N \times [0, T]$. This assumption can be verified either by making one more change of the unknown and creating a variational inequality which yields the result in combination with the rest of the proof (for such an argument I refer to [BSS]) or by some kind of boot-strap argument in combination again with the rest of the proof (for the details I refer to [KS2]).

4. Consider the upper- and lower-semicontinuous functions z^* and $z_* : \mathbb{R}^N \times [0, t_*) \to \mathbb{R}$, which are given by

$$z^*(x, t) = \limsup_{\substack{\epsilon \to 0 \\ (z, s) \to (x, t)}} z^\epsilon(y, s) \quad \text{and} \quad z_*(x, t) = \liminf_{\substack{\epsilon \to 0 \\ (y, s) \to (x, t)}} z^\epsilon(y, s).$$

It follows that

(3.12)
$$|Dz^*| \leq 1 \quad \text{and} \quad |Dz_*| \leq 1 \quad \text{in } \mathbb{R}^N \times [0, t_*).$$

5. Using once more (3.8), (3.9), (3.12) and standard arguments from the theory of viscosity solutions (see, for example, [Cr], [BSS], [Ba2], [BJ], etc.), one gets that z^* is a viscosity solution of

$$1 - |Dz| = 0 \quad \text{in } \{z < 0\},$$

and z_* is a viscosity solution of

$$|Dz| - 1 \quad \text{in } \{z > 0\}.$$

Finally, letting $\epsilon \to 0$ in (3.10) also yields that

$$z_t^* - \Delta z^* \leq 0 \quad \text{in } \{z^* < 0\}$$

and

$$z_{*t} - \Delta z_* \geq 0 \quad \text{in } \{z_* > 0\}.$$

The uniqueness results of Theorem 1.10 now yield that

$$z^* \leq d \text{ in } \{d < 0\} \text{ and } z_* \geq d \text{ in } \{d > 0\},$$

where d is the signed distance to Γ_t. In particular, if the evolution does not develop interior, it follows that

$$z^* = z_* = d.$$

6. The extension of the result in $\mathbb{R}^N \times [t_*, \infty)$ is now more or less immediate – I refer to [ESS] and [BSS] for the precise argument.

$\square$

The above proof is rather simple. It can also be extended to more general situations where some of the estimates mentioned, for example the gradient bound etc., are not so obvious. One then has to use one more step involving changing the z^ϵ's and hence z^* and z_* in such a way to produce an appropriate gradient bound. This argument was introduced in [BSS] and was employed rather successfully in [J] and [KS2].

The second proof of Theorem 3.2 follows along the lines of the original argument of [ESS] and its modification in [BSS] to cover more general problems, the basic argument being to use the signed distance to the evolving front to construct super- and sub-solutions to (3.4).

In preparation for the proof it is necessary to discuss the issue of existence and the properties of traveling waves for nonlinearities of the form $u \mapsto W'(u) - \epsilon a$, where W is a double well potential and ϵ and a are sufficiently small so that the equation

$$W'(u) - \epsilon a = 0$$

has exactly three zeroes $h_-^\epsilon < h_0^\epsilon < h_+^\epsilon$, which, of course, in the limit $\epsilon \to 0$ converge to the zeros of $W'(u)$, here $-1, 0$ and 1. Under some mild technical assumptions on W (see, for example, [AW], [FM], [BSS]) it turns out that for ϵ

and a sufficiently small there exists a unique pair $(c^\epsilon(a,\epsilon), q^{a,\epsilon})$ where $c^\epsilon(a,\epsilon)$ is a constant and $q^{a,\epsilon} : \mathbb{R} \to \mathbb{R}$ is such that

$$(3.13) \quad \begin{cases} c(a,\epsilon)\dot{q}^{a,\epsilon} + \ddot{q}^{a,\epsilon} = W'(q^{a,\epsilon}) - \epsilon a \text{ in } \mathbb{R}, \text{ and,} \\ q^{a,\epsilon}(\pm\infty) = h^{a,\epsilon}_\pm, \ q^{a,\epsilon}(0) = h^{a,\epsilon}_0 \text{ and } \dot{q}^{a,\epsilon} > 0 \text{ in } \mathbb{R}, \end{cases}$$

and, since W is assumed to have wells of the same depth,

$$(3.14) \qquad c(a,\epsilon) \to 0 \text{ and } \epsilon^{-1}c(a,\epsilon) \to c(a) \text{ as } \epsilon \to 0 \text{ and } c(a) \to 0 \text{ as } a \to 0.$$

Next for fixed $\delta, a > 0$, let $u^{\delta,a}$ be the solution of

$$(3.15) \quad \begin{cases} u^{\delta,a}_t = \text{tr}\left[(I - \frac{Du^{\delta,a} \otimes Du^{\delta,a}}{|Du^{\delta,a}|^2})D^2 u^{\delta,a}\right] - c(a)|Du^{\delta,a}| & \text{in } \mathbb{R}^N \times (0,\infty), \\ u^{\delta,a} = d_0 + \delta & \text{in } \mathbb{R}^N \times \{0\}, \end{cases}$$

where $c(a)$ is given by (3.14). Recall that the level sets of (3.15) move with normal velocity

$$V = \text{ mean curvature } - c(a)$$

and that if $d^{\delta,a}$ is the signed distance from the set $\{x : u^{\delta,a}(x,t) = 0\}$, then

$$(3.16) \qquad d^{\delta,a}_t \geq \Delta d^{\delta,a} - c(a)|Dd^{\delta,a}| \text{ and } |Dd^{\delta,a}| = 1 \text{ in } \{d^{\delta,a} > 0\}.$$

It is necessary to introduce an auxiliary function

$$(3.17) \qquad\qquad w^{\delta,a} = \eta_\delta(d^{\delta,a}) \text{ in } \mathbb{R}^N \times (0,\infty),$$

where $\eta_\delta : \mathbb{R} \to \mathbb{R}$ is a smooth function satisfying, for some constant C independent of δ,

$$(3.18) \quad \begin{cases} \eta_\delta(z) = -\delta & \text{if } z \leq \delta/4, \\ \eta_\delta(z) = z - \delta & \text{if } z \geq \delta/2, \\ \eta_\delta(z) \leq \delta/2 & \text{if } z \leq \delta/2, \\ 0 \leq \eta'_\delta(z) \leq C \text{ and } |\eta''_\delta| \leq C\delta^{-1} & \text{on } \mathbb{R}. \end{cases}$$

The following lemma follows from a straightforward combination of (3.16) and (3.18) – see, for example Lemma 3.1 of [ESS].

Lemma 3.3: *There exists a constant C, independent of δ and a, such that*

$$(3.19) \quad \begin{cases} (i) \ w^{\delta,a}_t - \Delta w^{\delta,a} + c(a)|Dw^{\delta,a}| \geq -\delta^{-1}C & \text{in } \mathbb{R}^N \times (0,t^*), \\ (ii) \ w^{\delta,a}_t - \Delta w^{\delta,a} + c(a)|Dw^{\delta,a}| \geq 0 & \text{in } \{d^{\delta,a} > \frac{\delta}{2}\}, \end{cases}$$

and

(3.20) $$|Dw^{\delta,a}| = 1 \quad in \ \{d^{\delta,a} > \delta/2\},$$

where t^* is the extinction time of $\{x : u^{\delta,a}(x,t) = 0\}$. $\square$

Finally define

(3.21) $$\Phi^\epsilon = q^{a,\epsilon}(\epsilon^{-1}w^{\delta,a}) \quad on \ \mathbb{R}^N \times [0,\infty).$$

Proposition 3.4: For every $a > 0$, Φ^ϵ is a supersolution of (3.4), if $\epsilon \le \epsilon_0(\delta,a)$ and $\delta \le \delta_0(a)$.

The proof of this proposition follows along the lines of Theorem 3.2 of [ESS] and Proposition 10.2 of [BSS] where I refer for more details as well as well more general statements. In the proof I present below, I will argue as if $w^{\delta,a}$ has actual derivatives, keeping in mind that actually everything has to be checked in the viscosity sense.

Proof of Proposition 3.4: 1. It follows from (3.13) and (3.21) that

$$\Phi^\epsilon_t - \Delta\Phi^\epsilon + \frac{1}{\epsilon^2}W'(\Phi^\epsilon) = \epsilon^{-1}\dot{q}^{a,\epsilon}(w^{\delta,a}_t - \Delta w^{\delta,a} + \epsilon^{-1}c(a.\epsilon)) - \ddot{\epsilon}^{-1}q^{a,\epsilon}(|Dw^{\delta,a}| - 1)^2 + \epsilon^{-1}a.$$

where $\dot{q}^{a,\epsilon}$ and $\ddot{q}^{a,\epsilon}$ are evaluated at $w^{\delta,a}/\epsilon$.

2. If $\frac{\delta}{2} < d^{\delta,a} < 2\delta$, (3.13), (3.19)(i), (3.20) and (3.18) yield

$$\Phi^\epsilon_t - \Delta\Phi^\epsilon + \epsilon^{-2}W'(\Phi^\epsilon) = \epsilon^{-1}\dot{q}^{a,\epsilon}(\epsilon^{-1}c(a,\epsilon) - c(a)) \ge 0$$

provided that ϵ is sufficiently small.

3. If $d^{\delta,a} \le \delta/2$ the choice of η_δ yields that

$$w^{\delta,a} \le -\delta/2.$$

Consequently the properties of the traveling waves and, in particular, the exponential decay of $\dot{q}$ and $\ddot{q}$ to 0 as $|\xi| \to \infty$ imply that there exists $K > 0$ such that

$$\epsilon^{-1}\dot{q}^{a,\epsilon}(\epsilon^{-1}w^{\delta,a}) + \epsilon^{-2}|\ddot{q}^{a,\epsilon}(\epsilon^{-1}w^{\delta,a})| \le Ke^{-K\delta/\epsilon}.$$

Using the fact that $|Dw^{\delta,a}| \le C$, which follows from (3.20) and $|Dd^{\delta,a}| \le 1$, and (3.19)(i) gives, in the limit $\epsilon \to 0$,

$$\Phi^\epsilon_t - \Delta\Phi^\epsilon + \epsilon^{-2}W'(\Phi^\epsilon) \ge Ke^{-K\delta/\epsilon}(-\delta^{-1} - c + o(1)) + \epsilon^{-1}a \to 0$$

as $\epsilon \to 0$.

4. If $d^{\delta,a} > 2\delta$, then $w^{\delta,a} > \delta$ and the argument is similar. $\square$

All the ground has now been prepared for

Second proof of Theorem 3.2: 1. In view of (3.7), Proposition 3.4 and the maximum principle, it follows that, for ϵ sufficiently small, given δ and a,

$$u^\epsilon \leq \Phi^\epsilon \quad \text{in } \mathbb{R}^N \times [0, t_*).$$

2. Let $(x_0, t_0) \in \mathbb{R}^N \times [0, t_*)$ be such that $u(x_0, t_0) = -\beta < 0$, where u is the solution of the mean curvature pde

$$\begin{cases} u_t = \text{tr}(I - \frac{Du \otimes Du}{|Du|^2})D^2u & \text{in } \mathbb{R}^N \times (0, \infty), \\ u = d_0 & \text{on } \mathbb{R}^N \times \{0\}. \end{cases}$$

The stability properties of viscosity solution yield the $u^{\delta,a} \to u$, as $\delta, a \to 0$, uniformly in (x, t). Hence, for sufficiently small a and δ, one has

$$u^{\delta,a}(x_0, t_0) < -\frac{\beta}{2} < 0.$$

But then $d^{\delta,a}(x_0, t_0) < 0$, hence

$$\limsup_{\epsilon \to 0} u^\epsilon(x_0, t_0) \leq \limsup_{\epsilon \to 0} \Phi^\epsilon(x_0, t_0) = -1.$$

Since, again by the maximum principle and (3.7), $u^\epsilon \geq -1$ it follows that

$$u^\epsilon \to -1 \quad \text{in } \bigcup_{t_* > t > 0} \{x : u(x, t) < 0\}.$$

3. A similar argument, which requires the analogue of Proposition 3.4 but for subsolutions and which follows along the lines of the aforementioned propositions, yields that

$$u^\epsilon \to 1 \quad \text{in } \bigcup_{0 < t < t_*} \{x : u(x, t) > 0\}.$$

4. For the fact that the result can be extended to $\mathbb{R}^N \times (0, \infty)$ instead of $\mathbb{R}^N \times (0, t_*)$, I refer to the related discussion in the first proof of Theorem 3.2. $\qquad\square$

The third proof of Theorem 3.2 relies on the abstract theory presented in Section 2, i.e. Theorem 2.1. It is necessary, however, to settle at this point the issue of the small time behavior. For the particular problem under consideration this was studied in [Ch]. The arguments of [Ch] can then be extended to apply to a more general situation. Below I only remark what follows from [Ch] and I instead refer to the actual paper for the precise statement.

To this end assume that Γ_0 is a smooth hypersurface in $\mathbb{R}^N$, and that, hence, propagates smoothly by mean curvature for small time t. Denoting, as usual, by d the signed distance to the front, [Ch] proves that under suitable regularity assumptions on u_0, there exists some $M > 0$ and $\tau > 0$ such that

$$(3.22) \quad \begin{cases} u_\epsilon(x, t) \geq 1 - \epsilon & \text{if } d(x, t) > M\epsilon|\ln \epsilon| \text{ and } \tau\epsilon^2|\ln \epsilon| \leq t \\ \text{and} \\ u_\epsilon(x, t) \leq -1 + \epsilon & \text{if } d(x, t) < -M\epsilon|\ln \epsilon| \text{ and } \tau\epsilon^2|\ln \epsilon| \leq t. \end{cases}$$

Third proof of Theorem 3.2. 1. Assumptions (H1), (H2) and (H3) of Theorem 2.1 are abviously satisfied in view of the choice of W and the maximum principle.

2. If $g = (1 - \delta)\mathbb{I}_{\{\phi \geq 0\}} - \mathbb{I}_{\{\phi < 0\}}$ in $\mathbb{R}^N$ is the initial datum for (3.4) where $0 < \delta < 1$ and ϕ is as in (H4), then (3.22) combined with the arguments in the second proof of Theorem 3.2 in the case where the motion is smooth, immediately yield that

$$\liminf_* S^\epsilon(h)[(1 - \delta)\mathbb{I}_{\{\phi \geq 0\}} - \mathbb{I}_{\{\phi > 0\}}](x) = 1$$

for h sufficiently small and $d(x, h) > 0$.

3. Actually the ϕ in (H4) does not satisfy the assumptions needed for the argument presented earlier. This can, however, be worked out by an easy sub- and supersolution argument using again (3.22) and observing this ϕ can be put between two others that have the appropriate form, using $q^{a,\epsilon}$ and $q^{-a,\epsilon}$ for a and ϵ sufficiently small. $\qquad\square$

I conclude this part remarking that the above ideas can be extended to treat for general reaction-diffusion equations of the form

$$u_t^\epsilon - \epsilon L u^\epsilon + \epsilon^{-1} W_\epsilon'(u^\epsilon, x, t) = 0 \quad \text{in } \mathbb{R}^N \times (0, \infty)$$

and

$$u_t^\epsilon - L u^\epsilon + \epsilon^{-2} W_\epsilon''(u^\epsilon, x, t) = 0 \quad \text{in } \mathbb{R}^N \times (0, \infty),$$

where L is now a general second-order elliptic operator with coefficients depending on (x, t) and $W_\epsilon(u, x, t)$ is a double-well potential for each (x, t), provided that the assumptions needed for the existence of traveling waves hold uniformly in (x, t) and the behavior as $\epsilon \to 0$ of W_ϵ is well understood. The aforementioned problems are studied in [BSS].

The arguments presented here can also be used to study the asymptotics of certain systems of reaction-diffusion equations like the FitzHugh-Nagumo-type systems. This was studied by Soravia and Souganidis [SorS] where the asymptotics of the following two systems were considered:

$$\begin{cases} u_t^\epsilon - \epsilon \Delta u^\epsilon + \epsilon^{-1} f^\epsilon(u^\epsilon, v^\epsilon) = 0 \\ v_t^\epsilon - (\Delta v^\epsilon - g^\epsilon(u^\epsilon, v^\epsilon)) = 0 \end{cases} \quad \text{in } \mathbb{R}^N \times (0, \infty)$$

and

$$\begin{cases} u_t^\epsilon - \Delta u^\epsilon + \epsilon^{-2} f^\epsilon(u^\epsilon, v^\epsilon) = 0 \\ v_t^\epsilon - (\Delta v^\epsilon - g^\epsilon(u^\epsilon, v^\epsilon)) = 0 \end{cases} \quad \text{in } \mathbb{R}^N \times (0, \infty),$$

with typical example of (f, g) given by

$$\begin{cases} f(u, v) = (u - \mu)(u^2 - 1) + v \\ g(u, v) = \sigma v - u. \end{cases}$$

3.2. Some new results.

Here I state without proof and, as usual, not all the technical assumptions some new results regarding the appearance of moving interfaces. These results are studied in [BS2] using the method described in Section 2.

The first result is about the asymptotics of

$$\begin{cases} u_t^\epsilon - \operatorname{div} A(Du^\epsilon) + \epsilon^{-2}W'(u^\epsilon) = 0 & \text{in } \mathbb{R}^N \times (0,\infty), \\ u^\epsilon = u_0 & \text{on } \mathbb{R}^N \times \{0\}, \end{cases}$$

where $A : \mathbb{R}^N \to \mathbb{R}^N$ is a positively homogeneous of degree one monotone vector field and W is a double well potential with two wells of equal depth. As usual, the results extend to the more general W's.

To state the theorem about the asymptotics of the above initial value problem define the matrix $a = ((a_{ij}))$ by

$$a_{ij} = A_{i,p_j}$$

and assume, without loss of generality, that it is symmetric.

Theorem 3.5: *Assume that* $\Gamma_0 = \partial\{x : u_0(x) > 0\} = \partial\{x : u_0(x) < 0\}$. *Let* $(\Gamma_t, \Omega_t^+, \Omega_t^-)$ *be the level set evolution of* $(\Gamma_0, \{u_0 > 0\}, \{u_0 < 0\})$ *with*

$$F(X,p) = tr\Big[A(\hat{p})X(I(A(\hat{p})p \cdot p)^{-1}A(\hat{p})p \otimes p) + 2(A(\hat{p})p \cdot p)X(I - \hat{p} \otimes \hat{p})DpA(\hat{p})p \cdot p\Big]$$

Then, as $\epsilon \to 0$,

$$u^\epsilon \to \begin{cases} 1 & \bigcup_{t>0}\Omega_t^+ \times \{t\}, \\ & \text{locally uniformly in} \\ -1 & \bigcup_{t>0}(\Omega_t^- \times \{t\}). \end{cases}$$

Formal results corresponding to the above theorem were obtained by Bellettini and Paolini [BP2].

The next result is about the asymptotics of problems like, for example,

$$\begin{cases} u^\epsilon - \epsilon\Delta u^\epsilon + b(\epsilon^{-1}x) \cdot Du^\epsilon + \epsilon^{-1}W'(u^\epsilon) = 0 & \text{in } \mathbb{R}^N \times (0,\infty), \\ u^\epsilon = u_0 & \text{on } \mathbb{R}^N \times (0,\infty), \end{cases}$$

where b is a periodic function and, as usual, W' is a double potential. It was proved by Xin [X] that under certain assumptions on b and W', the equation

$$u_t - \Delta u + b(x) \cdot Du + W'(u) = 0 \quad \text{in } \mathbb{R}^N \times (0,\infty)$$

admits multidimensional traveling wave solutions, i.e. special solutions of the form

$$u(x,t) = q(x \cdot k - c(k)t),$$

where $k \in S^{N-1}$, the unit sphere in $\mathbb{R}^N$. More precisely, [X] shows that for each direction $k \in S^{N-1}$ there exists a unique pair $(c(k), q)$ – the uniqueness of q being understood always up to translations – where $q : \mathbb{R} \times \mathbb{R}^N \to \mathbb{R}$ solves the problem

$$(3.23) \quad \begin{cases} c(k)\dot{q} + \Delta_x q + 2k\Delta_x \dot{q} + |k|^2 \ddot{q} = W'(q) \text{ in } \mathbb{R} \times \mathbb{R}^N \\ q(\pm\infty, x) = \pm 1, \ q(\xi, \cdot) \text{ periodic in } Q, \text{ and } \dot{q}(\xi, \cdot) > 0 \text{ in } \mathbb{R} \times Q. \end{cases}$$

The following theorem is proved in [BS2].

Theorem 3.6: *Assume that $\Gamma_0 = \partial\{x : u_0(x) > 0\} = \partial\{x : u_0(x) < 0\}$ and let $(\Gamma_t, \Omega_t^+, \Omega_t^-)$ be the level set evolution of $(\Gamma_0, \{u_0 > 0\}, \{u_0 < 0\})$. Then, as $\epsilon \to 0$,*

$$u^\epsilon \to \begin{cases} 1 & \bigcup_{t>0} \Omega_t^+ \times \{t\}. \\ & \text{locally uniformly in} \\ -1 & \bigcup_{t>0} (\Omega_t^- \times \{t\}. \end{cases}$$

$\square$

The next example is about the behavior of (3.4) set in a bounded domain with, for example, Neuman boundary conditions, i.e. the problem

$$\begin{cases} u_t^\epsilon - \Delta u^\epsilon + \epsilon^{-2} W'(u^\epsilon) = 0 \text{ in } \Omega \times (0, \infty), \\ \dfrac{\partial u^\epsilon}{\partial n} = 0 \text{ on } \partial\Omega \times (0, \infty), \\ u^\epsilon = u_0 \text{ on } \Omega \times \{0\}, \end{cases}$$

where, as usual for this scaling, W is a double-well potential with wells of the same depth. This problem was studied in [Ch] for short times and convex Ω and in Katsoulakis, Kossioris and Reitich [KKR] for general domains but short times and only for convex domains globally in time. The level set approach can be used to define motion by mean curvature, for example, in bounded domains with Neuman-type boundary conditions. This was developed by Giga and Sato [GS] where I refer for the exact definitions and details.

The following theorem is proved in [BS2].

Theorem 3.7: *Assume that $\Gamma_0 = \partial\{x : u_0(x) = 0\} = \partial\{x : u_0(x) > 0\}$ and let $(\Gamma_t, \Omega_t^+, \Omega_t^-)$ be the level set evolution of $(\Gamma_0, \{u_0 > 0\}, \{u_0 < 0\})$ by the mean curvature evolution with Neumann boundary. Then, as $\epsilon \to 0$,*

$$u^\epsilon \to \begin{cases} 1 & \bigcup_{t>0} \Omega_t^+ \times \{t\}, \\ & \text{locally uniformly in} \\ -1 & \bigcup_{t>0} (\Omega_t^- \times \{t\}. \end{cases}$$

$\square$

The last example I discuss here is about the asymptotics of the following fully nonlinear equation

$$u_t^\epsilon - \epsilon^{-2}(J^\epsilon * u^\epsilon - u^\epsilon) + \epsilon^{-2}W'(u^\epsilon) = 0 \quad \text{in } \mathbb{R}^N \times (0, \infty),$$

where W is the usual double-well potential and $J^\epsilon(x) = \epsilon^{-N}J(\epsilon^{-1}x)$, J being nonnegative, even, i.e. $J(z) = J(-z)$, and satisfying some integrability conditions. Such equations arise at the mesoscopic limit of stochastic Ising models with long-range interactions and Kac potentials, which are discussed in detail in Section 5 below.

To state the result I make the assumption that for each $n \in S^{N-1}$ there exists a unique standing wave solution $u(x, t) = q(x \cdot n, n)$ of

$$u_t + [u - J * u] + W'(u) = 0 \quad \text{in } \mathbb{R}^N \times (0, \infty),$$

i.e. a function $q : \mathbb{R} \times S^{N-1} \to \mathbb{R}$ such that

$$\begin{cases} \displaystyle\int J(y)q(\xi + n \cdot y, n)d\xi - q(\xi, n) = W'(q(\xi, n)) & \text{in } \mathbb{R}, \\ q(\pm\infty, n) = \pm 1 \quad \text{and} \quad \dot{q}(\xi, n) > 0 & \text{on } \mathbb{R}. \end{cases}$$

Notice that since J is not assumed to be isotropic q depends in a nonlinear way on the direction n. For a discussion about the existence of such q's I refer to Bates, Fife, Ren and Wang [BFRW] and Katsoulakis and Souganidis [KS3].

Next for any $n \in S^{N-1}$ consider the matrix $\theta(n)$ defined by

$$\theta(n) = \tfrac{1}{2}\left(\int \dot{q}^2(\xi, n)d\xi\right)^{-1} \times$$

$$\iint J(y)\dot{q}(\xi, n)[\dot{q}(\xi + n \cdot y, n)y \otimes y + D_n q(\xi + n \cdot y, n) \otimes y + y \otimes D_n q(\xi + n \cdot y, n)]dy d\xi.$$

The following is proved in [BS2].

Theorem 3.8: *Assume that $\Gamma_0 = \partial\{x : u_0(x) > 0\} = \partial\{x : u_0(x) > 0\}$. If $(\Gamma_t, \Omega_t^+, \Omega_t^-)$ is the level set evolution of $(\Gamma_0, \{u_0 > 0\}, \{u_0 < 0\})$ with normal velocity $V = -\mathrm{tr}[\theta(n)Dn]$, then, as $\epsilon \to 0$,*

$$u^\epsilon \to \begin{cases} 1 & \displaystyle\bigcup_{t>0}\Omega_t^+ \times \{t\}, \\[2mm] & \text{locally uniformly in} \\[2mm] -1 & \displaystyle\bigcup_{t>0}(\Omega_t^-) \times \{t\}. \end{cases}$$

At this point it is worth remarking about the qualitative difference between (3.4) and this last equation. The asymptotic behavior of (3.4) is always governed by mean curvature, as long as W is a double well potential with wells of equal depth, independently of the particular W. On the other hand this is not true anymore for the above model since the normal velocity depends on q which in turn depends non trivially on the specific form of W.

4. Asymptotics of reaction diffusion equations and systems of KPP-type.

Reaction-diffusion equations with KPP- (i.e. Kolmogorov, Petrovskii and Piskunov) type nonlinearties, a simple example being

$$(4.1) \qquad u_t - \Delta u + u(u - 1) = 0$$

arise in applications similar to the one's for equation of bistable-type, the main difference being on the physical assumptions regarding the chemistry. The study of the long time and large space asymptotics (4.1) and its generalizations, which after the usual (hyperbolic) scaling leads to asymptotic problems of the form

$$(4.2) \qquad u_t^\epsilon - \epsilon \Delta u^\epsilon + \epsilon^{-1} u^\epsilon (u^\epsilon - 1) = 0$$

have been the object of a great deal of research which began before the one for equations of bistable type and has been recently revisited in some work regarding turbulent flame propagation and combustion.

This section is divided into three parts. The first part is about the asymptotics, as $\epsilon \to 0$, of problems like (4.2), but with both more general nonlinearities and second-order operators. It also contains a discussion about the validity of Huygens principle and the connection with the asymptotics discussed in the previous section. In the second part, I discuss some results about KPP-type systems. In the third part, I present some more recent results regarding asymptotics of problems like (4.2) but with oscillatory coefficients and different scalings. This is related to turbulent flame propagation and combustion.

4.1. General Theory.

Here I consider the behavior, as $\epsilon \to 0$, of

$$(4.3) \quad \begin{cases} \text{(i) } u_t^\epsilon - a_{ij}(x)u_{x_i x_j}^\epsilon + b(x) \cdot Du^\epsilon + \epsilon^{-1} f(x, u^\epsilon) = 0 \quad \text{in } \mathbb{R}^N \times (0, \infty), \\[2mm] \text{(ii) } u^\epsilon = g \quad \text{on } \mathbb{R}^N \times \{0\}, \end{cases}$$

where the matrix $A = ((a_{ij})_{ij})$ is uniformly elliptic, the nonlinearity $f : \mathbb{R}^N \times \mathbb{R} \to \mathbb{R}$ is of KPP-type, i.e.

$$(4.4) \quad \begin{cases} f(x, u) < 0 \text{ for } u \in (0, 1) \text{ and } f(x, u) > 0 \text{ for } u \in (-\infty, 0) \cup (1, \infty) \\[2mm] \text{and} \\[2mm] c(x) = \dfrac{\partial f}{\partial u}(x, 0) = \inf_{0 < u \leq 1} \dfrac{f(x, u)}{u} < 0 \quad (x \in \mathbb{R}^N), \end{cases}$$

and the initial datum $g : \mathbb{R}^N \to \mathbb{R}$ is such that

$$(4.5) \qquad g \geq 0 \text{ and } \Omega_0 = \text{spt}(g) \text{ is bounded.}$$

It is necessary, of course, to make a number of technical assumptions on the dependence and regularity on x of the coefficients of (4.3) for which I refer to [ESo2], [MS], etc..

The asymptotic behavior of (4.3) was first studied, under some conditions, using probabilistic ideas from the theory of large deviations by Freidlin [Fr1,2], and later by Evans and Souganidis [ESo2], using ideas from the theory of viscosity solutions, without the restrictions of [Fr1,2]. More recently, Barles and Souganidis [BS3] revisited this issue and studied the rate of convergence as $\epsilon \to 0$ of the solution of u^ϵ to the equilibria.

The main result is:

Theorem 4.1: *Assume (4.4) and (4.5) and let $Z : \mathbb{R}^N \times [0, \infty) \to \mathbb{R}$ be the unique viscosity solution of the variational inequality*

$$(4.6) \quad \begin{cases} \min[Z_t - a_{ij}(x)Z_{x_i}Z_{x_j} + b(x) \cdot DZ + c(x), Z] = 0 \ \ in \ \mathbb{R}^N \times (0, \infty), \\ Z = \begin{cases} 0 & in \ \Omega_0 \times \{0\}, \\ -\infty & in \ (\mathbb{R}^N \backslash \overline{\Omega}_0) \times \{0\}. \end{cases} \end{cases}$$

Then, as $\epsilon \to 0$,

$$u^\epsilon \to \begin{cases} 0 & \quad\quad\quad\quad\quad\quad \{Z < 0\}, \\ & locally \ uniformly \ in \\ 1 & \quad\quad\quad\quad\quad\quad int\{Z = 0\}, \end{cases}$$

with exponential rate.

I continue with a sketch of the proof of the theorem and then discuss a number of interesting points and examples. In the proof below I make all the necessary assumptions to simplify the presentation and refer to [ESo2] and [BS3] for the details.

Proof of Theorem 4.1: 1. Assume, for simplicity, that $a_{ij} = \delta_{ij}$ and $b \equiv 0$. The general case follows in exactly the same way.

2. Assume for simplicity that $0 \le g \le 1$. Then the maximum principle yields that

$$0 < u < 1 \quad in \ \mathbb{R}^N \times [0, \infty).$$

3. Define $Z^\epsilon : \mathbb{R}^N \times [0, \infty) \to \mathbb{R}$ by

$$Z^\epsilon(x, t) = \epsilon \ln u^\epsilon(x, t).$$

A simple computation yields that

$$\begin{cases} Z_t^\epsilon - \epsilon \Delta Z^\epsilon - |DZ^\epsilon|^2 + (u^\epsilon)^{-1} f(u^\epsilon) = 0 \ \ in \ \mathbb{R}^N \times [0, \infty), \\ Z^\epsilon = \begin{cases} \epsilon \ln g & in \ \Omega_0 \times \{0\}, \\ -\infty & in \ (\mathbb{R}^N \backslash \overline{\Omega}_0) \times \{0\}. \end{cases} \end{cases}$$

4. Assume that for each compact K subset of $(\mathbb{R}^N \times (0, \infty)) \cup (\Omega_0 \times [0, \infty))$, there exists $C = C(K) > 0$ such that

$$\sup_{\epsilon > 0} \|Z^\epsilon\|_{L^\infty(K)} \le C.$$

Then

$$Z^*(x,t) = \limsup_{\substack{\epsilon \to 0 \\ (y,s) \to (x,t)}} Z^\epsilon(y,s) \quad \text{and} \quad Z_*(x,t) = \limsup_{\substack{\epsilon \to 0 \\ (y,s) \to (x,t)}} Z^\epsilon(y,s)$$

are well defined upper- and lower-semicontinuous functions respectively and in view of the bound in Step 2 above,

$$Z^*, Z_* \leq 0 \quad \text{in } \mathbb{R}^N \times [0,\infty).$$

5. Using (4.4) in the equation that Z^ϵ satisfies gives

$$Z_t^\epsilon - \epsilon \Delta Z^\epsilon - |DZ^\epsilon|^2 + c(x) \leq 0 \quad \text{in } \mathbb{R}^N \times [0,\infty).$$

Passing now the limit $\epsilon \to 0$ yields

$$\begin{cases} Z_t^* - |DZ^\epsilon|^2 + c(x) \leq 0 \quad \text{in } \mathbb{R}^N \times [0,\infty), \\ Z^* = \begin{cases} 0 & \text{in } \Omega_0 \times \{0\}, \\ -\infty & \text{in } (\mathbb{R}^N \backslash \overline{\Omega}_0) \times \{0\}. \end{cases} \end{cases}$$

6. It also turns out that Z_* is a supersolution of (4.6). Indeed, since $Z_* \leq 0$, it suffices to check that

$$Z_{*t} - |DZ_*|^2 + c(x) \geq 0 \quad \text{in } \{Z_* < 0\}.$$

But the relationship between Z^ϵ and u^ϵ and the KPP-assumption (4.4) on f immediately yield

$$\lim_{\epsilon \to 0} (u^\epsilon)^{-1} f(x, u^\epsilon) = c(x) \quad \text{in } \{Z_* < 0\},$$

hence the claim.

7. Since (4.6) admits a unique viscosity solution Z, it now follows that

$$Z^* = Z_* = Z \quad \text{in } \mathbb{R}^N \times [0,\infty).$$

The issue of existence and uniqueness of solutions to a large class of parabolic pde, with unbounded initial datum, (4.6) being one of them, was studied by Crandall, Lions and Souganidis [CLS], where I refer to for the details.

8. Since

$$u^\epsilon = e^{Z^\epsilon/\epsilon},$$

it is immediate that

$$u^\epsilon \to 0 \quad \text{exponentially fast in } \{Z < 0\}.$$

9. The local uniform convergence of u^ϵ to 1 in $\text{int}\{Z = 0\}$ is a bit more technical, hence I refer to [ESo2].

10. If $w^\epsilon : \mathbb{R}^N \times [0, \infty) \to \mathbb{R}$ is defined by

$$u^\epsilon = 1 - e^{w^\epsilon/\epsilon},$$

then w^ϵ solves

$$w_t^\epsilon - \epsilon \Delta w^\epsilon - |Dw|^2 + \frac{f(u^\epsilon, x)}{u^\epsilon - 1} = 0 \quad \text{in } \mathbb{R}^N \times (0, \infty).$$

It follows that, as $\epsilon \to 0$,

$$w^\epsilon \to w \text{ locally uniformly in int}\{Z = 0\},$$

where w is the unique viscosity solution of

$$\begin{cases} w_t - |Dw|^2 + d(x) = 0 & \text{in int}\{Z = 0\}, \\ w = 0 \text{ on } \partial\{Z = 0\} \cap \{t > 0\}, \\ w = \begin{cases} 0 & \text{in } \Omega_1 \times \{0\}, \\ -\infty & \text{in } (\mathbb{R}^N \backslash \Omega_1) \times \{0\}, \end{cases} \end{cases}$$

where

$$\Omega_1 = \{x \in \mathbb{R}^N : g(x) = 1\} \text{ and } d(x) = \frac{\partial f}{\partial u}(u, 1) > 0;$$

notice that Ω_1 may be empty.

The uniqueness of solutions of the above initial boundary value problem, which in turn yields the convergence of the w^ϵ's, is proved in [BS3].

11. Since $W < 0$ in int$\{Z = 0\}$, the proof is complete.

$\square$

It turns out (see [ESo2]) that there is an explicit but rather complicated formula for Z as the value function of a zero sum, two player differential game with stopping times. In some cases, however, this formula can be simplified as it is indicated below, without any proof. (See [ESo2] for details.)

Proposition 4.2: *Assume that $b \equiv 0$ and $c(x) = c$ in (4.3). Then*

$$Z = \overline{Z} \wedge 0 \quad \text{in } \mathbb{R}^N \times (0, \infty),$$

where $\overline{Z}$ is the solution of

$$\begin{cases} \overline{Z}_t - a_{ij}(x)\overline{Z}_{x_i}\overline{Z}_{x_j} + c = 0 & \text{in } \mathbb{R}^N \times (0, \infty), \\ \overline{Z} = \begin{cases} 0 & \text{in } \Omega_0 \times \{0\}, \\ -\infty & \text{in } (\mathbb{R}^N \backslash \overline{\Omega}_0) \times \{0\}, \end{cases} \end{cases}$$

which is given by the formula

$$\overline{Z}(x, t) = -\frac{d^2(x, \Omega_0)}{4t} - ct \quad ((x, t) \in \mathbb{R}^N \times (0, \infty)).$$

Here $d(x, \Omega_0)$ denotes the distance from x to Ω_0 in the Riemannian metric determined by A^{-1}.

An immediate corollary is:

Corollary 4.3: *Under the hypotheses of Proposition 4.2, as $\epsilon \to 0$,*

$$u^\epsilon(x, t) \to \begin{cases} 0 & d(x, \Omega_0) < 2(-c)^{1/2}t \\ & \text{exponentially fast locally uniformly in} \\ 1 & d(x, \Omega_0) < 2(-c)^{1/2}t. \end{cases}$$

Another interesting point is that under some relatively mild conditions on A and b, the front $\Gamma_t = \partial\{Z(\cdot, t) < 0\}$ is a Lipschitz continuous surface. This was proved by Freidlin and Lee [FL] using the formulae obtained in [ESo2]. (See also [MS] for a direct pde proof.)

Proposition 4.4: *Assume that $c(x) = c$ and, for simplicity, $A = I$. Then, if*

$$\|b\|_\infty < 2\sqrt{-c},$$

$t \mapsto \Gamma_t$ *is a Lipschitz continuous surface in $\mathbb{R}^N \times (0, \infty)$.* $\square$

A natural question at this point is why is it that Theorem 4.1 cannot be proved in the same way as the analogous theorems in Section 3. To answer this question it is necessary to recall (see [AW] for example) that nonlinearities of KPP-type do not admit a unique pair (α, q) with $\alpha \in \mathbb{R}$ and $q : \mathbb{R} \to \mathbb{R}$ (unique up to translations) such that -- here for simplicity I neglect the x-dependence of f,

$$\begin{cases} \alpha\dot{q} + \ddot{q} = f(q) & \text{in } \mathbb{R}, \\ q(-\infty) = 0, \ q(+\infty) = 1 \ \text{and} \ \dot{q} > 0 & \text{in } \mathbb{R}, \end{cases}$$

in contrast to the behavior of nonlinearities of bistable type -- but rather a whole continuum of velocities and, therefore, traveling waves! This rather intriguing fact, which is, of course, related to the different nature of the equilibria 0 and 1 of f, unstable and stable respectively for KPP - f's, rather than stable, stable as is the case for bistable f's, is the main reason about the difference in the analysis of the asymptotics of (3.3) and (4.3).

As a matter of fact this is also the source of some qualitative differences in the asymptotic behavior of (3.3) and (4.3) which goes well beyond the technical issue of how they are studied. This is related to the issue of whether some kind of Huygens-principle is valid for the fronts obtained by Theorem 4.1, which, in this context, is quantified as whether there is a geometric pde governing their evolution. The issue of whether there is a Huygens-principle was discussed at length in [Fr1,2], which contain a number of related examples. This problem was revised by Majda and Souganidis [MS], who gave some more examples and discussed the issue of existence or not of a geometric pde. The lack of such equations exhibited [MS] came to the real surprice to a number of researchers in the combustion community, where such equations, known as G-equations, were derived formally using nonlinearities which were not of KPP-type but "close" to that and were used freely in modeling and for computational purposes. Below I present some of the arguments and examples of [MS], for simplicity, however, I take

$$A = I.$$

It is immediate from (4.6) that Z satisfies

$$\min[Z_t + b(x) \cdot DZ - 2(-c)^{1/2}|DZ|, Z] \geq 0 \quad \text{in } \mathbb{R}^N \times [0, \infty),$$

which, in turn yields using the theory a generalized front propagation of Section 1, that

$$\{x : Z(x,t) < 0\} \subset \{x : W(x,t) < 0\},$$

where W solves the geometric pde

$$\begin{cases} W_t + b(x) \cdot DW = 2(-c(x))^{1/2}|DW| & \text{in } \mathbb{R}^N \times (0, \infty), \\ W = \begin{cases} d(\cdot, \Omega_0) & \text{in } \Omega_0, \\ -d(\cdot, \Omega_0) & \text{in } \mathbb{R}^N \backslash \overline{\Omega}_0. \end{cases} \end{cases}$$

The question now is whether $\Gamma_t = \partial\{Z(\cdot, t) < 0\}$ moves with normal velocity

$$V = 2(-c(x))^{1/2} + b(x) \cdot n.$$

It turns out (see [MS]) that this is indeed the case if $b(x) \equiv b$ and $c(x) \equiv c$ are x-independent. Actually, $A = I$ is not needed for this fact. This is, of course, related to Proposition 4.2 and Corollary 4.3. On the other hand if either c or b depend on x it turns out that there is no geometric pde for Γ_t. This actually can be exhibited by the following two examples.

The first example shows that the velocity of Γ_t depends nontrivially on the location of Γ_0, which means that one does not have a geometric motion in the sense discussed in Section 1. Indeed consider the asymptotic problem

$$\begin{cases} u_t^\epsilon - \epsilon u_{xx}^\epsilon + x u_x^\epsilon + \epsilon^{-1} u^\epsilon(u^\epsilon - 1) = 0 & \text{in } \mathbb{R} \times (0, \infty), \\ u^\epsilon = \chi_{(-\infty, a]} & \text{on } \mathbb{R} \times \{0\}. \end{cases}$$

Since the solution Z of

$$\begin{cases} \min(Z_t - (Z_x)^2 + x Z_x - 1, Z) = 0 & \text{in } \mathbb{R} \times (0, \infty), \\ Z = \begin{cases} 0 & \text{in } (-\infty, a] \times \{0\}, \\ -\infty & \text{in } (a, \infty) \times \{0\}, \end{cases} \end{cases}$$

is given by

$$Z(x,t) = \left(t - \frac{(x - e^t a)_+^2}{2(e^{2t} - 1)}\right) \wedge 0 \qquad ((x,t) \in \mathbb{R} \times (0, \infty)),$$

where $x_+ = \max(x, 0)$, as one can verify by direct computation. It follows that the front Γ_t is the curve

$$x(t) = e^t a + (2t(e^{2t} - 1))^{1/2},$$

which clearly does not satisfy the ode

$$\dot{\tilde{x}} = 2 + \tilde{x},$$

which is the geometric motion derived earlier. Moreover the velocity $\dot{x}(t)$ of $x(t)$ depends on the location a of Γ_0 and it is not universal.

The second example shows that although the front, in view of Proposition 4.4, can be characterized as a Lipschitz continuous function $t : \mathbb{R} \to \mathbb{R}$, i.e. $t = t(x)$, nevertheless there exist b such that the map $x \mapsto t(x)$ is not a monotone function x. This, in turn, yields that the front does not propagate according to Huygen's princple, since the lack of monotonicity of $t(x)$ implies that at some points ahead of the front "new sources" appear. Since this example is rather tedious, I only state it below and I refer to [MS] for the details.

Consider again the one-dimensional problem

$$\begin{cases} u_t^\epsilon - \epsilon u_{xx}^\epsilon + b(x) u_x^\epsilon + \epsilon^{-1} u^\epsilon(u^\epsilon - 1) = 0 & \text{in } \mathbb{R} \times (0, \infty) \\ u^\epsilon = \chi_{(-\infty, 0]} \text{ on } \mathbb{R} \times \{0\}, \end{cases}$$

where now $b : \mathbb{R} \to \mathbb{R}$ is smooth and decreasing. It follows that if b is within a small distance from the piecewise constant function

$$\bar{b}(x) = \begin{cases} \lambda b & \text{if } x < \bar{x}, \\ b & \text{if } x > \bar{x}, \end{cases}$$

for some $\bar{x} > 0$, where λ and b are chosen so that

$$1 < \lambda, \ 0 < b < 2, \ 0 < \lambda b < 2 \ \text{and} \ 1 < (\lambda^2 - \frac{1}{2} - \lambda\sqrt{\lambda^2 - 1})b,$$

then $x \mapsto t(x)$ is not a monotone function of x.

4.2. Some results about systems.

Here I present result regarding the asymptotics of KPP-type systems of reaction-diffusion equations coupled only through the zeroeth-order terms. Partial results in this direction were first obtained in [Fr1,2]. Evans and Souganidis [ESo3] obtained some related results for linear systems. The first general result about the asymptotics of KPP-systems were obtained by Barles, Evans and Souganidis [BES]. Extensions of these results were later obtained in [FrL].

The problem considered in [BES] is the scaled reaction-diffusion system

(4.7)
$$\begin{cases} u_{k,t}^\epsilon - \epsilon d_k \Delta u_k^\epsilon + \epsilon^{-1} f_k(u^\epsilon) = 0 \text{ in } \mathbb{R}^N \times (0, \infty), \\ u_k^\epsilon = g_k \text{ on } \mathbb{R}^N \times \{0\} \quad (k = 1, \dots m). \end{cases}$$

Here the positive constants $d_k > 0$ are given and the smooth, bounded functions $g : \mathbb{R}^N \to \mathbb{R}^m$ are assumed to satisfy

$$(4.8) \qquad g_k \geq 0 \text{ and } \Omega_0 = \{g_k > 0\} \text{ is bounded } (k = 1, \ldots, m),$$

As for the smooth and bounded vector field $f : \mathbb{R}^N \to \mathbb{R}^m$, the essential assumptions are:

$$(4.9) \qquad f(0) = 0$$

and if

$$C \equiv Df(0) = \left(\left(\frac{\partial f_k}{\partial u_\ell}(0) \right) \right)$$

then

$$(4.10) \qquad \begin{cases} (i) c_{u\ell} < 0 & 1 \leq k, \ell \leq m \\ (ii) f_k(u) \geq c_{u\ell} u_\ell & (k = 1, \ldots m). \end{cases}$$

I refer to [BES] for one more structural assumption on f which is necessary to guarantee that some uniform L^∞-bounds on the $u^\epsilon : \mathbb{R}^N \to \mathbb{R}^m$ exist.

Given $p \in \mathbb{R}^N$ define now the $m \times m$-matrix

$$B(p) \equiv \text{diag}(\ldots, -d_k|p|^2, \ldots)$$

and then set

$$A(p) \equiv B(p) + C.$$

Since the matrix $A(p)$ has negative entries, the Perron-Frobenius theory asserts $A(p)$ possesses a simple, real eigenvalue $\lambda^0 = \lambda^0(A(p))$ satisfying Re $\lambda > \lambda^0$ for all other eigenvalues λ of $A(p)$. Define the Hamiltonian

$$H(p) = \lambda^0(A(p)) \qquad (p \in \mathbb{R}^N),$$

and consider the unique viscosity solution Z

$$\begin{cases} Z_t + H(DZ) = 0 & \text{in } \mathbb{R}^N \times (0, \infty) \\ Z = \begin{cases} 0 & \text{on } \Omega_0 \times \{0\} \\ -\infty & \text{in int}(\mathbb{R}^N \backslash \Omega_0) \times \{0\}. \end{cases} \end{cases}$$

Theorem 4.5: *Assume (4.8), (4.9) and (4.10). Then, as $\epsilon \to 0$,*

$$u^\epsilon \to 0 \quad \text{locally uniformly in } \{Z > 0\}$$

and

$$\liminf_{\epsilon \to 0} u_k^\epsilon > 0 \quad \text{uniformly on compact subsets of } \{Z < 0\}. \qquad \square$$

4.9. Front dynamics for turbulent reaction-diffusion equations.

Recently Majda and Souganidis [MS] developed simplified effective equations for the large scale front propagation of turbulent reaction-diffusion equations in the simplest prototypical situation involving advection for turbulent velocity fields with two separated scales. In addition to developing a rigorous theory [MS] studied the issue of the validity of the Huygens principle, discussed earlier in this section, and provided elementary but also sharp bounds on the velocity of the propagating front. In subsequent papers, Embid, Majda and Souganidis [EMS1,2] studied the special case of steady incompressible velocity field consisting of a mean flow plus a small-scale periodic shear both analytically and computationally. The results of [EMS1] give information about,among others, the dependence of the turbulent flame speed on the turbulence intensity. [EMS2] considers the comparison between the results of [MS] and the theory of the G-equation and shows that although in some regimes of the ratio between the transverse component of the mean flow and the shear the agreement is excellent, in some other regimes the G-theory underestimates the flame speeds computed through the nonlinear averaging theory of [MS]. This is, of course, related to the issue of failure of the Huygens principle.

Here I discuss the following asymptotic problem, which is similar to the one considered in [MS]

$$(4.11) \quad \begin{cases} u_t^\epsilon - \epsilon \Delta u^\epsilon + V(x, \delta^{-1}x) \cdot Du^\epsilon + \frac{1}{\epsilon}u^\epsilon(u^\epsilon - 1) = 0 & \text{in } \mathbb{R}^N \times (0, \infty), \\ u^\epsilon = g & \text{on } \mathbb{R}^N \times \{0\}. \end{cases}$$

The main assumption on V is that

$$(4.12) \qquad (x,y) \mapsto V(x,y) \text{ is periodic with respect to } y \text{ in } Q.$$

As before the assumption on g is that

$$(4.13) \qquad 0 \le g \text{ and } \Omega_0 = \{g > 0\} \text{ is bounded.}$$

Given $p, x \in \mathbb{R}^N$ and $a \in [0, \infty)$ consider next the issue of solving the "cell-problem"

$$(4.14) \qquad -a\Delta w - |p + Dw|^2 + V(x,y) \cdot (p + Dw) = -H_a(p,x) \text{ in } Q$$

i.e. the problem of finding a unique $H_a(p,x)$ and a unique up to constant periodic, in Q, $w : \mathbb{R}^N \to \mathbb{R}$ solving (4.14). It turns out that the homogenization theory for Hamilton-Jacobi equations developed by Lions, Papanicolaou and Varadhan [LPV] (see also Evans [E1]) applies to this case and yields the existence of such a unique $H_a(p,x)$. Notice that when $\alpha \ne 0$, then $-H_\alpha(p,x)$ is the principal eigenvalue of the non-divergence form operator

$$-a\Delta \quad - (2p - V(x,y)) \cdot D - |p|^2 + V(x,y) \cdot p \quad \text{in } Q.$$

Finally, consider the variational inequality

$$(4.15) \quad \begin{cases} \min[Z_t^a - H_a(DZ^a, x) - 1, Z^a] = 0 & \text{in } \mathbb{R}^N \times (0, \infty), \\ Z^a = \begin{cases} 0 & \text{in } \Omega_0 \times \{0\}, \\ -\infty & \text{in int}(\mathbb{R}^N \backslash \overline{\Omega}_0) \times \{0\}. \end{cases} \end{cases}$$

The asymptotic behavior of (4.11) is characterized by the following theorem, which in [MS] is studied when $\delta = \epsilon^\alpha$ with $\alpha \in (0, 1]$ and for general f's of KPP-type.

Theorem 4.6: *Assume (4.12) and (4.13) and let Z^a be the solution of (4.15). If $\delta^{-1}\epsilon \to a \in [0, \infty)$, then*

$$\lim_{\substack{\epsilon \to 0 \\ \delta \to 0}} u^\epsilon(x, t) = \begin{cases} 0 & \text{in } \{Z^a < 0\}, \\ 1 & \text{in } int\{Z^a = 0\}. \end{cases}$$

$\square$

The proof follows along the lines of the proof of Theorem 4.5 appropriately modified to take care of the nonlinear averaging, which can be accomplished using the "perturbed-test function" method of Evans [E1]. Instead of writing all the technicalities below I only present a formal argument and refer to [MS] for the details.

To this end observe that the usual transformation $u^{\epsilon,\delta} = \exp \epsilon^{-1} Z^{\epsilon,\delta}$ yields the equation

(4.16) $\qquad Z_t^{\epsilon,\delta} - \epsilon \Delta Z^{\epsilon,\delta} - |DZ^{\epsilon,\delta}|^2 + V(x, \delta^{-1}x) \cdot DZ^{\epsilon,\delta} + u^{\epsilon,\delta} - 1 = 0.$

For simplicity neglect the $u^{\epsilon,\delta}$ term above, which is of course the source of the variational inequality, and consider the following expansion for $Z^{\epsilon,\delta}$:

$$Z^{\epsilon,\delta} = Z^0(x, t) + \delta Z^1(\delta^{-1}x) + O(\delta^2).$$

A simple computation yields that the following equation should hold

$$Z_t^0 - \delta^{-1}\epsilon \Delta Z^0 - |DZ^0 + DZ^1|^2 + V(x, \delta^{-1}x) \cdot (DZ^0 + DZ^1) - 1 = 0$$

which, of course, explains the role played by $H^a(p, x)$ in Theorem 4.6.

I conclude this section with a discussion of a specific example from the homogenization theory of Hamilton-Jacobi equations, which yields a formula for the solution $H_0(p, x)$ of the cell problem. Having such a formula is important in the context of turbulent flame propagation, because it provides a model which permits to validate numerical algorithms to compute the effective flame velocity.

To this end, consider the case $N = 2$, $a = 0$, $V(x, y) = \bar{v}(x) + (v(y_2), 0)$ and $\langle v \rangle = 0$. In view of the uniqueness of $H_0(p, x)$, the cell problem reduces to finding $H_0(p, x)$ and a $Q \cap \mathbb{R}$-periodic $w : \mathbb{R} \to \mathbb{R}$ such that

$$|w' + (p_2 - \frac{1}{2}\bar{v}_2)|^2 = H_0(p, x) - p_1^2 + \frac{1}{4}(\bar{v}_2)^2 + (\bar{v}_1 + v(y))p_1 \quad \text{in } \mathbb{R},$$

where, denotes differentiation with respect to $y \in \mathbb{R}$.

A simple argument (see [PLV] and [MS]) based on the observation that if w is a $Q \cap \mathbb{R}$-periodic solution of $|w' + \xi|^2 = F(y)$ and $F(y) > 0$ in $Q \cap \mathbb{R}$, then

$$|\xi| = \langle F(y)^{1/2} \rangle,$$

yields that

$$H_0(p, x) = \begin{cases} p_1^2 - \frac{1}{4}\bar{v}_2^2(x) - \bar{v}_1(x)p_1 + M & \text{if } |p_2 - \frac{\bar{v}_2(x)}{2}| \leq \langle (M + v(y_2)p_1)^{1/2} \rangle \\ p_1^2 - \frac{1}{4}\bar{v}_2^2(x) - \bar{v}_1(x)p_1 + M + A & \text{if } |p_2 - \frac{\bar{v}_2(x)}{2}| = \langle (M + v(y_2)p_1 + A)^{1/2} \rangle, \end{cases}$$

where

$$M = \max_Q(-v(y)p_1).$$

5. Macroscopic (hydrodynamic) limits of stochastic Ising models with long range interactions.

One of the most striking applications of the theory of the generalized front propagation and viscosity solutions has been the development of a rigorous theory about macroscopic limits of stochastic Ising model and the rigorous justification of the appearance of moving interfaces in the limit. Such results are important both from the mathematical and statistical mechanics point of view. They also provide a theoretical justification, from the microscopic point of view, for the phenomenological theories of phase transitions, which are obtained with arguments from continuum mechanics, as well as for the numerical computations performed by physicists to compute evolving fronts.

In the first part of this section I present very briefly some of the key facts from the equilibrium theory of Ising models. In the second part I discuss the dynamic theory and derive the appropriate mean field equations. Finally in the third part I present the results about the macroscopic limits.

5.1 Equilibrium theory.

Stochastic Ising models are the canonical Gibbsian models used in statistical mechanics to describe phase transitions. Describing in detail such models is well beyond the scope of these notes. Instead below, abusing, if needed, the mathematical rigor as well as the actual meaning of the terms I use, I present a brief summary of these models and refer to the monograph of DeMasi and Presutti [DP], the books by Spohn [Sp1] and Liggett [Lig], the papers by DeMasi, Orlandi, Presutti and Triolo [DOPT1,2,3] and the references therein for the complete and rigorous theory.

To this end one considers the lattice $\mathbb{Z}^N$, the spin $\sigma(x) = \pm 1$ at $x \in \mathbb{Z}^N$, the configuration (sample) space $\Sigma = \{-1, 1\}^{\mathbb{Z}^N}$ and the Gibbs (equilibrium) measures μ^β on Σ which depend on the inverse temperature $\beta > 0$ and the Hamiltonian (energy) function H, which is given by

$$H(\sigma) = -\sum_{x \neq y} J(x, y)\sigma(x)\sigma(y) - h\sum_x \sigma(x).$$

Here $J \geq 0$ is the interaction potential – the sign condition on J means that one deals with ferromagnetic models – and h is the external magnetization field. It should be noted that in the way that H is defined above, the sums on the right-hand side may diverge. This is corrected in the rigorous setting (see, for example, [DP]) by considering the model in some finite set of size Λ and then by letting

$\Lambda \to \infty$. This is the kind of rigor that, as mentioned above, will be not considered at least in this part of Section 5.

It turns out (see, for example, [Lig]) that for any $\beta > 0$, as long as $h \neq 0$, there exists a unique Gibbs measure. On the other hand, if $h = 0$ there exists β_c such that for $\beta < \beta_c$ there still exists a unique Gibbs measure, but for $\beta > \beta_c$ there exist at least two probability measures $\mu_\pm$ on Σ such that any linear combination $\alpha\mu_- + (1-\alpha)\mu_+$, where $\alpha \in [0,1]$, is also Gibbs measure. The appearance of many Gibbs measures is a manifestation that phase transition occurs for $\beta > \beta_c$.

5.2. Dynamic (Nonequilibrium) Theory Mesoscopic limit.

Studying the phase transitions from the dynamic point of view for $\beta > \beta_c$ amounts to introducing some dynamics, i.e. a Markov process on Σ, which have the Gibbs measures as invariant measures and to analyzing the way this process evolves any initial distribution (measure) to the equilibria measures. Convenient quantities (order parameters) to analyze in this context are the moments of the evolving measure, the first one being the total magnetization m, which, of course, will develop an interface for large times. The shape and evolution of this interface is of great interest both theoretically and in the applications.

A very general example of such dynamics which have the Gibbs measures as invariant measures is the spin-flip dynamics, which, loosely speaking, is a sequence of flips σ^x, where

$$\sigma^x(y) = \begin{cases} \sigma(y) & \text{if } y \neq x, \\ -\sigma(y) & \text{if } y = x, \end{cases}$$

with rate

$$c(x, \sigma) = \Psi(\Delta_x H),$$

for an appropriate Ψ, where $\Delta_x H$ is the energy difference due to a spin flip at x. The only restriction on Ψ, besides some obvious continuity assumptions, which is related to the requirement that the Gibbs measures are invariant for the dynamics, is that it satisfies the balance law

$$\Psi(r) = \Psi(-r)e^{-r}.$$

Classical choices of Ψ are $(1 + e^{-2r})$, which corresponds to the Glauber dynamics and $e^{-r/2}$, which is the Metropolis dynamics.

More precisely the spin-flip dynamics is a Markov jump process on Σ with generator given by

$$Lf(\sigma) = \sum_x c(x, \sigma)[f(\sigma^x) - f(\sigma)],$$

acting on cylindrical functions on Σ – see [Lig] for the precise definition. The solution of the equation

$$\frac{d}{dt}f_t = L_\gamma f_t, \qquad f_0 = f$$

is then given by

$$f_t(\sigma) = \sum_{\sigma'} e^{L_\gamma t}(\sigma, \sigma')f(\sigma'),$$

where $e^{L_\gamma t}(\sigma, \sigma')$ are the transition probabilities of the process.

The full-stochastic jump process σ_t is constructed as follows: The initial configurations σ^0 are randomly distributed according to some measure μ on Σ. Given a σ^0, $\sigma_t = \sigma^0$ for an exponentially distributed waiting time with rate $\sum_y c(y, \sigma^0)$, then σ_t jumps to a new configuration $\sigma^1 = \sigma^x$ with probability $c(x, \sigma^0)/\sum_y c(y, \sigma^0)$. Then $\sigma_t = \sigma^1$ for another exponentially distributed waiting time with rate $\sum_y c(y, \sigma^1)$, etc. Notice that, in view of the positivity of J, the probability of a spin flip at x is higher when the spin at x is different from that of most of its neighbors than it is when the spin agrees with most of its neighbors. Thus the system prefers configurations in which the spins tend to be aligned with one another. This property in the language of statistical mechanics, is refered to as ferromagnetism.

In view of the previous discussion one is interested in the behavior of the system as $t \to \infty$. Another classical limit, known as the Lebowitz-Penrose limit [LP], is to study the behavior of the system also as the interaction range tends to infinity. In this limit, known in the physics literature as grain coarsening, there is a law of large numbers effect that dampens the oscillations and causes the whole collection to evolve deterministically. An important question is whether these limits commute and, if not, whether there is a particular scaling or scalings for which one can study both.

These issues are addressed by Katsoulakis and Souganidis [KS3] for the general dynamics described above with long-range interactions, with rate

$$c_\gamma(x, \sigma) = \Psi(\Delta_x H_\gamma),$$

where γ^{-1} is the interaction range,

$$H_\gamma(x) = -\sum_{y \neq x} J_\gamma(x, y)\sigma(x)\sigma(y)$$

and

$$J_\gamma(x, y) = \gamma^N J(\gamma(x - y)).$$

Throughout the remaining of this section it is assumed that

(5.1) $\qquad \begin{cases} J : \mathbb{R}^N \to [0, \infty) \text{ satisfies some integrability conditions} \\\\ \text{and is even, i.e. } J(z) = J(-z). \end{cases}$

The behavior of the model under the mesoscopic scaling $(x, t) \to (\gamma x, t)$ in the limit $\gamma \to 0$ was studied in [DOPT1] for isotropic J's and in [KS3] in the general setting of (5.1).

To state the result it is necessary to introduce for each $n \in \mathbb{Z}^+$ the notation

$$\mathbb{Z}_n^N = \{\underline{x} = (x_1, \ldots, x_n) \in (\mathbb{Z}^N)^n : x_1 \neq \cdots \neq x_n\},$$

and to consider the fully nonlinear integral differential equation

(5.2) $\qquad m_t + \Phi(\beta(J * m))[m - \tanh \beta J * m] = 0 \quad \text{in } \mathbb{R}^N \times (0, \infty),$

where $J * m$ denotes the usual convolution and Φ is given by

(5.3) $$\Phi(r) = \Psi(-2r)(1 + e^{-2r})^{-1}.$$

The result, which is proved in [DOPT1] and [KS3] under some simplified assumption on the initial measure μ^γ is:

Theorem 5.1: *Assume that that intial measure μ^γ is a product measure such that*

$$\mathbb{E}_{\mu^\gamma}(\sigma(x)) = m_0(\gamma x) \qquad (x \in \mathbb{Z}^N).$$

Then, for each $n \in \mathbb{Z}^+$,

$$\lim_{\gamma \to 0} \sup_{\underline{x} \in \mathbb{Z}_n^N} \left| \mathbb{E}_{\mu^\gamma}\left(\prod_{i=1}^n \sigma_t(x_i) \right) - \prod_{i=1}^n m(\gamma x_i, t) \right| = 0,$$

where m is the unique solution of (5.3) with initial datum m_0. $\square$

5.3. Macroscopic behavior of the mean field equation.

Here I discuss the asymptotic behavior for large r and t of the solution of (5.2). To avoid confusion I use x to denote points on the lattice and r to denote points in $\mathbb{R}^N$.

Assuming that

(5.4) $$\beta \hat{J} = \beta \int J(y)dy > 1,$$

it is easy to see that (5.2) admits three steady solutions $-m_\beta$, 0 and m_β such that $-m_\beta < 0 < m_\beta$. It then becomes a natural question, as is was for the reaction-diffusion problems, to try to understand the interface generated when, as $t \to \infty$, m approaches $\pm m_\beta$.

The appropriate scaling to consider is, as for the reaction-diffusion equation with a double-well potential with wells of the same depth, $(r, t) \mapsto (\lambda r, \lambda^2 t)$, which leads to singular problem

(5.5) $$m_t^\lambda + \lambda^{-2}\Phi(\beta J^\lambda * m^\lambda)[m^\lambda - \tanh \beta J^\lambda * m^\lambda] = 0 \text{ in } \mathbb{R}^N \times (0, \infty),$$

where

$$m^\lambda(r, t) = m(\lambda^{-1}r, \lambda^{-2}t)$$

and

$$J^\lambda(r) = \lambda^{-N}J(\lambda^{-1}r).$$

An important step in the analysis to follow is the existence of traveling, in this case standing, wave solutions for (5.2), i.e. solutions of the form

$$m(r, t) = q(r \cdot n, n),$$

where $n \in S^{N-1}$, and such that $q(\pm\infty, n) = \pm m_\beta$.

It turns out that q solves the equation

$$(5.6) \qquad q(\xi, n) = \tanh \beta \int J(y) q(\xi + n \cdot y, n) dy.$$

Notice that this equation is independent of Φ and hence Ψ, i.e. independent of the dynamics assumed for the underlying spin models, and that, since J is not assumed to be isotropic, q must depend on a nontrivial way on the direction n. For the existence and properties of such q's I refer to [KS3]. In what follows I will use $\dot{q}$ and $D_n q$ to denote the derivative of q with respect to ξ and n respectively.

For each $n \in S^{N-1}$ consider the symmetric matrix

$$\theta(n) = \left[\int \frac{\dot{q}^2(\xi, n)}{\Phi(\beta(J * q))(1 - q^2(\xi, n))} d\xi \right]^{-1} \times$$

$$(5.7) \qquad \times \frac{\beta}{2} \left[\iint J(y) \dot{q}(\xi, n) [\dot{q}(\xi + n \cdot y, n) y \otimes y + D_n q(\xi + n \cdot y, n) \otimes y \right.$$

$$\left. + y \otimes D_n q(\xi + n \cdot y, n)] dy d\xi \right].$$

The result about the asymptotic behavior, in the limit $\lambda \to 0$, of the solution m^λ of (5.5) is stated, without all the technical conditions, in the following theorem.

Theorem 5.2: *Let* $\Omega_0 = \{x : m^\lambda(x, 0) > 0\}$, *assume that* $\Gamma_0 = \partial \Omega_0 = \partial \{x : m^\lambda(x, 0) < 0\}$ *and consider the evolution* $(\Gamma_t, \Omega_t^+, \Omega_t^-)$ *of* $(\Gamma_0, \Omega_0, \mathbb{R}^N \backslash \overline{\Omega}_0)$ *with normal velocity* $V = -\operatorname{tr}[\theta(n) Dn]$, *with* $\theta(n)$ *given by (5.8). Then, as* $\lambda \to 0$,

$$m^\lambda \to \begin{cases} +m_\beta & \\ & \textit{locally uniformly in} \\ -m_\beta & \end{cases} \qquad \begin{aligned} & \bigcup_{t>0} \Omega_t^+ \times \{t\}, \\ & \bigcup_{t>0} \Omega_t^- \times \{t\}. \end{aligned}$$

This theorem, whose proof is based on the ideas developed in Section 2, is proved in [KS3]. Earlier related results were obtained by Katsoulakis and Souganidis [KS2], who studied (5.5) when J is isotropic, i.e. $J(r) = J(|r|)$ and for Glauber dynamics and Jerrard [J], who studied a local version of (5.5) again in the isotropic case. Finally [DOPT1] studied again the isotropic problem and Glauber dynamics under the assumption that the moving interface is smooth.

Below I present a formal argument to explain why it is that the asymptotics of (5.5) are governed by the motion $V = -\operatorname{tr}[\theta(n) Dn]$. To this end let $d(r, t)$ be the signed distance to the evolving front Γ_t and try to find an asymptotic expansion for m^λ of the form

$$(5.8) \qquad m^\lambda(r, t) = q \quad {}^{-1} d(r, t), Dd(r, t)) + \lambda Q(\lambda^{-1} d(r, t), Dd(r, t)) + O(\lambda^2),$$

where Q is to be chosen.

Next insert (5.8) in (5.5) to find

$$\lambda^{-1}\dot{q}(\xi,n)d_t + \lambda^{-2}\Phi(\beta J^\lambda * m^\lambda)[q(\xi,n) + \lambda Q(\xi,n)-$$

$$- \tanh\left[\beta\int J(y)[q(\frac{d(r+y)}{\lambda}, Dd(r+\lambda y)) + \lambda Q(\frac{d(r+\lambda y)}{\lambda}, Dd(r+\lambda y))dy] + O(\lambda^2)]\right.$$

$$= O(1),$$

where

$$\xi = \lambda^{-1}d(r,t) \quad \text{and} \quad n = Dd(r,t).$$

Expanding now the expression involving the tanh around $\tanh\beta\int J(y)q(\xi + n\cdot y,n)dy$, using (5.6) as well as the fact that $(\tanh u)' = 1 - (\tanh u)^2$, one finds that in order for the coefficient of $\frac{1}{\lambda}$ to be zero Q must satisfy the equation

$$(5.9) \quad \frac{1}{1-q^2(\xi,n)}Q(\xi,n) - \int J(y)Q(\xi + n\cdot y,n) =$$

$$= \frac{\dot{q}(\xi,n)}{\Phi(\beta(J*q))}d_t - \frac{\beta}{2}\int J(y)[\dot{q}(\xi + y\cdot n,n)(D^2dy,y) + (D_nq(\xi + y\cdot n,n), D^2dy)]dy$$

where d_t and D^2d are evaluated at (r,t).

If the operator L is defined by

$$LQ = \frac{1}{1-q^2}Q - \int J(y)Q(\xi + n\cdot y,n)dy,$$

differentiating (5.6) with respect to ξ yields that

$$L\dot{q} = 0.$$

Then, by a Fredholm alternative type of argument, one finds that in order for Q satisfying (5.9) to exist, it is necessary for the right-hand side of (5.9) to be orthogonal to $\dot{q}(\xi,e)$, i.e. it is necessary to have the compatibility condition

$$\left(\int \frac{\dot{q}^2(\xi,n)}{\Phi(\beta(J*q))(1-q^2(\xi,n))}d\xi\right)d_t$$

$$= \frac{\beta}{2}\iint J(y)\dot{q}(\xi,n)[\dot{q}(\xi + y\cdot n,n)(D^2dy,y) + D_nq(\xi + y\cdot n,n)D^2dy]dyd\xi.$$

After some elementary manipulations this last expression yields

$$d_t = \text{tr}(\theta(Dd)D^2d),$$

with $\theta(n)$ as in (5.7). This concludes the formal argument about the validity of Theorem 5.2.

I conclude the discussion about the macroscopic behavior of (5.5) with few remarks about the form of $\theta(n)$. Indeed observe that $\theta(n)$ is given in (5.7) as a product of a constant, which can be identified as the inverse of the mobility, and a matrix which can be identified as the surface tension of the interface. Notice that the surface tension depends on J and the mobility depends on the dynamics in terms of Φ and the energy via J.

5.4. Macroscopic (hydrodynamic) limit for the stochastic Ising model.

The results presented in the last two parts of this section suggest that if one could perhaps scale the particle system first $(x,t) \to (\gamma x, t)$ and then $(\gamma x, t) \to (\lambda \gamma x, \lambda^2 t)$, then the moving interface given by Theorem 5.2 will also govern the asymptotic behavior of the spin system. This is of course not obvious because it is not clear that two aforementioned limits commute and, moreover, one does not have and will not have, in view of the lack of regularity for the moving front, enough smoothness to perform a good enough asymptotic expansion to study this issue. On the other hand, these two limits suggest that it may be possible to scale the particle system like $(x,t) \mapsto (\gamma\lambda(\gamma), \lambda^2(\gamma)t)$ where $\lambda(\gamma) \to 0$, as $\gamma \to 0$, so that, in the limit $\gamma \to 0$, one obtains a moving interface. This actually happens provided that one chooses $\lambda(\gamma)$ which converge to 0 relatively slowly.

The result, again stated in a rather informal way, is contained in the following theorem. The statement is however admitedly cumbersome due to the fact that the particle system and the evolving front are, of course, defined at a different scale.

Theorem 5.3: *There exists $\rho^* > 0$ such that for any $\lambda(\gamma) \to 0$ slower than γ^{ρ^*} (i.e. $\lambda(\gamma)\gamma^{-\rho^*} \to +\infty$ as $\gamma \to 0$) the following hold: Let Ω_0 be an open subset of $\mathbb{R}^N$ such that $\Gamma_0 = \partial\Omega_0 = \partial(\mathbb{R}^N\backslash\overline{\Omega}_0)$ and*

$$\mathbb{E}_{\mu^\gamma} \sigma(x) > 0 \quad \textit{iff} \quad \gamma\lambda(\gamma)x \in \Omega_0$$

and

$$\mathbb{E}_{\mu_\gamma} \sigma(x) < 0 \quad \textit{iff} \quad \gamma\lambda(\gamma)x \in \mathbb{R}^N\backslash\overline{\Omega}_0.$$

If $(\Gamma_t, \Omega_t^+, \Omega_t^-)$ is the evolution of $(\Gamma_0, \Omega_0, \mathbb{R}^N\backslash\overline{\Omega}_0)$ with normal velocity $V = -tr[\theta(n)Dn]$, where $\theta(n)$ is given by (5.7), then

$$\sup_{\underline{x}\in\mathbb{Z}_n^N} \left| \mathbb{E}_{\mu^\gamma} \prod_{i=1}^n \sigma_{t\lambda(\gamma)^{-2}}(x_i) - m_\beta^n \prod_{x_i\in\mathcal{N}_t^\gamma} (-1) \right| = O(\lambda),$$

with the limit local uniform in t, where

$$N_t^\gamma = \{x \in \mathbb{Z}^N : (\gamma\lambda(\gamma)x, t) \in \Omega_t^-\}. \qquad \square$$

Theorem 5.3 is proved in [KS3]. A similar theorem but for the case of isentropic interactions and Glauber dynamics was obtained earlier in [KS2] for all times and in [DOPT1] as long as the resulting motion is smooth and for a particular choice of $\lambda(\gamma)$. An analogous result put for nearest neighbor interaction Ising models in two dimensions, where the curvature motion is always smooth, was obtained by Spohn [Sp2]. When the dynamics of nearest neighbor model are coupled with Kawasaki dynamics at infinite temperature, which delocalize the interaction, the macroscopic limit, which turns out to be motion by mean curvature, was studied by Bonaventura [Bo] for smooth motions, i.e., for small time and a particular $\lambda(\gamma)$, and by Katsoulakis and Souganidis [KS1] in the generality of Theorem 5.3 and even when fattening occurs. It should be noted that the mean field equation corresponding to the mesoscopic limit of Ising models with nearest neighbor spin flip Glauber dynamics and simple exchange (stirring) Kawasaki dynamics at infinite termpature is the Allen-Cahn equation (3.4) for an appropriate double-well potential W, as it was proved by DeMasi, Ferrari and Lebowitz [DFL].

As mentioned earlier besides its theoretical interest Theorem 5.3 provides a justification, from microscopic considerations, to phenomenological theories of phase transitions like sharp-interface models derived by thermodynamic arguments ([Gu]) and sometimes by scaling of a Landau-Ginzburg model, i.e. reaction-diffusion equations. The theorem can also be thought of as providing a theoretical justification of the validity of some Monte Carlo-type methods, which have long been implemented in the physics literature, to approximate evolving fronts at any time. Notice that in this context the stochastic spin dynamics are unaffected by the possible appearance of singularities in the flow. A final remark about the importance of Theorem 5.3 is related to the discussion at the end of Part 4 of this section and the form of $\theta(n)$. Indeed this result provides by microscopic considerations the "correct" form for the mobility of the interface, which depends, of course on the dynamics but also on the equilibrium considerations via J, something which is occasionally neglected in the phenomenological theories, where the choice of the mobility appears to be some times arbitrary.

I conclude the discussion about Theorem 5.3 with few words about its proof. At first glance it may appear that Theorem 5.3 follows by straightforward combinations of Theorem 5.1 and 5.2. This is far from being true since Theorem 5.1 does not provide any information at the appropriate time scale. The needed argument can be described as follows:

A careful study of the particle system involving a version of the local central limit theorem yields that the evolving spins are close, with respect to some γ-dependent seminorm and for short times, to the solution of (5.2), the error being, however, algebraic in γ. Unfortunately, this last observation eliminates the possibility of a straightforward discretization in time and then application of Theorem 5.2, since the errors add up! Instead one needs to consider two different fronts which move with velocities slightly larger and smaller than the desired ones and then to use these fronts as appropriate barriers to control the errors at each time step. It is exactly this choice of fronts and the need to control the errors that leads to the restriction on the scale $\lambda(\gamma)$.

References

[AC] S. M. Allen and J. W. Cahn, A macroscopic theory for antiphase boundary motion and its application to antiphase domain coarsening, *Acta Metal.* **27** (1979), 1085–1095.

[AGLM] L. Alvarez, F. Guichard, P.-L. Lions and J.-M. Morel, Axioms and fundamental equations of image processing, *Arch. Rat. Mech. Anal.* **123** (1992), 199–257.

[ACI] S. Angenent, D. L. Chopp and T. Ilmanen, A computed example of nonuniqueness of mean curvature flow is $\mathbb{R}^3$, *Comm PDE*, **20** (1995), 1937–1958.

[AIV] S. Angenent, T. Ilmanen and J. L. Velazquez, Nonuniqueness in geometric heat flows, in preparation.

[AW] D. G. Aronson and H. Weinberger, Multidimensional nonlinear diffusion arising in population genetics, *Adv. Math.* **30** (1978), 33–76.

[Ba1] G. Barles, Remark on a flame propagation model, Rapport INRIA 464 (1985).

[Ba2] G. Barles, Discontinuous viscosity solutions of first-order Hamilton-Jacobi equations: a guided visit, *Nonlinear Analysis TMA*, in press.

[BBS] G. Barles, L. Bronsard and P. E. Souganidis, Front propagation for reaction-diffusion equations of bistable type, *Anal. Nonlin.* **9** (1992), 479–506.

[BG] G. Barles and C. Georgelin, A simple proof of an approximation scheme for computing motion by mean curvature, *SIAM J. Num. Anal* **32** (1995), 484–500.

[BES] G. Barles, L. C. Evans and P. E. Souganidis, Wavefront propagation for reaction-diffusion systems of PDE, *Duke Math. J.* **61** (1990), 835–858.

[BP] G. Barles and B. Perthame, Discontinuous solutions of deterministic optimal stopping time problems, *M2AN* **21** (1987), 557–579.

[BSS] G. Barles, H. M. Soner and P. E. Souganidis, Front propagation and phase field theory, *SIAM J. Cont. Opt.* **31** (1993), 439–469.

[BS1] G. Barles and P. E. Souganidis, Convergence of approximation schemes for full nonlinear equations, *Asymptotic Analysis* **4** (1989), 271–283.

[BS2] G. Barles and P. E. Souganidis, A new approach to front propagation: Theory and Applications, *Arch. Rat. Mech. Anal.*, to appear.

[BS3] G. Barles and P. E. Souganidis, A remark on the asymptotic behavior of the solution of the KPP equation, *C. R. Acad. Sci. Paris* **319** Série I (1994), 679–684.

[BJ] E. N. Barron and R. Jensen, Semicontinuous viscosity solutions for Hamilton-Jacobi Equations with convex Hamiltonians, *Comm. in PDE*, in press.

[BFRW] P. Bates, P. Fife, X. Ren and X. Wang, Traveling waves in a convolution model for phast transitions, preprint.

[BP1] G. Bellettini and M. Paolini, Two examples of fattening for the mean curvature flow with a driving force, *Mat. App.* **5** (1994), 229–236.

[BP2] G. Bellettini and M. Paolini, Some results on minimal barriers in the sense of DeGiorgi to driven motion by mean curvature. preprint.

[BP3] G. Bellettini and M. Paolini, Anisotropic motion by mean curvature in the context of Finsler geometry, preprint.

[Bo] L. Bonaventura, Motion by curvature in an interacting spin system, preprint.

[Br] K. A. Brakke, *The motion of a surface by its Mean Curvature*, Princeton University Press, Princeton, NJ, (1978).

[BrK] L. Bronsard and R. Kohn, Motion by mean curvature as the singular limit of Ginzburg-Landau model, *J. Diff. Eqs.* **90** (1991), 211–237.

[Ca1] G. Caginalp, An analysis of a phase field model of a free boundary, *Arch. Rat. Mech.* **92** (1986), 205–245.

[Ca2] G. Caginalp, Mathematical models of phase boundaries, in *Material Instabilities in continuum Mechanics and Related Mathematical Problems*, ed. J. Ball, Clarendon Press, Oxford (1988), 35–52.

[Ch] X.-Y. Chen, Generation and propagation of the interface for reaction-diffusion equation, *J. Diff. Eqs.* **96** (1992), 116–141.

[CGG] Y.-G. Chen, Y. Giga and S. Goto, Uniqueness and existence of viscosity solutions of generalized mean curvature flow equations, *J. Diff. Geom.* **33** (1991), 749–786.

[Cr] M. G. Crandall, CIME Lectures.

[CIL] M. G. Crandall, H. Ishii and P.-L. Lions, User's guide to viscosity solutions of second order partial differential equations, *Bul. AMS* **27** (1992), 1–67.

[CLS] M. G. Crandall, P.-L. Lions and P. E. Souganidis, Universal bounds and maximal solutions for certain evolution equations, *Arch. Rat. Mech. Anal.* **105** (1989), 163–190.

[D] E. DeGiorgi, Some conjectures on flow by mean curvature, *Proc. Capri Workshop*, 1990, Benevento-Bruno-Sbardone editors.

[DFL] A. DeMasi, P. Ferrari and J. Lebowitz, Reaction-diffusion equations for interacting particle systems, *J. Stat. Phys.* **44** (1986), 589–644.

[DOPT1] A. DeMasi, E. Orlandi, E. Presutti and L. Triolo, Glauber evolution with Kač potentials: I. Mesoscopic and macroscopic limits, interface dynamics, *Nonlinearity* **7** (1994), 633–696; II. Fluctuations, preprint. III. Spinodal decomposition, preprint.

[DOPT2] A. DeMasi, E. Orlandi, E. Presutti and L. Triolo, Motion by curvature by scaling non local evolution equations, *J. Stat. Phys.* **73** (1993), 543–570.

[DOPT3] A. DeMasi, E. Orlandi, E. Presutti and L. Triolo, Stability of the interface in a model of phase separation, *Proc. Royal Soc. Edinb.* **124A** (1994), 1013–1022.

[DP] A. DeMasi and E. Presutti, *Mathematical Methods for Hydrodynamic Limits*, Lecture Notes in Mathematics, Springer-Verlag, Berlin, 1991.

[DS] P. DeMottoni and M. Schatzman, Development of interfaces in $\mathbb{R}^N$, *Proc. Royal Soc. Edinb.* **116A** (1990), 207–220.

[EMS1] P. F. Embid, A. Majda and P. E. Souganidis, Effective geometric front dynamics for premixed turbulent combustion with separated velocity scales, *Combust. Sci. Tech.* **103** (1994), 85–115.

[EMS2] P. F. Embid, A. Majda and P. E. Souganidis, Comparison of turbulent flame speeds from complete averaging and the G-equation, *Physics Fluids* **7** (1995), 2052–2060.

[E1] L. C. Evans, The perturbed test function method for viscosity solutions of nonlinear PDE, *Proc. Royal Soc. Edinb.* **111A** (1989), 359–375.

[E2] L. C. Evans, CIME Lectures.

[ESS] L. C. Evans, H. M. Soner and P. E. Souganidis, Phase transitions and generalized motion by mean curvature, *Comm. Pure Appl. Math.* **45** (1992), 1097–1123.

[ESo1] L. C. Evans and P. E. Souganidis, Differential games and representation formulae for solutions of Hamilton-Jacobi-Isaacs equations, *Ind. Univ. Math. J.* **33** (1984), 773–797.

[ESo2] L. C. Evans and P. E. Souganidis, A PDE approach to geometric optics for certain reaction-diffusion equations, *Ind. Univ. Math. J.* **38** (1989), 141–172.

[ESo3] L. C. Evans and P. E. Souganidis, A PDE approach to certain large deviation problems for systems of parabolic equations, *Ann. Inst. H. Poincaré, Anal. Non Lineáire* **6** (Suppl.) (1994), 229-258.

[ESp] L. C. Evans and J. Spruck, Motion of level sets by mean curvature I, *J. Diff. Geom.* **33** (1991), 635–681.

[Fi] P. C. Fife, Nonlinear diffusive waves, CMBS Conf., Utah 1987, CMBS Conf. Series (1989).

[FM] P. C. Fife and B. McLeod, The approach of solutions of nonlinear diffusion equations to travelling solutions, *Arch. Rat. Mech. An.* **65** (1977), 335–361.

[Fr1] M. Freidlin, *Functional Integration and Partial Differential Equations*, Ann. Math. Stud. **109**, Princeton, NJ (1985), Princeton University Press.

[Fr2] M. Freidlin, Limit theorems for large deviations of reaction-diffusion equation, *Ann. Prob.* **13** (1985), 639–675.

[FL] M. Freidlin and Y. T. Lee, Wave front propagation for a class of space non-homogeneous reaction-diffusion systems, preprint.

[Ga] J. Gärtner, Bistable reaction-diffusion equations and excitable media, *Math. Nachr.* **112** (1983), 125–152.

[GGi] M.-H. Giga and Y. Giga, Geometric evolution by nonsmooth interfacial energy, Hokkaido Univ., preprint.

[GGo] Y. Giga and S. Goto, Motion of hypersurfaces and geometric equations, *J. Math. Soc. Japan* **44** (1992), 99–111.

[GGIS] Y. Giga, S. Goto, H. Ishii and M. H. Sato, Comparison principle and convexity preserving properties for singular degenerate parabolic equations on unbounded domains, *Ind. Univ. Math. J.* **40** (1992), 443–470.

[GS] Y. Giga and M.-H. Sato, Generalized interface evolution with Neumann boundary condition, *Proceedings Japan Acad.* **67 Ser. A** (1991), 263–266.

[GT] D. Gilbarg and N. S. Trüdinger, *Elliptic partial differential equations of second-order*, Springer-Verlag, New York (1983).

[Go] S. Goto, Generalized motion of hypersurfaces whose growth speed depends superlinearly on curvature tensor, *J. Diff. Int. Eqs.* **7** (1994), 323–343.

[Gu] M. E. Gurtin, Multiphase thermodynamics with interfacial structure. 1. Heat conduction and the capillary balance law, *Arch. Rat. Mech. Anal.* **104** (1988), 185–221.

[GSS] M. E. Gurtin, H. M. Soner and P. E. Souganidis, Anisotropic motion of an interface relaxed by the formation of infinitesimal wrinkles, *J. Diff. Eqs.* **119** (1995), 54–108.

[Il1] T. Ilmanen, The level-set flow on a manifold, *Proc. Symp. in Pure Math.* **54** (1993), 193–204.

[Il2] T. Ilmanen, Convergence of the Allen-Cahn equation to Brakke's motion by mean curvature, *J. Diff. Geom.* **38** (1993), 417–461.

[Is1] H. Ishii, Hamilton-Jacobi equations with discontinuous Hamiltonians on arbitrary open sets, *Bull. Fac. Sci. Eng. Chuo Univ.* **26** (1985), 5–24.

[Is2] H. Ishii, Parabolic pde with discontinuities and evolution of interfaces, preprint.

[IPS] H. Ishii, G. Pires and P. E. Souganidis, Threshold dynamics and front propagation, preprint.

[IS] H. Ishii and P. E. Souganidis, Generalized motion of noncompact hypersurfaces with velocity having arbitrary growth on the curvature tensor, *Tohoku Math. J.* **47** (1995), 227–250.

[JLS] R. Jensen, P.-L. Lions and P. E. Souganidis, A uniqueness result for viscosity solutions of second-order fully nonlinear pde's, *Proc. AMS* **102** (1988), 975–978.

[J] R. Jerrard, Fully nonlinear phase transitions and generalized mean curvature motions, *Comm. PDE* **20** (1995), 223–265.

[KKR] M. Katsoulakis, G. Kossioris and F. Reitich, Generalized motion by mean curvature with Neumann conditions and the Allen-Cahn model for phase transitions, *J. Geom. Anal.*, to appear.

[KS1] M. Katsoulakis and P. E. Souganidis, Interacting particle systems and generalized mean curvature evolution, *Arch. Rat. Mech. Anal.* **127** (1994), 133–157.

[KS2] M. Katsoulakis and P. E. Souganidis, Generalized motion by mean curvature as a macroscopic limit of stochastic Ising models with long range interactions and Glauber dynamics, *Comm. Math. Phys* **169** (1995), 61–97.

[KS3] M. Katsoulakis and P. E. Souganidis, Stochastic Ising models and anisotropic front propagation, *J. Stat. Phys.*, in press.

[LP] J. Lebowitz and O. Penrose, Rigorous treatment of the Van der Waals Maxwell theory of the liquid vapour transition, *J. Math. Phys.* **98** (1966), 98–113.

[Lig] T. Liggett, *Interacting Particle Systems*, Springer-Verlag, New York, 1985.

[Lio] P.-L. Lions, Axiomatic derivation of image processing models, preprint.

[LPV] P.-L. Lions, G. Papanicolaou and S. R. S. Varadhan, Homogenization of the Hamilton-Jacobi equations, preprint.

[MS] A. Majda and P. E. Souganidis, Large scale front dynamics for turbulent reaction-diffusion equations with separated velocity scales, *Nonlinearity* **7** (1994), 1–30.

[NPV] R. H. Nochetto, M. Paolini and C. Verdi, Optimal interface error estimates for the mean curvature flow, *Ann. Sc. Norm. Sup. Pisa* **21** (1994), 193–212.

[OhS] M. Ohnuma and M. Sato, Singular degenerate parabolic equations with applications to geometric evolutions, *J. Dif. Int. Eqs* **6** (1993), 1265–1280.

[OJK] T. Ohta, D. Jasnow and K. Kawasaki, Universal scaling in the motion of random intervaces, *Phys. Rev. Lett.* **49** (1982), 1223-1226.

[OsS] S. Osher and J. Sethian, Fronts moving with curvature dependent speed: Algorithms based on Hamilton-Jacobi equations, *J. Comp. Phys.* **79** (1988), 12–49.

[RSK] J. Rubinstein, P. Sternberg and J. B. Keller, Fast reaction, slow diffusion and curve shortening, *SIAM J. Appl. Math.* **49** (1989), 116–133.

[Son1] H. M. Soner, Motion of a set by the curvature of its boundary, *J. Diff. Eqs.* **101** (1993), 313–372.

[Son2] H. M. Soner, Ginzburg-Landau equation and motion by mean curvature, I: Convergence, *J. Geom. Anal.*, in press, II: *J. Geom. Anal.*, in press.

[SonS] H. M. Soner and P. E. Souganidis, Uniqueness and singularities of rotationally symmetric domains moving by mean curvature, *Comm. PDE* **18** (1993), 859–894.

[Sor] P. Soravia, Generalized motion of a front along its normal direction: A differential games approach, preprint.

[SorS] P. Soravia and P. E. Souganidis, Phase field theory for FitzHugh-Nagumo type systems, *SIAM J. Math. Anal.*, **42** (1996), 1341–1359.

[Sp1] H. Spohn, *Large Scale Dynamics of Interacting Particles*, Springer-Verlag (1991), New York.

[Sp2] H. Spohn, Interface motion in models with stochastic dynamics, *J. Stat. Phys.* **71** (1993), 1081–1132.

[X] J. X. Xin, Existence and nonexistence of traveling waves and reaction-diffusion front propagation in periodic media, in press.

C.I.M.E. Session on "Viscosity Solutions and Applications"

List of participants

O. ARENA, Istituto di Matematica, Facoltà di Architettura, Via dell'Agnolo 14, 50122 Firenze, Italy
M. ARISAWA, CEREMADE, Univ. Paris-Dauphine, Place Maréchal de Lattre de Tassigny, 75775 Paris Cedex 16, France
F. BAGAGIOLO, Dip.to di Matematica, Università di Trento, 38050 Povo, Trento, Italy
G. BECCHERE, Dip.to di Matematica, Via F. Buonarroti 2, 56127 Pisa, Italy
G. BELLETTINI, Istituto di Matematiche Appl., Fac. di Ing., Via Bonanno 25b, 56126 Pisa, Italy
D. BERTACCINI, Dip.to di Matematica "U. Dini", Viale Morgagni, 67/A, 50134 Firenze, Italy
I. BIRINDELLI, Dip.to di Matematica, Univ. "La Sapienza", P.le A. Moro 2, 00185 Roma, Italy
S. BOTTACIN, Dip.to di Matematica, Via Belzoni 7, 35131 Padova, Italy
A. BRIANI, Dip.to di Matematica, Via F. Buonarroti 2, 56127 Pisa, Italy
A. CUTRI', Dip.to di Matematica, Univ. di Roma "Tor Vergata", Via della Ricerca Scientifica, 00185 Roma, Italy
F. DA LIO, Dip.to di Matematica, Via Belzoni 7, 35131 Padova, Italy
L. ESPOSITO, Dip.to di Matematica e Appl., Via Cintia, 86126 Napoli, Italy
S. FAGGIAN, Via Napoli 17/A, 30172 Mestre, Italy
M. FALCONE, Dip.to di Matematica, Univ. "La Sapienza", P.le A. Moro 2, 00185 Roma, Italy
S. FINZI VITA, Dip.to di Matematica, Univ. "La Sapienza", P.le A. Moro 2, 00185 Roma, Italy
L. FREDDI, Dip.to di Matematica e Informatica, Via delle Scienze 206, 33100 Udine, Italy
U. GIANAZZA, Dip.to di Matematica, Via Abbiategrasso 215, 27100 Pavia, Italy
P. GOATIN, Via Ardigò 18, 35126 Padova, Italy
S. KOIKE, CEREMADE, Univ. Paris-Dauphine, Place du Maréchal de Lattre de Tassigny, 75775 Paris Cedex 16, France
G. KOSSIORIS, Dept. of Math., Univ. of Crete, 71409 Heraklion, Crete, Greece
F. LASCIALFARI, Dip.to di Matematica, P.zza di Porta S. Donato 5, 40137 Bologna, Italy
S. LIGABUE, Dip.to di Matematica, Univ. di Roma "Tor Vergata", Via della Ricerca Scientifica, 00185 Roma, Italy
P. LORETI, IAC-CNR, Viale del Policlinico 137, 00161 Roma, Italy
R. MAGNANINI, Dip. di Matematica "U. Dini", Viale Morgagni 67/A, 50134 Firenze, Italy
P. MARCATI, Dip.to di Matematica, Via Vetoio, 67010 Coppito, L'Aquila, Italy
M. MARINUCCI, Dip.to di Matematica, Univ. "La Sapienza", P.le A. Moro 2, 00185 Roma, Italy
A. MARTA, Dip.to di Matematica, Univ. "La Sapienza", P.le A. Moro 2, 00185 Roma, Italy
G. MINGIONE, Dip.to di Matematica e Appl., Via Cintia, 86126 Napoli, Italy
A. MONTANARI, Dip.to di Matematica, Piazza di Porta S. Donato 5, 40127 Bologna, Italy
M. MOTTA, Dip.to di Matematica, Via Belzoni 7, 35131 Padova, Italy
F. NARDINI, Dip.to di Matematica, Piazza di Porta S. Donato 5, 40127 Bologna, Italy
M. NEDELJKOV, Inst. of Math., Fac. of Sci., Univ. of Novi Sad, 21000 Novi Sad, Yugoslavia
H. NGUYEN, Fac. of Maths and Inf., Delft University., 2600 GA Delft, The Netherlands
G. OMEL'YANOV, Moscow State Inst. of Electr. and Math., B.Vuzovski 3/12, 109028 Moscow
G. E. PIRES, R.Jose Mello e Castro 13, 3dto, 1700 Lisboa, Portugal
E ROUY, Scuola Normale Superiore, Piazza dei Cavalieri 7, 56126 Pisa,Italy
B. RUBINO, Dip.to di Matematica. pura ed appl., 67010 Coppito, L'Aquila, Italy
M. SAMMARTINO, Dip.to di Matematica e Appl., Via Archirafi 34, 90123 Palermo, Italy
C. SARTORI, Dip.to di Metodi e Modelli Matematici, Via Belzoni 7, 35100 Padova, Italy
G. SAVARE', IAN-CNR, Via Abbiategrasso 209, 27100 Pavia,Italy
R. SCHATZLE, School of Math. and Phys. Sci., Univ. of Sussex, GB-Falmer, Brighton, U.K.

A. SICONOLFI, Dip.to di Matematica, Univ. "La Sapienza", P.le A. Moro 2, 00185 Roma, Italy

C. SINESTRARI, Dip.to di Matematica, Univ. di Roma "Tor Vergata", Via della Ricerca
        Scientifica, 00185 Roma, Italy

I. STRATIS, Dept. of Math., Univ. of Athens, Panepistimiopolis, GR 15784 Athens, Greece

N. TADDIA, Dip.to di Matematica, Via Machiavelli 35, 44100 Ferrara

G. TALENTI, Dip.to di Matematica "U. Dini", V.le Morgagni 67/A, 50134 Firenze, Italy

V.M. TORTORELLI, Dip.to di Matematica, Via F. Buonarroti 2, 56127 Pisa, Italy

A. TOURIN, CEREMADE, Univ. Paris IX Dauphine, Place de Lattre de Tassigny
        75775 Paris Cedex, France

P. TREBESCHI, Dip.to di Matematica, Via F. Buonarroti 2, 56127 Pisa, Italy

FONDAZIONE C.I.M.E.
CENTRO INTERNAZIONALE MATEMATICO ESTIVO
INTERNATIONAL MATHEMATICAL SUMMER CENTER

"Mathematics Inspired by Biology"

is the subject of the First 1997 C.I.M.E. Session.

The session, sponsored by the Consiglio Nazionale delle Ricerche (C.N.R), the Ministero dell'Università e della Ricerca Scientifica e Tecnologica (M.U.R.S.T.), and European Community will take place, under the scientific directions of Professors VINCENZO CAPASSO (Università di Milano) and ODO DIEKMANN (University of Utrecht) in Martina Franca (Taranto), **from 13 to 20 June, 1997.**

Courses

a) **Dynamics of physiologically structured populations** (6 lectures in English)
   Prof. Odo DIEKMANN (University of Utrecht)

Outline:
1. Formulation and analysis of general linear models. The connection with multi-type branching processes. The definition of the basic reproduction ration R_0 and the Malthusian parameter r. Spectral analysis and asymptotic large time behaviour.
2. Nonlinear models: density dependence through feedback via environmental interaction variables. Stability boundaries in parameter space. Numerical bifurcation methods.
3. Evolutionary considerations. Invasibility and ESS (Evolutionarily Stable Strategies). Relationship with fitness measures.
4. Case studies:
   - cannibalism
   - Daphnia feeding on algae
   - reproduction strategy of Salmon

References

- J.A.J. Metz & O. Diekmann (eds.), Dynamics of Physiologically Structured Populations, Lect.Notes in Biomath. 68, 1986, Springer.
- O. Diekmann, M. Gyllenberg, J. A. J. Metz & H. R. Thieme, On the formulation and analysis of general deterministic structured population models, I. Linear theory, preprint.
- P. Jagers, The growth and stabilization of populations, Statistical Science 6 (1991), 269-283.
- P. Jagers, The deterministic evolution of general branching populations, preprint.
- O. Diekmann, M. A. Kirkilonis, B. Lisser, M. Louter-Nool, A. M. de Roos & B. P. Sommeijer, Numerical continuation of equilibria of physiologically structured population models, preprint.
- S. D. Mylius & O. Diekmann, On evolutionarily stable life histories, optimization and the need to be specific about density dependence, OIKOS 74 (1995), 218-224.
  J. A. J. Meetz, S. M. Mylius & O. Diekmann, When does evolution optimise? preprint.
- A. M. de Roos, A gentle introduction to physiologically structured population models, preprint.
- F. van den Bosch, A. M. de Roos & W. Gabriel, Cannibalism as a life boat mechanism, J. Math. Biol. 26 (1988), 619-633.
- V. Kaitala & W. W. Getz, Population dynamics and harvesting of semel-parous species with phenotypic and genotypic variability in reproductive age, J. Math. Biol. 33 (1995), 521-556.

b) **When is space important in modelling biological systems?** (6 lectures in English).
Prof. Rick DURRETT (Cornell University)

Outline:
In these lectures I will give an introduction to stochastic spatial models (also called cellular automata or interacting particle systems) and relate their properties to those of the ordinary differential equations that result if one ignores space and assumes instead that all individuals interact equally. In brief one finds that if the ODE has an attracting fixed point then both approaches (spatial and non-spatial) agree that coexistence will occur. How if the ODE has two or more locally stable equilibria or periodic orbits? Then the two approaches can come to radically different predictions. We will illustrate these principles by the study of a number of different examples

- competition of Daphnia species in rock pools
- allelopathy in E. coli
- spatial versions of Prisoner's dilemma in Maynard Smith's evolutionary games framework
- a species competition model of Silvertown et al. and how it contrasts with a three species ODE system of May and Leonard

References

- R. Durrett (1955) Ten lectures on Particle Systems, pages 97-201 in Springer Lecture Notes in Math. 1608.
- R. Durrett and S. Levin (1994) Stochastic spatial models: a user's guide to ecological applications. Phil. Trans. Roy. Soc. B 343, 329-350.
- R. Durrett and S. Levin (1994) The importance of being discrete and spatial. Theor. Pop. Biol. 46 (1994), 363-394.
- R. Durrett and S. Levin (1996) Allelopathy in spatially distributed populations. Preprint
- R. Durrett and C. Neuhauser (1996) Coexistence results for some competition models.

c) **Random walk systems modeling spread and interaction** (6 lectures in English).
Prof. K. P. HADELER (University of Tübingen)

Outline:
Random walk systems are semilinear systems of hyperbolic equations that describe spatial spread and interaction of species. From a modeling point of view they are similar to reaction diffusion equations but they do not show infinitely fast propagation. These systems comprise correlated random walks, certain types of Boltzmann equations, the Cattaneo system, and others. Mathematically, they are closely related to damped wave equations. The aim of the course is the derivation of such systems from biological modeling assumptions, exploration of the connections to other approaches to spatial spread (parabolic equations, stochastic processes), application to biological problems and presentation of a qualitative theory, as far as it is available

1. The problems and their history
Reaction diffusion equations; velocity jump processes and Boltzmann type equations; correlated random walks; nonlinear interactions; Cattaneo problems; reaction telegraph equations; boundary conditions.
2. Linear theory
Explicit representations; operator semigroup theory; positivity properties; spectral properties; compactness properties.
3. Semilinear systems
Invariance and monotonicity; Lyapunov functions; gradient systems; stationary points; global attractors.
4. Spatial spread
Travelling fronts and pulses; spread of epidemics; bifurcation and patterns; Turing phenomenon; free boundary value problems.
5. Comparison of hyperbolic and parabolic problems.

References

- Goldstein, S, On diffusion by discontinuous movements and the telegraph equation. Quart. J. Mech. Appl. Math. 4 (1951), 129-156.
- Kac, M., A stochastic model related to the telegrapher's equation. (1956), reprinted Rocky Math. J. 4 (1974), 497-509.
- Dunbar, S., A branching random evolution and a nonlinear hyperbolic equation. SIAM J. Appl. Math. 48 (1988), 1510-1526.
- Othmer, H. G., Dunbar, S., Alt, W., Models of dispersal in biological systems. J. Math. Biol. 26 (1988), 263-298.
- Hale, J. K., Asymptotic behavior of dissipative systems. Amer. Math. Soc., Providence, R.I. 1988.
- Temam, R., Infinite dimensional dynamical systems in mechanics and physics. Appl. Math. Series 68, Springer 1988.

- Hadeler, K. P., Reaction telegraph equations and random walks. Canad. Appl. Math. Quart. 2 (1994), 27-43.
- Hadeler, K. P., Reaction telegraph equations and random walk systems. 31 pp. In: S. van Strien, S. Verduyn Lunel (eds.), "Dynamical systems and their applications in science". Royal Academy of the Netherlands, North Holland 1966.

d) Mathematical Modelling in Morphogenesis (6 lectures in English).
Prof. Philip MAINI (Oxford University)

Outline:

A central issue in developmental biology is the formation of spatial and spatio temporal patterns in the early embryon, a process known as morphogenesis. In 1952, Turing published a seminal paper in which he showed that a system of chemicals, reacting and diffusing, could spontaneously generate spatial patterns in chemical concentrations, and he termed this process "diffusion-driven instability". This paper has generated a great deal of research into more general reaction diffusion systems, consisting of nonlinear coupled parabolic partial differential equations (Turing's model was linear), both from the viewpoint of mathematical theory, and from the application to diverse patterning phenomena in development. In these lectures I shall discuss the analysis of these equations and show how this analysis carries over to other models, for example, mechanochemical models. Specifically, I shall focus on linear analysis and nonlinear bifurcation analysis. This analysis will be extended to non-standard problems.

References

- G. C. Cruywagen, P.K. Maini, J. D. Murray, Sequential pattern formation in a model for skin morphogenesis, IMA J. Math. Appl. Med. & Biol., 9 (1992), 227-248.
- R. Dillon, P.K. Maini, H. G. Othmer, Pattern formation in generalised Turing systems. I. Steady-state patterns in systems with mixed boundary conditions, J. Math. Biol., 32 (1994), 345-393.
- P. K. Maini, D. L. Benson, J. A. Sherratt, Pattern formation in reaction diffusion models with spatially inhomogeneous diffusion coefficients, IMA J. Math. Appl. Med. & Biol.,) (1992), 197-213.
- J. D. Murray, Mathematical biology, Springer Verlag, 1989.
- A. M. Turing, The chemical basis of morphogenesis, Phil. Trans. Roy. Soc. Lond., B 237 (1952), 37-72.

e) The Dynamics of Competition (6 lectures in English).
Prof. Hal L. SMITH (Arizona State University, Tempe)

Outline:

Competitive relations among populations of organisms are among the most studied by ecologists and there is a vast literature on mathematical modeling of competition. Quite recently, there have been a number of breakthroughs in the mathematical understanding of the dynamics of competitive systems, that is, of those systems of ordinary, delay, difference and partial differential equations arising in the modeling of competition. In these lectures I intend to describe some of the new results and consider their applications to microbial competition for nutrients in a chemostat.

References

- deMottoni, P. and Schiaffino, A., Competition systems with periodic coefficients: a geometric approach, J. Math. Biology, 11 (1982), 319-335.
- Hess, P. and Lazer, A. C., On an abstract competition model and applications, Nonlinear Analysis T.M.A., 16 (1991), 917-940.
- Hess, P., Periodic-parabolic boundary value problems and positivity, Longman Scientific and Technical, New York, 1991.
- Hirsch, M., Systems of differential equations that are competitive or cooperative. VI: A local C' closing lemma for 3-dimensional systems, Ergod. Th. Dynamical Sys., 11 (1991), 443-454.
- Smith, H. L., Monotone dynamical systems: an introduction to the theory of competitive and cooperative systems, Math. Surveys and Monographs, 41, Amer. Math. Soc. 1995.
- Smith, H. L., A discrete, size-structured model of microbial growth and competition in the chemostat, to appear, J. Math. Biology.
- Smith, H. L. and Waltman, P., The theory of the chemostat, dynamics of microbial competition, Cambridge Studies in Math. Biology, Cambridge University Press, 1995.
- Hsu, S.-B., Smith, H. L. and Waltman, P., Competitive exclusion and coexistence for competitive systems in ordered Banach spaces, to appear, Trans. Amer. Math. Soc.
- Hsu, S. B., Smith, H. L. and Waltmanm P., Dynamics of Competition in the unstirred chemostat, Canadian Applied Math. Quart., 2 (1994), 461-483.

FONDAZIONE C.I.M.E.
CENTRO INTERNAZIONALE MATEMATICO ESTIVO
INTERNATIONAL MATHEMATICAL SUMMER CENTER

"Advanced Numerical Approximation of Nonlinear Hyperbolic Equations"

is the subject of the Second 1997 C.I.M.E. Session.

The session, sponsored by the Consiglio Nazionale delle Ricerche (C.N.R), the Ministero dell'Università e della Ricerca Scientifica e Tecnologica (M.U.R.S.T.), and European Community will take place, under the scientific direction of Professors ALFIO QUARTERONI (Politecnico di Milano) at Grand Hotel San Michele, Cetraro (Cosenza), **from 23 to 28 June, 1997.**

Courses

a) **Discontinuous Galerkin Methods for Nonlinear Conservation Laws** (5 lectures in English).
Prof. Bernardo COCKBURN (University of Minnesota, Minneapolis)

Outline:
Lecture 1:
The nonlinear scalar conservation law as paradigm. The continuous dependence structure associated with the vanishing viscosity method for nonlinear scalar conservation laws. The entropy inequality and the entropy solution. The DG method as a vanishing viscosity method.

Lecture 2:
Features of the DG method. The use of two-point monotone numerical fluxes as an artificial viscosity associated with interelement edges: Relation with monotone schemes. Slope limiting as an artificial viscosity associate with the interior of the elements: Relation with the streamline-diffusion method.

Lecture 3:
Theoretical properties of the DG method. Stability properties and a priori and a posteriori error estimates.

Lecture 4:
Computational results for the scalar conservation law.

Lecture 5:
Extension to general multidimensional hyperbolic systems and computational results for the Euler equations of gas dynamics.

b) **Adaptive methods for differential equations and application to compressible flow problems** (6 lectures in English)
Prof. Claes JOHNSON (Chalmers University of Technology, Göteborg)

Outline:
We present a general methodology for adaptive error control for Galerkin methods for differential equations based on a posteriori error estimates involving the residual of the computed solution. The methodology is developed in the monograph Computational Differential Equations by Eriksson, Estep, Hansbo and Johnson, Cambridge University Press, 1996, and the companion volume Advanced Computational Differential Equations, by Eriksson, Estep, Hansbo, Johnson and Levenstam (in preparation), and is realized in the software femlab available on Internet http://www.math.chalmers.se/femlab. The a posteriori error estimates involve stability factors which are

estimated through auxiliary computation solving dual linearized problems. The size of the stability factor determines if the the specific problem considered is computable in the sense that the error in some norm can be made sufficiently small with the available computational resources. We present applications to a variety of problems including heat conduction, compressible and incompressible fluid flow, reaction-diffusion problems, wave propagation, elasticity/plasticity, and systems of ordinary differential equations.

Lecture 1: Introduction. A posteriori error estimates for Galerkin methods
Lecture 2: Applications to diffusion problems
Lecture 3: Applications to compressible and incompressible flow
Lecture 4: Applications to reaction-diffusion problems
Lecture 5: Applications to wave propagation, and elasticity/plasticity.
Lecture 6: Applications to dynamical systems.

e) Essentially Non-Oscillatory (ENO) And Weighted, Essentially Non-Oscillatory (WENO) Schemes For Hyperbolic Conservation Laws (6 lectures in English)
Prof. Chi-Wang SHU (Brown University, Providence)

Outline:
In this mini-course we will describe the construction, analysis, and application of ENO (Essentially Non-Oscillatory) and WENO (Weighted Essentially Non-Oscillatory) schemes for hyperbolic conservation laws. ENO and WENO schemes are high order accurate finite difference schemes designed for problems with piecewise smooth solutions with discontinuities. The key idea lies at the approximation level, where a nonlinear adaptive procedure is used to automatically choose the smoothest local stencil, hence avoiding crossing discontinuities in the interpolation as much as possible. The talk will be basically self-contained, assuming only the background of hyperbolic conservation laws which will be provided by the first lectures of Professor Tadmor.

Lecture 1: ENO and WENO interpolation
    We will describe the basic idea of ENO interpolation, starting from the Newton form of one dimensional polynomial interpolations. We will show how the local smoothness of a function can be effectively represented by its divided or undivided differences, and how this information can help in choosing the stencils in ENO or the weights in WENO. Approximation results will be presented.

Lecture 2: More on interpolation
    It will be a continuation of the first. We will discuss some advanced topics in the ENO and WENO interpolation, such as different building blocks, multi dimensions including general triangulations, sub-cell resolutions, etc.

Lecture 3: Two formulations of schemes for conservation laws
    We will describe the conservative formulations for numerical schemes approximating a scalar, one dimensional conservation law. Both the cell averaged (finite volume) and point value (finite difference) formulations will be provided, and their similarities, differences, and relative advantages will be discussed. ENO and WENO interpolation procedure developed in the first two lectures will then be applied to both formulations. Total variation stable time discretization will be discussed.

Lecture 4: Two dimensions and systems
    Is a continuation of the third. We will discuss the generalization of both formulations of the scheme to two and higher dimensions and to systems. Again the difference and relative advantages of both formulations will be discussed. For systems, the necessity of using local characteristic decompositions will be illustrated, together with some recent attempts to make this part of the algorithm, which accounts for most of the CPU time, cheaper.

Lecture 5: Practical issues
    Practical issues of the ENO and WENO algorithms, such as implementation for workstations, vector and parallel supercomputers, how to treat various boundary conditions, curvilinear coordinates, how to use artificial compression to sharpen contact discontinuities, will be discussed. ENO schemes applied to Hamilton-Jacobi equations will also be discussed.

Lecture 6: Applications to computational physics
    Application of ENO and WENO schemes to computational physics, including compressible Euler and Navier-Stokes equations, incompressible flow, and semiconductor device simulations, will be discussed.

d) High Resolution Methods for the Approximate Solution of Nonlinear Conservation Laws and Related Equations (5 lectures in English)
Prof. Eitan TADMOR (UCLA and Tel Aviv University)

Outline:
The following issues will be addressed.
Conservation laws: Scalar conservation laws, One dimensional systems; Riemann's problem, Godunov, Lax-Friedrichs and Glimm schemes, Multidimensional systems
Finite Difference Methods: TVD Schemes, three- and five-points stencil scheme, Upwind vs. central schemes, TVB approximations. Quasimonotone schemes; Time discretizations
Godunov-Type Methods and schemes
Finite Volume Schemes and error estimates
Streamline Diffusion Finite Element Schemes: The entropy variables
Spectral Viscosity and Hyper Viscosity Approximations; Kinetic Approximations

Lecture 1: Approximate solutions of nonlinear conservation laws - a general overview
During the recent decade there was an enormous amount of activity related to the construction and analysis of modern algorithms for the approximate solution of nonlinear hyperbolic conservation laws and related problems.
To present the successful achievements of this activity we discuss some of the analytical tools which are used in the development of the convergence theories associated with these achievements. In particular, we highlight the issues of compactness, compensated compactness, measure-valued solutions, averaging lemma,...while we motivate our overview of finite-difference approximations (e.g. TVD schemes), finite-volume approximations (e.g. convergence to measure-valued solutions on general grids), finite-element schemes (e.g., the streamline-dissusion method), spectral schemes (e.g. spectral viscosity method), and kinetic schemes (e.g., BGK-like and relaxation schemes).

Lecture 2: Approximate solutions of nonlinear conservation law - nonoscillatory central schemes
We discuss high-resolution approximations of hyperbolic conservation laws which are based on t *centraldifferencing*. The building block of such schemes is the use of staggeredgrids. The main advantage is simplicity, since no Riemann problems are involved. In particular, we avoid the time-consuming field-by-field decompositions required by (approximate) Riemann solvers of upwind difference schemes.
Typically, staggering suffers from excessive numerical dissipation. Here, excessive dissipation is compensated by using modern, high-resolution, non-oscillatory reconstructions.
We prove the non-oscillatory behavior of our central procedure in the scalar framework: For both the second- and third-order schemes we provide total-variation bounds, one-sided Lipschitz bounds (which in turn yield precise error estimates), as well as entropy and multidimensional L^{∞} stability estimates.
Finally, we report on a variety of numerical experiments, including second -and third-order approximations of one-dimensional problems (Euler and MHD equations), as well as two-dimensional systems (including compressible and incompressible equations). The numerical experiments demonstrate that these central schemes offer simple, robust, Riemann-solver-free approximations for the solution of one and two-dimensional systems. At the same time, these central schemes achieve the same quality results as the high-resolution upwind schemes.

Lecture 3: Approximate solutions of nonlinear conservation laws - convergence rate estimates
Convergence analysis of approximate solutions to nonlinear conservation laws is often accomplished by BV or compensated-compactness arguments, which lack convergence rate estimates. An L^1-error estimate is available for monotone approximations.
We present an alternative convergence rate analysis. As a stability condition we assume Lip^+-stability in agreement with Oleinik's E-condition.
We show that a family of approximate solutions, v^{eps}, which is Lip^+-stable, satisfies

$$\| v^{eps}(.,t) - u(.,t) \|_{Lip} \leq C \| v^{eps}(.,0) - u(.,0) \|_{Lip}$$

Consequently, familiar L^p and new pointwise error estimates are derived. We demonstrate these estimates for viscous and kinetic approximations, finite-difference schemes, spectral methods, coupled systems with relaxation...

Lecture 4: Approximate solutions of nonlinear conservation laws - the spectral viscosity method
Numerical tests indicate that the convergence of spectral approximations to nonlinear conservation laws may --- and in some cases we prove it must --- fail, with or without post-processing the numerical solution. This failure is related to the global nature of spectral methods. Since nonlinear conservation laws exhibit spontaneous shock discontinuities, the spectral approximation pollutes unstable Gibbs oscillations overall the computational domain, and the lack of entropy dissipation prevents convergence in these cases.

The Spectral Viscosity (SV) method attempts to stabilize the spectral approximation by augmenting the latter with high frequency viscosity regularization, which could be efficiently implemented in the spectral, rather than the physical space. The additional SV is small enough to retain the formal spectral accuracy of the underlying approximation; yet, the SV is shown to be large enough to enforce a sufficient amount of entropy dissipation, and hence, by compensated compactness arguments, to prevent unstable spurious oscillations. Recent convergence results for the SV approximations of initial- and initial-boundary value problems will be surveyed. Numerical experiments will be presented to confirm that by post-processing the SV solution, one recovers the exact entropy solution within spectral accuracy.

Lecture 5: Approximate solutions of nonlinear conservation laws: entropy, kinetic formulations and regularizing effects

We present a new formulation of multidimensional scalar conservation laws and certain 2x2 one-dimensional systems (including the isentropic equations), which includes both the equation and the entropy criterion. This formulation is a kinetic one involving an additional variable called velocity by analogy. We also give some applications of this formulation to new compactness and regularity results for entropy solutions based upon the velocity averaging lemmas. Finally, we show that this kinetic formulation is in fact valid and meaningful for more general classes of equations like equations involving nonlinear second-order terms.

Seminars

A number of seminars will be offered during the Session.

Applications

Those who want to attend the Session should fill in an application to the Director of C.I.M.E at the address below, **not later than April 15, 1997.**

An important consideration in the acceptance of applications is the scientific relevance of the Session to the field of interest of the applicant.

Applicants are invited to submit, along with their application, a letter of recommendation.

Participation will only be allowed to persons who have applied in due time and have had their application accepted.

CIME will be able to partially support some of the youngest participants. Those who plan to apply for support have to mention explicitely in the application form.

Attendance

No registration fee is requested.
Lectures will be held at Grand Hotel San Michele, Cetraro (Cosenza) on June 23, 24, 25, 26, 27, 28.
Participants are requested to register at the Grand Hotel San Michele on June 22, 1997.

Site and lodging

The Session will be held at Grand Hotel San Michele, Cetraro (Cosenza, Italy).

Prices for full pensions (bed, meals) are roughly 140.000 lires p.p. a day in a single room, 120.000 lires in a double room. Cheaper arrangements for multiple lodging in a residence are available. More detailed information may be obtained from the Direction of the Hotel (tel +39-982-91012; telefax +39-982-91430)

Lecture Notes

Lecture notes will be published as soon as possible after the Session.

ROBERTO CONTI
Director, C.I.M.E.

PIETRO ZECCA
Secretary, C.I.M.E.

Fondazione C.I.M.E.
c/o Dipartimento di Matematica "U. Dini"
Viale Morgagni, 67/A - 50134 FIRENZE (Italy)
Tel. +39-55-434975 / +39-55-4237123
FAX +39-55-434975 / +39-55-4222695
E-mail CIME@UDINI.MATH.UNIFI.IT

Informations, programs and application form can be obtained on http: //www.math.unifi.it/CIME

FONDAZIONE C.I.M.E.
CENTRO INTERNAZIONALE MATEMATICO ESTIVO
INTERNATIONAL MATHEMATICAL SUMMER CENTER

"Arithmetical Theory of Elliptic Curves"

is the subject of the Third 1997 C.I.M.E. Session.

The session, sponsored by the Consiglio Nazionale delle Ricerche (C.N.R), the European Community under the "Training and Mobilty of Researchers" Programme and the Ministero dell'Università e della Ricerca Scientifica e Tecnologica (M.U.R.S.T.), will take place, under the scientific direction of Professor CARLO VIOLA (Università di Pisa) at Grand Hotel San Michele, Cetraro (Cosenza), **from 12 to 19 July, 1997.**

Courses

a) Iwasawa Theory for Elliptic Curves Without Complex Multiplication. (6 lectures in English).
Prof. John COATES (Cambridge University)

Outline:
Let E be an elliptic curve over $\mathbf{Q}$ without complex multiplication, and let p be a prime number. A considerable amount is now known about the Iwasawa theory of E over the cyclotomic $\mathbf{Z}_p$ extension of $\mathbf{Q}$, and some of this material will be discussed in the courses by Greenberg and Rubin. On the other hand, very little is known about the Iwasawa theory of E over the field F_∞ which is obtained by adjoining all p-power division points on E to $\mathbf{Q}$. If G_∞ denotes the Galois group of F_∞ over Q, then G_∞ is an open subgroup of $GL_2(\mathbf{Z}_p)$ by a theorem of Serre, since E is assumed not to have complex multiplication. The course will begin by discussing some of the basic properties of the non-abelian Iwasawa algebra $\Lambda = \mathbf{Z}_p[[G_\infty]]$. It will then give our present fragmentary knowledge of various basic Iwasawa modules over Λ, which arise when one studies the arithmetic of E over F_∞. Much of the course will be concerned with posing open questions which seem to merit further study.

References

The course will only assume basic results about elliptic curves and their Galois cohomology, most of which are contained in

- J. Silverman, "*The Arithmetic of Elliptic Curves*", GTM 106, (1986), Springer.

b) Iwasawa Theory for Elliptic Curves. (6 lectures in English)
Prof. Ralph GREENBERG (University of Washington, Seattle)

Outline:
This course will present some of the basic results and conjectures concerning the behavior of the Selmer group of an elliptic curve defined over a number field F in a tower of cyclotomic extensions of F. We will mostly discuss the case where the elliptic curve

has good ordinary reduction at the primes of F dividing a rational prime p and the cyclotomic extensions are obtained by adjoin p-power roots of unity of F. One of the main results we will prove is Mazur's "Control Theorem". The proof will depend on a rather simple description of the Selmer group which also will allow us to study specific examples. We will also discuss the "Main Conjecture" which was formulated by Mazur and its relationship to the Birch and Swinnerton-Dyer Conjecture. If time permits, we will discuss generalizations of this Main Conjecture.

References
1) J. Silverman, The Arithmetic of Elliptic Curves, GTM 106, Springer.
2) S. Lang, Cyclotomic Fields I and II, GTM 121, Springer.
3) L. Washington, Introduction to Cyclotomic Fields, GTM 83, Springer.
4) J. P. Serre, Cohomologie galoisienne, LNM 5, Springer.
5) J. Tate, Duality Theorems in Galois Cohomology over Number Fields, Proceedings of the ICM, Stockholm, 1962, pp. 288-295.

In reference 1, one should become familiar with the Tate module of an elliptic curve and reduction modulo a prime p. In references 2 or 3 one can find an introduction to the structure of finitely generated modules over the Iwasawa algebra. It may be useful to consult references 4 and 5 to become familiar with Galois cohomology.

c) **Two-dimensional representations of Gal $(\overline{\mathbf{Q}}/\mathbf{Q})$**. (6 lectures in English)
Prof. Kenneth A. RIBET (University of California, Berkeley)

Outline:
The general theme of the course is that two-dimensional representations of $Gal\left(\overline{\mathbf{Q}}/\mathbf{Q}\right)$ can be shown to have large images in appropriate circumstances. The lectures will touch on topics to be selected from the following list:

1. *Recent work by Darmon and Merel on modular curves* associated to normalizers of non-split Cartan Subgroups of $\mathbf{GL}(2, \mathbf{Z}/p\mathbf{Z})$. A preprint by these authors proves, among other things, that the Fermat-like equations $x^p + y^p = 2z^p$ has only the trivial non-zero solutions in integers x, y and z when p is an odd prime number! Technically, the main theorem of the paper is an analogue of a result pertaining to split Cartan subgroups of $\mathbf{GL}(2, \mathbf{Z}/p\mathbf{Z})$ which appears in B. Mazur's 1978 paper *"Rational isogenies of prime degree"*. The connection with two-dimensional representations of $Gal(\overline{\mathbf{Q}}/\mathbf{Q})$ comes about because Merel has used the Darmon-Merel theorem to prove that the *mod p* representations of $Gal(\overline{\mathbf{Q}}/\mathbf{Q})$ arising from non-CM Frey elliptic curves have large images for $p > 2$. [To read about this work prior to the conference, download Darmon-Merel paper (http://www.math.mcgill.ca/darmon/pub/Winding/paper.html) and follow the references given in that article]

.2. *Semistable representations.* In his 1972 article on Galois properties of torsion points of elliptic curves, J.-P. Serre proved that the *mod p* representation of $Gal(\overline{\mathbf{Q}}/\mathbf{Q})$ associated to a semistable elliptic curve over $\mathbf{Q}$ is either reducible or surjective, provided that the prime p is at least 7. In his 1995 Bourbaki seminar on the work of Wiles and Taylor-Wiles, Serre extended his theorem to cover the primes $p = 3,5$. Using group-theoretic results of L. E. Dickson, one may replace $\mathbf{Z}/p\mathbf{Z}$ by a finite etale $\mathbf{Z}/p\mathbf{Z}$ algebra. [See (ftp://math.berkeley.edu/pub/Preprints/Ken-Ribet/semistable.tex).].

3. *Adelic representations.* A recent theorem of R. Coleman, B. Kaskel and K. Ribet concerns point P on the modular curve $X_0(37)$ for which the class of the divisor $(P) - (P_\infty)$ has finite order. (Here P_∞ is the "cusp at infinity" on $X_0(37)$.) The theorem states that the only such points are the two cusps on $X_0(37)$. Our proof relies heavily on results of B. Kaskel, who calculated the image of the "adelic" representation $Gal(\overline{\mathbf{Q}}/\mathbf{Q}) - GL(2, R)$ which is defined by the action of $Gal(\overline{\mathbf{Q}}/\mathbf{Q})$ on the torsion points on the Jacobian of $X_0(37)$. Here R is the ring of pairs $(a, b) \in \widehat{Z} \times \widehat{Z}$ for which a and b have the same

parity. In my lectures, I would like to explain how Kaskel arrives at this results. [Although neither Kaskel's work nor the Coleman-Kaskel-Ribert is available at this time, one might look at 6 of Part I of the Lang-Trotter book on "Frobenius distributions" (Lecture Notes in Math, **504**) to get a feel for this kind of study].

4. *l-adic representation attached to modular forms.* Let $f_1, \ldots, f_p$ be newforms of weights $k_1, \ldots, k_p$ possibly of different levels and characters. Let $E = E_1 \times \cdots \times E_t$ be the product of the number fields generated by the coefficients of the different forms. For each prime , the l-adic representations attached to the different f_i furnish a product representation $\rho_l : Gal(\overline{\mathbf{Q}}/\mathbf{Q}) \to \mathbf{GL}\,(2, E \otimes Q_l)$. In principle, we know how to calculate the image of ρ "up to finite groups", i.e., the Lie algebras of $\rho Gal(\overline{\mathbf{Q}}/\mathbf{Q})$. I hope to explain the answer - and how we know that it is correct. [One might consult my article in Lecture Notes in Math. **601** and my article in Math. Annalen **253** (1980)].

d) Elliptic curves with complex multiplication and the conjecture of Birch and Swinnerton-Dyer. (6 lecture in English)
Prof. Carl RUBIN (Ohio State University)

Outline:
The conjecture of Birch and Swinnerton-Dyer relates the arithmetic of an elliptic curve with the behavior of its L-function. These lectures will give a survey of the state of our knowledge of this conjecture in the case of elliptic curves with complex multiplication, where the results are strongest. Specific topics will include:
1) Elliptic curves with complex multiplication.
2) Descent (following Coates and Wiles).
3) Elliptic units.
4) Euler systems and ideal class groups.
5) Iwasawa theory and the "main conjecture".

References
- Shimura, G., Introduction to the arithmetic theory of automorphic functions, Princeton: Princeton Univ. Press (1971), Chapter 5.
- Silverman, J., Advanced topics in the arithmetic of elliptic curves, Graduate Texts in Math. 151, New York: Springer-Verlag (1994), Chapter II.
- de Shalit. E., Iwasawa theory of elliptic curves with complex multiplication, Perspectives in Math. 3, Orlando: Academic Press (1987).
- Coates, J., Wiles, A., On the conjecture of Birch and Swinnerton-Dyer, Invent. math. 39 (1977) 223-251.
- Rubin, K., The main conjecture. Appendix to: Cyclotomic fields I and II, S. Lang, Graduate Texts in Math. 121, New York: Springer-Verlag (1990) 397-419.
- Rubin, K., The 'main conjectures' of Iwasawa theory for imaginary quadratic fields, Invent. Math. 103 (1991) 25-68.

LIST OF C.I.M.E. SEMINARS   Publisher

1954 - 1. Analisi funzionale   C.I.M.E.
        2. Quadratura delle superficie e questioni connesse   "
        3. Equazioni differenziali non lineari   "

1955 - 4. Teorema di Riemann-Roch e questioni connesse   "
        5. Teoria dei numeri   "
        6. Topologia   "
        7. Teorie non linearizzate in elasticità, idrodinamica,aerodinamica   "
        8. Geometria proiettivo-differenziale   "

1956 - 9. Equazioni alle derivate parziali a caratteristiche reali   "
       10. Propagazione delle onde elettromagnetiche   "
       11. Teoria della funzioni di più variabili complesse e delle
           funzioni automorfe   "

1957 - 12. Geometria aritmetica e algebrica (2 vol.)   "
       13. Integrali singolari e questioni connesse   "
       14. Teoria della turbolenza (2 vol.)   "

1958 - 15. Vedute e problemi attuali in relatività generale   "
       16. Problemi di geometria differenziale in grande   "
       17. Il principio di minimo e le sue applicazioni alle equazioni
           funzionali   "

1959 - 18. Induzione e statistica   "
       19. Teoria algebrica dei meccanismi automatici (2 vol.)   "
       20. Gruppi, anelli di Lie e teoria della coomologia   "

1960 - 21. Sistemi dinamici e teoremi ergodici   "
       22. Forme differenziali e loro integrali   "

1961 - 23. Geometria del calcolo delle variazioni (2 vol.)   "
       24. Teoria delle distribuzioni   "
       25. Onde superficiali   "

1962 - 26. Topologia differenziale   "
       27. Autovalori e autosoluzioni   "
       28. Magnetofluidodinamica   "

1972 - 59. Non-linear mechanics   "
        60. Finite geometric structures and their applications   "
        61. Geometric measure theory and minimal surfaces   "

1973 - 62. Complex analysis   "
        63. New variational techniques in mathematical physics "
        64. Spectral analysis   "

1974 - 65. Stability problems   "
        66. Singularities of analytic spaces   "
        67. Eigenvalues of non linear problems   "

1975 - 68. Theoretical computer sciences   "
        69. Model theory and applications   "
        70. Differential operators and manifolds   "

1976 - 71. Statistical Mechanics   Ed Liguori, Napoli
        72. Hyperbolicity   "
        73. Differential topology   "

1977 - 74. Materials with memory   "
        75. Pseudodifferential operators with applications   "
        76. Algebraic surfaces   "

1978 - 77. Stochastic differential equations   "
        78. Dynamical systems   Ed Liguori, Napoli and Birhäuser Verlag

1979 - 79. Recursion theory and computational complexity   "
        80. Mathematics of biology   "

1980 - 81. Wave propagation   "
        82. Harmonic analysis and group representations   "
        83. Matroid theory and its applications   "

1981 - 84. Kinetic Theories and the Boltzmann Equation (LNM 1048) Springer-Verlag
        85. Algebraic Threefolds   (LNM 947) "
        86. Nonlinear Filtering and Stochastic Control (LNM 972) "

1982 - 87. Invariant Theory   (LNM 996) "
        88. Thermodynamics and Constitutive Equations (LN Physics 228) "
        89. Fluid Dynamics   (LNM 1047) "

1993 –	117.	Integrable Systems and Quantum Groups	(LNM 1620) Springer-Verlag
	118.	Algebraic Cycles and Hodge Theory	(LNM 1594)
	119.	Phase Transitions and Hysteresis	(LNM 1584) "
1994 –	120.	Recent Mathematical Methods in Nonlinear Wave Propagation	(LNM 1640) "
	121.	Dynamical Systems	(LNM 1609) "
	122.	Transcendental Methods in Algebraic Geometry	(LNM 1646) "
1995 –	123.	Probabilistic Models for Nonlinear PDE's	(LNM 1627) "
	124.	Viscosity Solutions and Applications	(LNM 1660) "
	125.	Vector Bundles on Curves. New Directions	(LNM 1649) "
1996 –	126.	Integral Geometry, Radon Transforms and Complex Analysis	to appear "
	127.	Calculus of Variations and Geometric Evolution Problems	to appear "
	128.	Financial Mathematics	(LNM 1656) "
1997 –	129.	Mathematics Inspired by Biology	to appear "
	130.	Advanced Numerical Approximation of Nonlinear Hyperbolic Equations	to appear "
	131.	Arithmetic Theory of Elliptic Curves	to appear "
	132.	Quantum Cohomology	to appear "